Surgical Anatomy of the Infratemporal Fossa

Surgical Anatomy of the Infratemporal Fossa

John D. Langdon

Professor and Head of Department
Department of Oral and Maxillofacial Surgery
King's College London, UK

Barry K.B. Berkovitz

Reader in Anatomy
Division of Anatomy, Cell and Human Biology
King's College London, UK

Bernard J. Moxham

Professor of Anatomy and Head of Teaching in Biosciences
Cardiff School of Biosciences
Cardiff University, UK

MARTIN DUNITZ

First published in the United Kingdom in 2003
by Martin Dunitz, Taylor & Francis Group plc, 11 New Fetter Lane, London EC4P 4EE

Tel.: +44 (0) 20 7583 9855
Fax.: +44 (0) 20 7842 2298
E-mail: info@dunitz.co.uk
Website: http://www.dunitz.co.uk

A CIP record for this book is available from the British Library.

ISBN 1–899066–79–9

Distributed in the USA by
Fulfilment Center
Taylor & Francis
10650 Tobben Drive
Independence, KY 41051, USA
Toll Free Tel.: +1 800 634 7064
E-mail: taylorandfrancis@thomsonlearning.com

Distributed in Canada by
Taylor & Francis
74 Rolark Drive
Scarborough, Ontario M1R 4G2, Canada
Toll Free Tel.: +1 877 226 2237
E-mail: tal_fran@istar.ca

Distributed in the rest of the world by
Tomson Publishing Services
Cheriton House
North Way
Andover, Hampshire SP10 5BE, UK
Tel.: +44 (0)1264 332424
E-mail: salesorder.tandf@tomsonpublishingservices.co.uk

Project management: Top Draw Design
Medical illustrators: Philip Wilson FMAA, RMIP
Debbie Maizels CBiol, MIBiol

Printed in China by Imago

Contents

Contributors

James Allibone
Consultant Spinal Neurosurgeon
Royal National Orthopaedic Hospital
Stanmore, Middlesex, UK

Barry K.B. Berkovitz
Reader in Anatomy
Division of Anatomy, Cell and Human Biology
GKT School of Biomedical Sciences
King's College London
London, UK

Luigi Clauser
Professor and Head of Department of Craniomaxillofacial Surgery and Centre for Orbital Surgery and Pathology
St Anna Hospital and University
Ferrara, Italy

Camillo Curioni
Professor of Maxillofacial Surgery
University of Ferrara
Ferrara, Italy

Barrie T. Evans
Consultant Oral and Maxillofacial Surgeon
Southampton University Hospitals
Southampton, UK

John D. Langdon
Professor and Head of Department
Department of Oral and Maxillofacial Surgery
GKT Dental Institute
King's College London
London, UK

Dennis T. Lanigan
Head, Division of Oral and Maxillofacial Surgery
Department of Biological, Diagnostic, and Surgical Sciences
College of Dentistry
University of Saskatchewan
Saskatoon, Saskatchewan, Canada

Bernard J. Moxham
Professor of Anatomy and Head of Teaching in Biosciences
Cardiff School of Biosciences
Cardiff University
Cardiff, UK

J. Phillip Rood
Professor, Department of Oral and Maxillofacial Surgery
GKT Dental Institute
King's College London
London, UK

Colin Shieff
Consultant Neurosurgeon
Department of Neurosurgery
Royal Free Hospital
London, UK

Ram M. Tiwari
Head, Neck, Skull Base and Reconstructive Surgeon
Bangalore Institute of Oncology
and
Chief, Division of Surgical Oncology and Head and Neck Surgery
Manipal Hospital
Bangalore, India

David Wiesenfeld
Oral and Maxillofacial Surgeon
Head of Oral and Maxillofacial Surgery Unit
The Royal Melbourne Hospital
and
Oral and Maxillofacial Surgeon
Peter MacCallum Cancer Institute
Senior Associate, Department of Surgery and School of Dental Science
The University of Melbourne
Melbourne, Australia

Preface

Specialisation in the medical and dental professions has always been an important issue. With the development of new undergraduate and postgraduate training schemes, together with the concomitant reduction in the teaching of basic science, particularly anatomy, the issue is assuming even greater significance. This book takes cognisance of educational, and political, concerns and also takes account of the rapid clinical and surgical developments that have taken place in recent years.

The infratemporal fossa is of particular importance to the dentist and to the oral and maxillofacial surgeon. However, as its anatomical boundaries impinge onto other regions of clinical interest, it is also relevant to other medical specialities. For example, an infratemporal approach is now used for certain neurosurgical procedures. The main aim of this book is to bring together descriptions of the anatomy of this area with coverage of the main aspects of clinical and surgical relevance, topics not normally found in a single text. In relation to the anatomy of the infratemporal fossa, attention has also been given to the temporomandibular joint, parotid gland and pterygopalatine fossa. Clinical aspects that are covered concern local anaesthesia, spread of infection, surgical approaches to the infratemporal fossa and to the facial nerve, tumours and tumour-like disorders in the region and maxillofacial trauma and orthognathic surgery. The scope and specialisation of the subject matter requires contributions from experts in a variety of disciplines. Inevitably, there will be some overlap between chapters. The editors regard this as a positive feature, giving the reader the opportunity to hear the views of different experts in the field of head and neck surgery.

We hope that our attempts to provide a guide to the infratemporal fossa by correlating anatomical, clinical and surgical subject matter, and our efforts to provide a visual commentary by the use of numerous colour illustrations, will prove beneficial to our readers whatever their level of specialisation.

J.D. Langdon
B.K.B. Berkovitz
B.J. Moxham

Chapter 1

Regional and sectional anatomy of the infratemporal fossa

B.J. MOXHAM AND B.K.B BERKOVITZ

THE BOUNDARIES OF THE INFRATEMPORAL FOSSA

The infratemporal fossa is the space located deep to the ramus of the mandible. Further reading of the anatomy of this region is available from Hollingshead,[1] Berkovitz and Moxham,[2] *Gray's Anatomy*[3] and Lang.[4] Together with the temporal fossa, pterygoid processes and maxillary tuberosity, the infratemporal fossa has been thought of by some anatomists as part of a 'masticatory muscle compartment' or 'masticatory space'.[4]

The fossa is bounded anteriorly by the posterior surface of the maxilla and posteriorly by the styloid apparatus, carotid sheath and deep part of the parotid gland. Medially lies the lateral pterygoid plate and the superior constrictor muscle of the pharynx. Laterally lies the ramus of the mandible. The roof is formed by the infratemporal surface of the greater wing of the sphenoid. The infratemporal fossa has no anatomical floor, being continuous with tissue spaces in the neck.

The infratemporal fossa communicates with the temporal fossa deep to the zygomatic arch. It also communicates with the pterygopalatine fossa through the pterygomaxillary fissure (Fig. 1.1). At the base of the cranium, the foramen ovale, the foramen spinosum and the sphenoidal emissary foramen (of Vesalius)

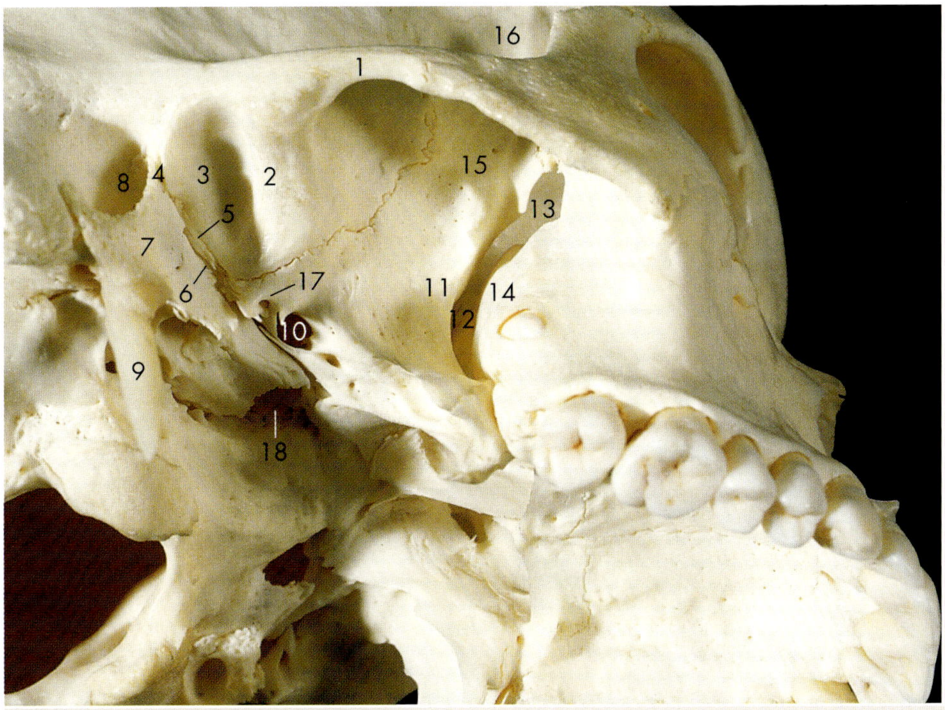

1 Zygomatic arch
2 Articular eminence
3 Mandibular fossa
4 Squamotympanic fissure
5 Petrotympanic fissure
6 Petrosquamous fissure
7 Tympanic plate
8 External acoustic meatus
9 Styloid process
10 Foramen ovale
11 Lateral pterygoid plate
12 Pterygomaxillary fissure leading into pterygopalatine fossa
13 Inferior orbital fissure
14 Posterior surface of maxilla
15 Infratemporal crest of sphenoid bone
16 Greater wing of sphenoid bone
17 Foramen spinosum
18 Foramn lacerum

Figure 1.1 Skull showing osteology related to the infratemporal fossa.

enter the fossa through the sphenoid bone. The foramen lacerum and the petrotympanic, squamotympanic and petrosquamous fissures are also found close to the infratemporal fossa. On the medial surface of the ramus of the mandible is the mandibular foramen.

THE CONTENTS OF THE INFRATEMPORAL FOSSA

The major structures that occupy the infratemporal fossa are:

- The lateral and medial pterygoid muscles
- The mandibular division of the trigeminal nerve
- The chorda tympani branch of the facial nerve
- The otic parasympathetic ganglion
- The maxillary artery and branches
- The pterygoid venous plexus
- The deep 'lobe' of the parotid gland.

The key to understanding the relationships of structures within the infratemporal fossa is the lateral pterygoid muscle. This lies in the roof of the fossa, running anteroposteriorly in a horizontal plane from the region

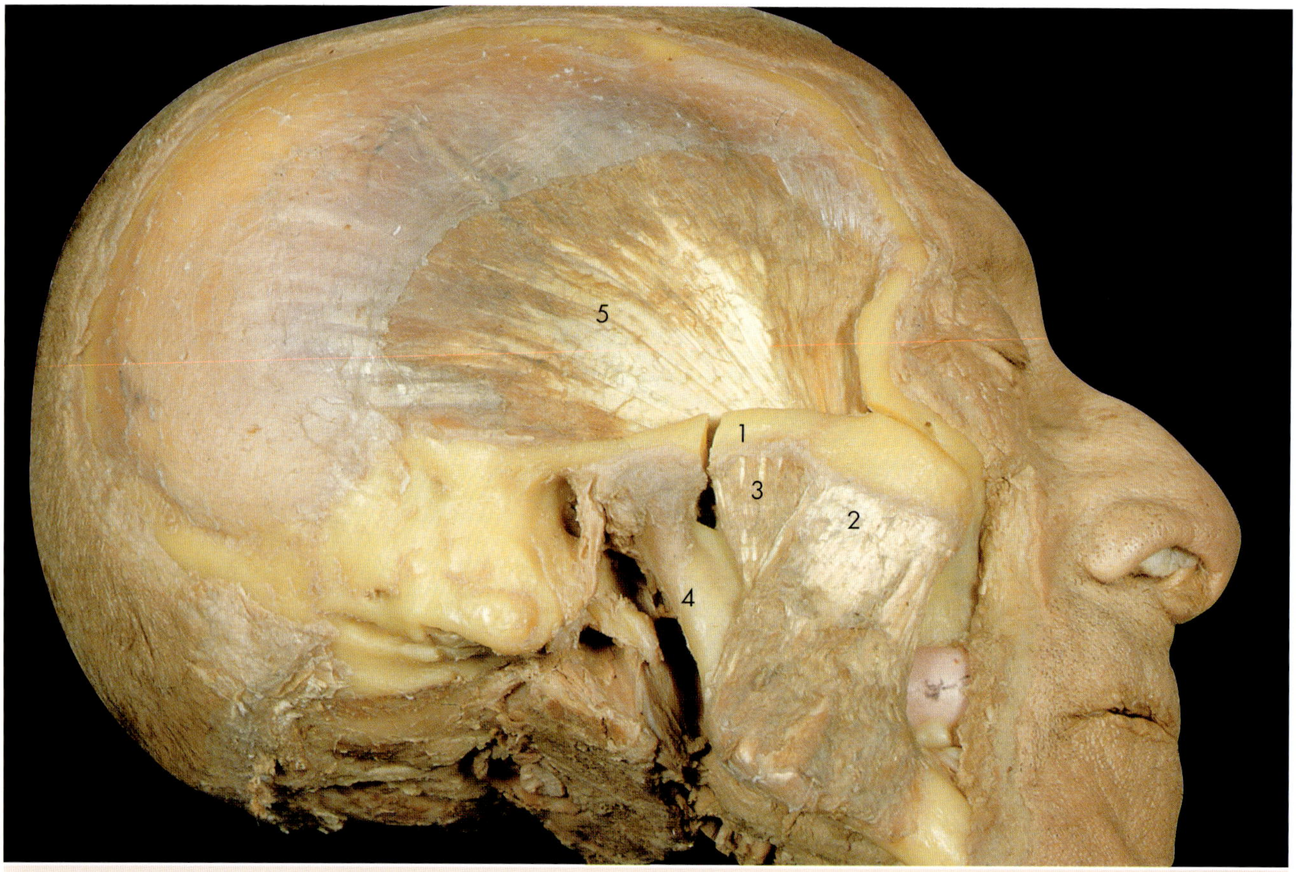

1 Zygomatic arch
2 Masseter muscle (superficial head)
3 Masseter muscle (deep head)
4 Posterior border of ramus of mandible
5 Temporalis muscle

Figure 1.2 The masseter muscle. (Courtesy of Professor S. Standring, GKT School of Biomedical Sciences, London.)

of the pterygoid plates to the mandibular condyle. Deep to the muscle arise the branches of the mandibular nerve and the main origin of the medial pterygoid muscle. The maxillary artery generally passes superficial to the lower head of the lateral pterygoid. The buccal branch of the mandibular nerve passes between the two heads that comprise the lateral pterygoid muscle. Emerging below the inferior border of the muscle are the medial pterygoid muscle and the lingual and inferior alveolar nerves. At the upper border emerge the deep temporal nerves and vessels. Concentrated around and within the lateral pterygoid muscle lies a venous network, the pterygoid venous plexus.

THE MASTICATORY MUSCLES

The four primary masticatory muscles are the masseter, temporalis, lateral and medial pterygoid muscles. Being derived from the first branchial arch, the muscles are supplied by the mandibular nerve. The pterygoid muscles lie within the infratemporal fossa. The masseter muscle arises from the zygomatic arch and is attached to the lateral surface of the ramus of the mandible (Fig. 1.2). The temporalis muscle arises from the floor of the temporal fossa and the overlying temporal fascia, passes behind the zygomatic arch, and is attached to the anterior and medial surface of the coronoid process (Fig. 1.3).

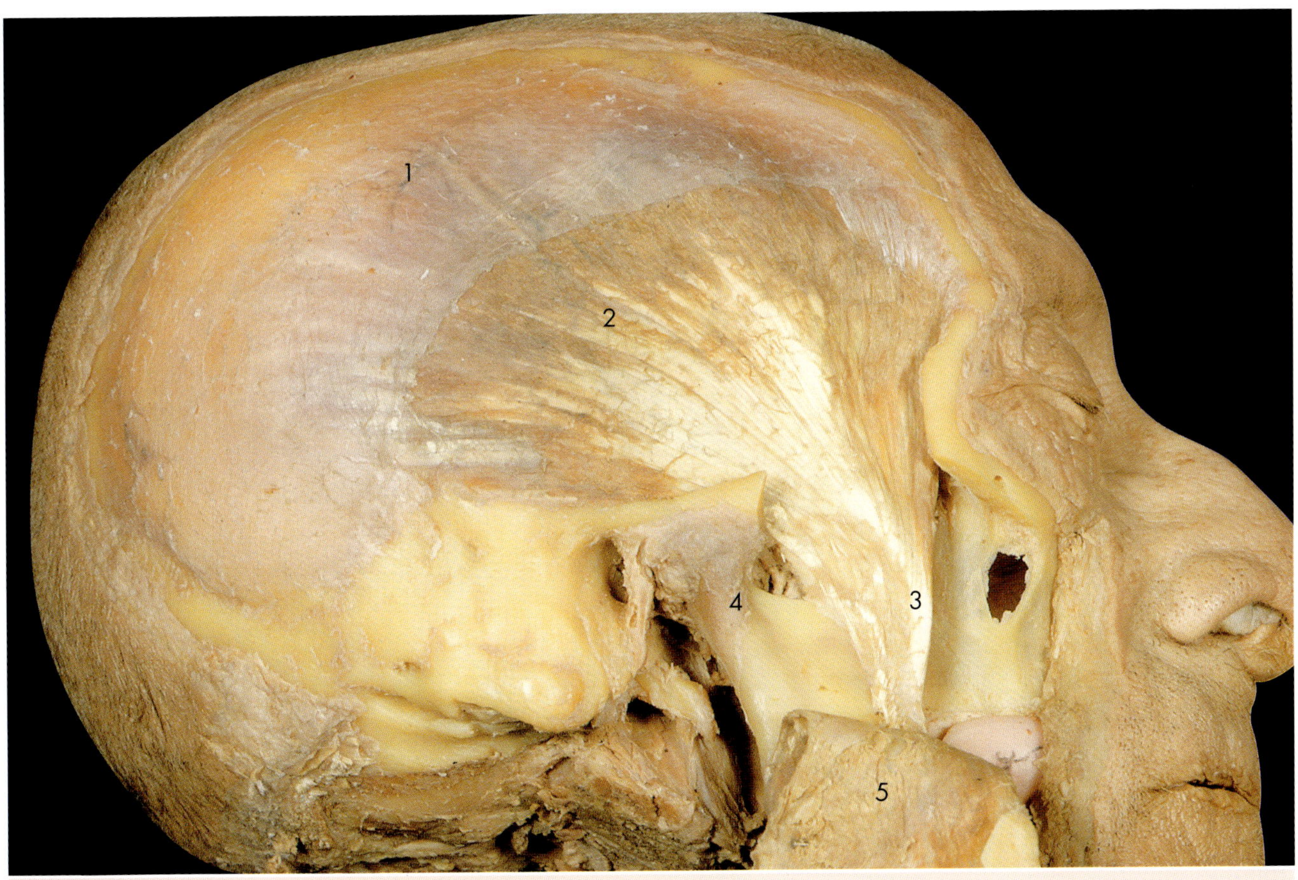

1 Temporal fascia arising from superior temporal line
2 Temporalis muscle arising from inferior temporal line
3 Attachment of temporalis muscle to coronoid process of mandible
4 Mandibular condyle
5 Reflected part of masseter muscle revealed following reflection of the zygomatic arch

Figure 1.3 The temporalis muscle, revealed following reflection of the zygomatic arch. (Courtesy of Professor S. Standring, GKT School of Biomedical Sciences, London.)

THE LATERAL PTERYGOID MUSCLE[4–7] (FIGS 1.4–1.6)

This muscle has two separate and distinct heads.[5,8–10] The larger, lower head is sometimes referred to as the pterygoid head. The smaller, upper head has been termed the infratemporal head. Some anatomists claim that the lateral pterygoid has three heads – the upper head having two slips of muscle.[7]

Attachments

The bulk of the muscle is formed by its lower head. This arises mainly from the lateral surface of the lateral pterygoid plate of the sphenoid bone, although some fibres

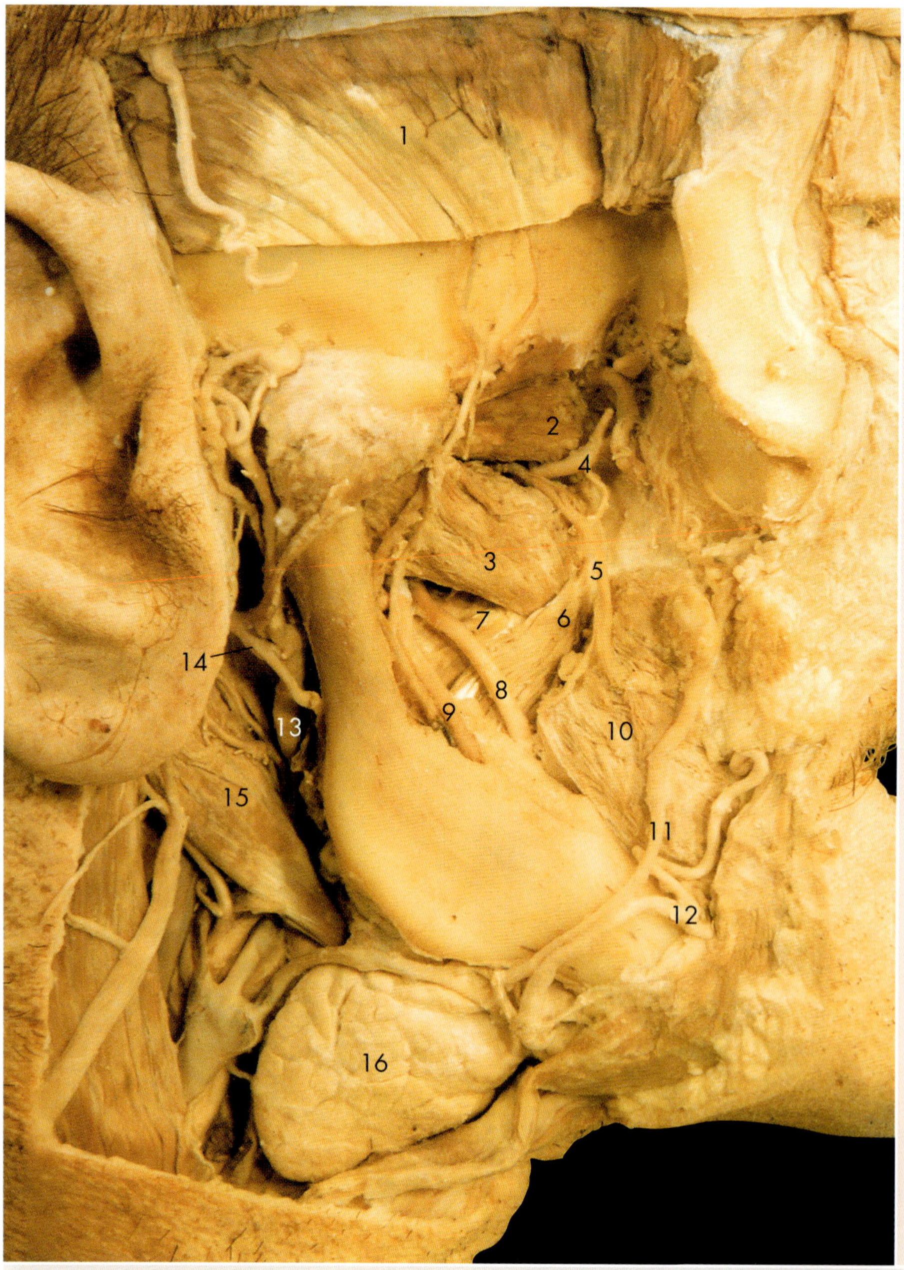

1 Temporalis muscle
2 Lateral pterygoid muscle (upper head)
3 Lateral pterygoid muscle (lower head)
4 Maxillary artery
5 Buccal branch of mandibular nerve
6 Medial pterygoid muscle (superficial head)
7 Medial pterygoid muscle (deep head)
8 Lingual nerve
9 Inferior alveolar nerve and vessel
10 Buccinator muscle
11 Facial vein
12 Facial artery
13 External carotid artery
14 Facial nerve
15 Digastric muscle (posterior belly)
16 Submandibular gland

Figure 1.4 The pterygoid muscles viewed laterally. (Courtesy of Professor S. Standring, GKT School of Biomedical Sciences, London.)

may arise from the maxillary surface of the pterygoid plate. The most superior fibres of the lower head run more horizontally than the more inferior fibres. The length and thickness of the lower head varies considerably from site to site and from individual to individual.[4] The smaller upper head takes origin from the infratemporal surface of the greater wing of the sphenoid (usually medial to the infratemporal crest) and, under cover of the temporalis muscle, runs in a groove-like depression in the infratemporal roof. The two heads converge near the point of insertion (about 1 cm anterior to the neck of the mandibular condyle). The fibres of the upper head are said to insert primarily into the capsule and articular disc of the temporomandibular joint.[11] Indeed,

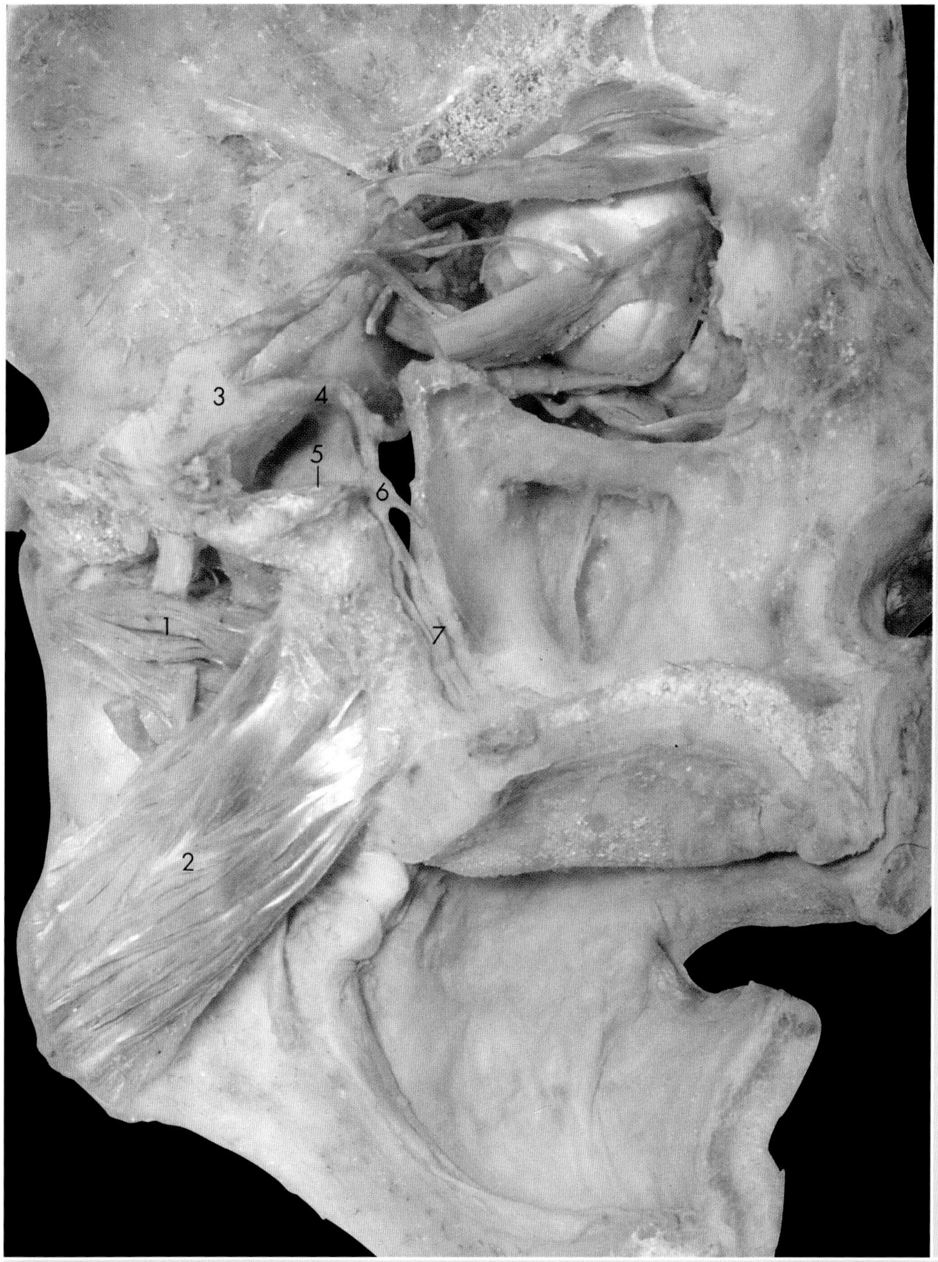

1 Lateral pterygoid muscle
2 Medial pterygoid muscle
3 Trigeminal ganglion
4 Maxillary nerve
5 Nerve of pterygoid canal
6 Pterygopalatine ganglion
7 Palatine nerves

Figure 1.5 The pterygoid muscles viewed medially. (Courtesy of Professor L. Garey, Anatomy Department, Imperial College Medical School, London.)

the articular disc has sometimes been thought of as a (cartilaginous) tendon of the upper head of the lateral pterygoid muscle.[12] The fibres from the lower head insert into the pterygoid fovea on the mandibular condyle.[13–15] Some anatomists have reported that there are no or few muscle fibres inserting into the articular disc of the temporomandibular joint.[16] The attachment of the lateral pterygoid muscle is discussed further on pages 35–36.

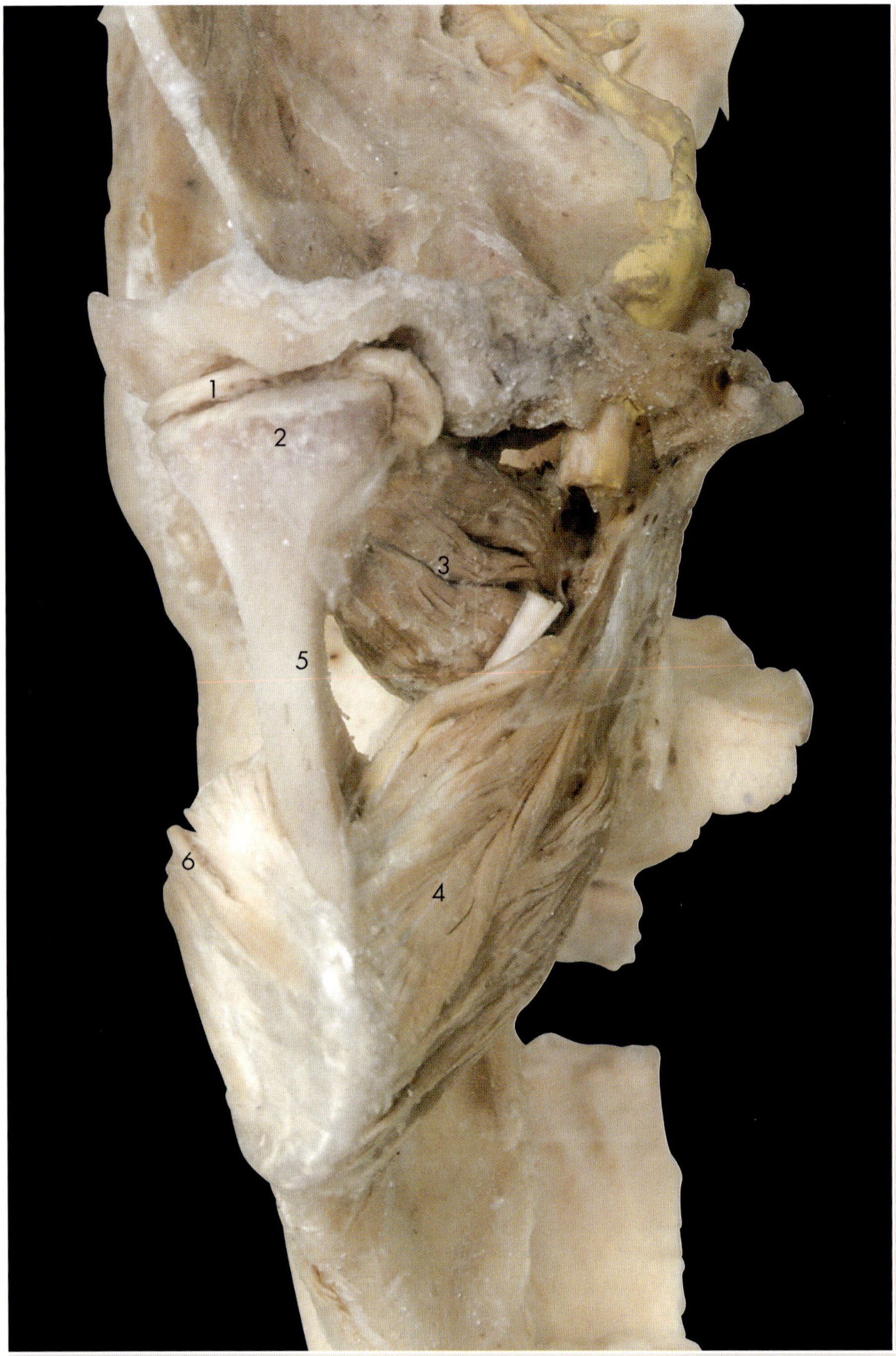

1 Articular disc of temporomandibular joint
2 Mandibular condyle
3 Lateral pterygoid muscle
4 Medial pterygoid muscle
5 Posterior border of ramus of mandible
6 Cut surface of masseter muscle

Figure 1.6 The pterygoid muscles viewed posteriorly. (Courtesy of Professor L. Garey, Anatomy Department, Imperial College Medical School, London.)

Innervation

The nerves to the lateral pterygoid (one for each head) arise from the anterior trunk of the mandibular nerve, deep to the muscle. The upper head and the lateral part of the lower head receive their innervation from a branch of the buccal nerve. However, the medial part of the lower head has a branch arising directly from the anterior trunk of the mandibular division of the trigeminal nerve.[4,17]

Vasculature

The arterial supply is derived from the maxillary artery (pterygoid branches) as it crosses the lateral pterygoid muscle and from the ascending palatine artery (a branch of the facial artery).[4]

Actions

The main action of the muscle is to assist in opening the jaws by pulling the mandibular condyle and the articular disc of the temporomandibular joint forwards and downwards along the posterior slope of the articular eminence. In addition, the muscle is involved in protrusion and in lateral movements of the mandible. The muscle is involved in stabilising the disc–condyle complex. From EMG findings, it has also been reported that the upper head of the lateral pterygoid muscle is inactive during jaw opening and might help to elevate the mandible. Furthermore, this head can restrain backward movements of the articular disc of the temporomandibular joint. EMGs also suggest that the lower head is a synergist of the suprahyoid muscles and the upper head is an antagonist.[6,9] It has been estimated that the upper head of the lateral pterygoid muscle can exert a tensile force of about 40 N and the lower head a greater force of approximately 130 N.[18]

THE MEDIAL PTERYGOID MUSCLE[4,6]

Attachments

This muscle is the deepest of the four muscles of mastication. It consists of two heads. The bulk of the muscle arises as a deep head from the medial surface of the lateral pterygoid plate. Thus, the lateral pterygoid plate of the sphenoid bone gives rise to both pterygoid muscles. A common mistake is the belief that the medial pterygoid muscle arises from the medial pterygoid plate. However, the medial pterygoid plate gives origin only to a small proportion of the superior constrictor muscle of the pharynx. The smaller, superficial head of the medial pterygoid muscle (sometimes called the tuberal head) originates from the maxillary tuberosity and the neighbouring part of the palatine bone (pyramidal process).[4,6] From these sites, the fibres pass downwards and backwards to insert into the roughened surface of the angle of the mandible on its medial aspect.

In terms of the muscle's dimension, the medial pterygoid muscle has on average a cross-section of 1.5 cm^2, being larger in males than females.[4] In relation to the long axis of the ramus of the mandible, the muscle shows an obliquity of approximately 30° on the medial side. At its origin on the pterygoid plate, the width of the muscle can vary between 20 mm and 35 mm. At its insertion, the muscle varies between 10 mm and 18 mm. The length of the main deep head of the muscle varies from 43 mm to 50 mm, although the most anterior fibres are usually shorter (32–50 mm). Within the muscle can be discerned up to six tendinous plates. The intramuscular tendinous plates form a multipennate structure.[6,18]

Following analysis by CT scanning, the maximum cross-sectional area of the medial pterygoid was centred at the level of the mandibular foramen. The maximum cross-sectional area was also found to be highly correlated with volume.[19] Other work [20] showed that the cross-sectional area of the medial pterygoid significantly decreases with age, with a greater decrease in edentulous subjects. Furthermore, there is a significant decrease in the density of the muscle with increasing age (previously interpreted to indicate an increase in fat and fibrous tissue).

Innervation

The nerve to the medial pterygoid muscle arises from the mandibular nerve (deep to the lateral pterygoid muscle), before the nerve divides into anterior and posterior trunks.

Vasculature

Like the lateral pterygoid muscle, the medial pterygoid derives its arterial supply from the maxillary artery.

Actions

The medial pterygoid muscle is an elevator of the mandible. It assists in lateral and protrusive movements. The medial pterygoid muscle is synergistic to the masseter muscle. In addition, the medial pterygoid muscle and the masseter muscle together provide a 'sling' to support the angle of the mandible.[4] It has been estimated that the elevating force provided by this pterygoid masseter sling can be as great as 420 N.[18] It has been reported[21] that similar bite-force efficiencies for the medial pterygoid muscle can be found in persons with disparate facial features.

THE PTERYGOID HIATUS[4,14]

This is the space bounded superiorly by the inferior margin of the lower head of the lateral pterygoid muscle and inferiorly by the posterior margin of the medial pterygoid muscle. In this space run the lingual and inferior alveolar nerves, the first part of the maxillary artery, part of the pterygoid venous plexus and the sphenomandibular ligament. Its dimensions are of some importance to maxillofacial surgeons, the superior margin being approximately 14–21 mm, the posterior border 14–31 mm and the frontal dimension 3–8 mm.

THE SPHENOMANDIBULAR MUSCLE[22]

It has been reported in the scientific and medical literature that a previously unknown muscle, termed the 'sphenomandibular muscle', exists within the infratemporal fossa. It has been proposed that this muscle is a fifth member of the 'muscles of mastication'. It appears to take origin from the greater wing of the sphenoid bone (at the base of the temporal fossa) and extends downwards and backwards to be inserted onto the inner and anterior aspect of the mandibular coronoid process and the anterior edge of the mandibular ramus. It would appear from its orientation that the muscle aids elevation (and perhaps protrusion) of the mandible. An alternative explanation for the muscle is that it is a previously unidentified component of a known muscle. Indeed, it may therefore be linked to the medial pterygoid muscle or be considered part of the temporalis muscle.

THE MANDIBULAR NERVE (FIGS 1.7–1.9)

This is the largest division of the trigeminal nerve and is the only one to contain motor as well as sensory fibres. Developmentally, it is the nerve of the first branchial arch and is thus responsible for supplying structures derived from it. Its sensory fibres supply the mandibular teeth and their supporting structures, the mucosa of the anterior two-thirds of the tongue and the floor of the mouth, the skin of the lower part of the face (including the lower lip) and parts of the temporal region and auricle. Its motor fibres supply the four 'muscles of mastication' and the mylohyoid, anterior belly of digastric, tensor veli palatini and tensor tympani muscles.

The mandibular nerve is formed in the infratemporal fossa by the union of the sensory and motor roots immediately after they leave the skull at the foramen ovale. Within the foramen ovale, the motor root (or roots) lie posteromedially to the sensory root and these roots are accompanied by emissary veins, the lesser petrosal nerve (from the glossopharyngeal nerve) going to the otic ganglion and by the accessory meningeal artery. As the mandibular nerve leaves the foramen ovale, it lies on the tensor veli palatini muscle and is covered laterally by the upper head of the lateral pterygoid muscle (slightly anterior to the neck of the mandible). After a short course, the nerve divides into a smaller anterior trunk and a larger posterior trunk. Before this division, the main trunk gives off two branches – the meningeal branch and the nerve to medial pterygoid. The anterior trunk of the mandibular nerve is mainly motor, the posterior trunk mainly sensory.

BRANCHES

- Meningeal branch (nervus spinosus)
- Nerve to medial pterygoid
- Anterior trunk:
    - Masseteric nerve
    - Deep temporal nerves
    - Nerve to lateral pterygoid
    - Buccal nerve
- Posterior trunk:
    - Auriculotemporal nerve
    - Lingual nerve
    - Inferior alveolar nerve

The meningeal branch of the mandibular nerve (nervus spinosus)

This arises from the main trunk of the mandibular nerve. It is a 'recurrent nerve' as it runs back into the middle cranial fossa through the foramen spinosum. It supplies the dura mater lining the middle and anterior cranial fossae and the mucosa of the mastoid antrum and mastoid air cells.

The nerve to the medial pterygoid muscle (Figs 1.8, 1.9, 1.12)

This enters the deep surface of the muscle and also gives slender branches that pass uninterrupted through the otic ganglion to supply the tensor tympani and tensor veli palatini muscles.

The masseteric nerve

This is usually the first branch of the anterior trunk of the mandibular nerve. It passes above the upper border of the lateral pterygoid muscle (accompanying the posterior deep temporal nerve) and then crosses the mandibular notch (between the condylar and coronoid processes) to be distributed into the masseter muscle. It also gives an articular branch to the temporomandibular joint. The nerve enters the masseter muscle as two branches. The upper branch is smaller and runs to the deeper layers of the muscle. The larger, lower trunk innervates the more superficial layers of the masseter muscle.

The deep temporal nerves (Fig. 1.5)

These nerves also pass above the lateral pterygoid muscle. Anatomists have provided varying descriptions for them. Anterior, middle and posterior deep temporal nerves may be recognised.

The nerve to the lateral pterygoid muscle[17]

This may arise separately or may run with the buccal nerve before entering the deep surface of the lateral pterygoid muscle.

The buccal branch of the mandibular nerve[23]

This is the only sensory branch of the anterior trunk of the mandibular nerve. On emerging between the upper and lower heads of the lateral pterygoid muscle (Figs 1.4, 1.7), it passes downwards and forwards across the lower head to contact the medial surface of the temporalis muscle as it inserts onto the coronoid process of the mandible (Figs 1.4, 1.7–1.9). It then clears the ramus of the mandible to lie on the lateral surface of the buccinator muscle in the cheek. At this point, it is close to the retromolar fossa of the mandible. It now gives branches to the skin of the cheek before piercing the buccinator to supply its lining mucosa, the buccal sulcus and the buccal gingiva related to the mandibular molar and premolar teeth. It may also carry secretomotor fibres to minor salivary glands in the buccal mucosa, these being postganglionic fibres from the otic ganglion. The buccal branch of the mandibular nerve may be seen to 'anastomose' with the buccal branches of the facial nerve.

The auriculotemporal nerve[24–26]

This is the first branch of the posterior trunk of the mandibular nerve. It is essentially sensory but it also distributes autonomic fibres to the parotid gland derived from the otic ganglion. It usually arises as two roots (approx. 75% of cases) that encircle the middle meningeal artery and unite behind the artery (Figs 1.8, 1.12). The nerve then runs backwards under the lateral pterygoid muscle to lie beneath the mandibular condyle (between the condyle and the sphenomandibular ligament) (Fig.1.9). On entering the parotid region, it turns to emerge superficially between the temporomandibular joint and the external acoustic meatus (Fig. 1.8). From the upper surface of the parotid gland, the auriculotemporal nerve ascends on the side of the head with the superficial temporal vessels (Figs 8.3, 8.19), passing over the posterior part of the zygomatic arch. It gives several branches along its course:

- *Ganglionic branches* which communicate with the otic ganglion.
- *Articular branches* which enter the posterior part of the temporomandibular joint; these carry proprioceptive information important in mastication.
- *Parotid branches* which convey parasympathetic secretomotor fibres and sympathetic fibres to the parotid gland; these fibres are related to the otic ganglion. Sensory fibres from the auriculotemporal nerve supply the gland (with the exception of the capsule, which is innervated by the great auricular nerve).
- *Auricular branches* (usually two) which supply the tragus and crus of the helix of the auricle, part of the external acoustic meatus, and the outer (lateral) surface of the tympanic membrane.
- *Superficial temporal branches* which are cutaneous nerves supplying part of the skin of the temple.

The lingual nerve (Figs 1.4, 1.7–1.9, 1.12)
This is the second branch of the posterior trunk of the mandibular nerve. It is essentially a sensory nerve but, following union with the chorda tympani branch of the facial nerve, it also contains parasympathetic fibres. Initially, the nerve lies on the tensor veli palatini muscle deep to the lateral pterygoid muscle. Here, the chorda tympani nerve (which has entered the infratemporal fossa via the petrotympanic fissure and passed over the spine of the sphenoid bone) joins the posterior surface

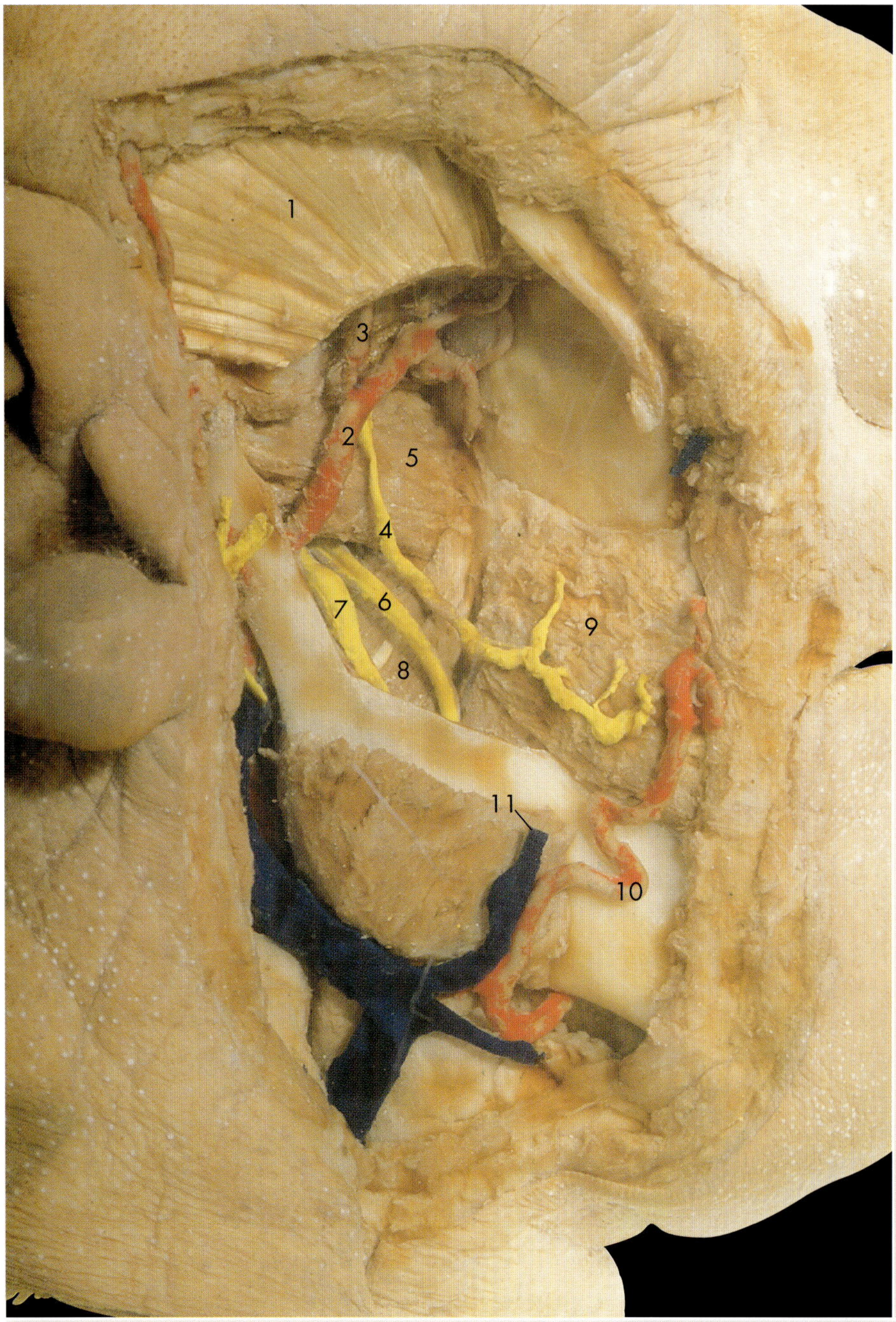

1 Temporalis muscle
2 Maxillary artery
3 Deep temporal artery
4 Buccal branch of mandibular nerve
5 Lateral pterygoid muscle (lower head)
6 Lingual nerve
7 Inferior alveolar nerve
8 Medial pterygoid muscle
9 Buccinator muscle
10 Facial artery
11 Cut end of facial vein

Figure 1.7 The mandibular nerve viewed buccally. (Courtesy of Professor L. Garey, Anatomy Department, Imperial College Medical School, London.)

of the lingual nerve (Figs 1.8, 1.12). Emerging from the inferior border of the lateral pterygoid muscle, the lingual nerve curves downwards and forwards in the space between the ramus of the mandible and the medial pterygoid muscle (pterygomandibular space) (Fig. 1.9). At this level, it lies anterior to, and slightly deeper than, the inferior alveolar nerve. The lingual nerve then leaves the infratemporal fossa, passing downwards and forwards to lie close to the lingual alveolar plate of the mandibular third molar. Before curving forwards into

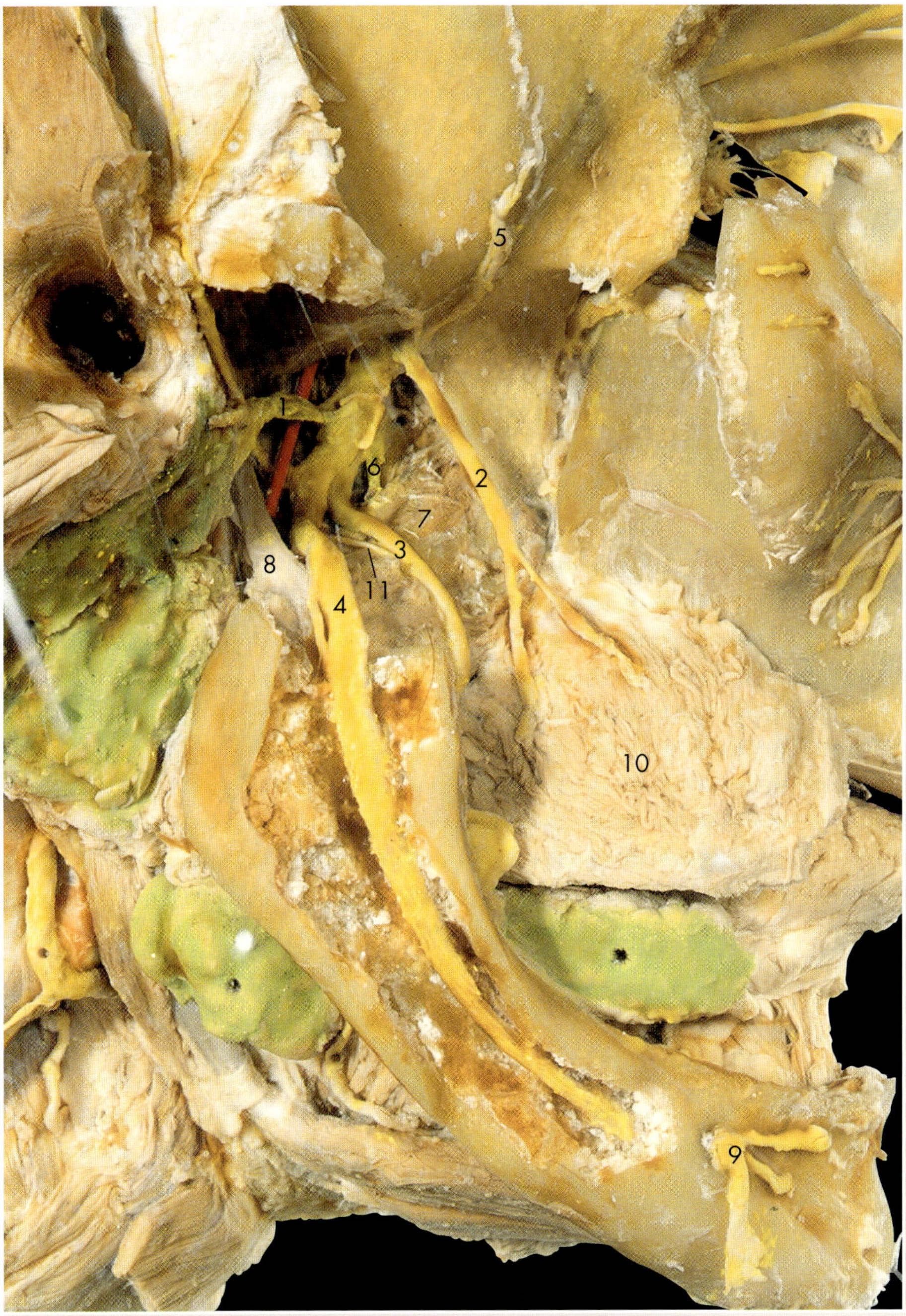

1 Roots of auriculotemporal nerve surrounding middle meningeal artery
2 Buccal branch of mandibular nerve
3 Lingual nerve
4 Inferior alveolar nerve with mylohyoid branch
5 Deep temporal nerve
6 Nerve to medial pterygoid muscle
7 Medial pterygoid muscle
8 Sphenomandibular ligament
9 Mental nerve
10 Buccinator muscle
11 Chorda tympani nerve

Figure 1.8 The mandibular nerve with the lateral pterygoid muscle removed. (Courtesy of Professor C. Dean, Department of Anatomy and Developmental Biology, University College London.)

the tongue, the nerve is found above the origin of the mylohyoid muscle and lateral to the hyoglossus muscle.

The close relationship of the lingual nerve to the third molar tooth makes the nerve susceptible to damage during removal of the tooth. In addition, in about one in seven cases, the lingual nerve is actually located above the lingual bony plate in the third molar region and is liable to damage during surgery.[27–29]

The lingual nerve supplies the mucosa covering the anterior two-thirds of the dorsum of the tongue, the ventral surface of the tongue, the floor of the mouth and the lingual gingivae of the mandibular teeth. The chorda tympani fibres travelling with the lingual nerve are of two types: sensory and parasympathetic. The sensory fibres are associated with taste for the anterior two-thirds of the dorsum of the tongue. The parasympathetic fibres are preganglionic fibres that pass to the submandibular ganglion. Postganglionic fibres are distributed to the submandibular and sublingual salivary glands.

The chorda tympani branch of the facial nerve (Figs 1.8, 1.12, 2.19)

This is distributed through the lingual nerve and has two types of fibres. Sensory fibres are associated with taste to the anterior two-thirds of the tongue. Parasympathetic fibres are preganglionic to the submandibular

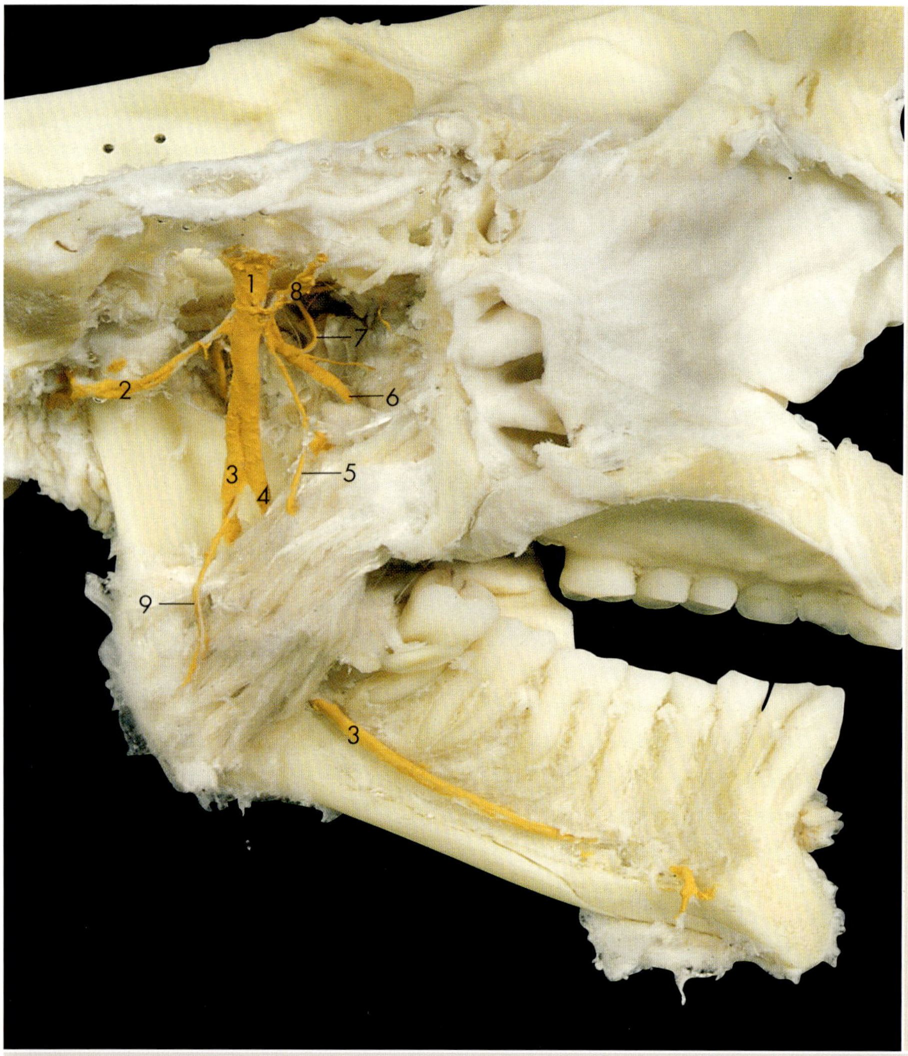

1 Mandibular nerve trunk
2 Auriculotemporal nerve
3 Inferior alveolar nerve
4 Lingual nerve
5 Nerve to medial pterygoid muscle
6 Buccal branch of mandibular nerve
7 Nerve to lateral pterygoid muscle
8 Otic ganglion
9 Mylohyoid nerve

Figure 1.9 The mandibular nerve viewed medially. (Courtesy of Professor C. Dean, Department of Anatomy and Developmental Biology, University College London.)

ganglion (Fig. 1.12). Postganglionic fibres are secretomotor to the submandibular and sublingual glands.

The inferior alveolar nerve (Figs 1.4, 1.7–1.9, 1.12) This is the largest branch of the mandibular division of the trigeminal nerve. It is the third branch of the posterior trunk of the mandibular nerve. Although it is essentially a sensory nerve, it also carries motor fibres which are given off as the mylohyoid nerve. Indeed, the mylohyoid nerve contains all the motor fibres of the posterior trunk of the mandibular nerve. The inferior alveolar nerve descends deep to the lateral pterygoid muscle, posterior to the lingual nerve in the pterygoid hiatus. Here, it is crossed by the maxillary artery. On emerging at the inferior border of the muscle, it passes between the sphenomandibular ligament and the ramus of the mandible to enter the mandibular foramen. It is accompanied in its course by inferior alveolar blood vessels.

The mylohyoid nerve is given off just before the mandibular foramen (Figs 1.8, 2.23). It pierces the sphenomandibular ligament and runs in a groove (the mylohyoid groove) which lies immediately below the mandibular foramen. The mylohyoid nerve supplies the mylohyoid muscle and the anterior belly of the digastric. The mylohyoid nerve may also contain sensory fibres that supply the skin of the chin and medial parts of the submandibular triangle in the suprahyoid region.

The main distribution of the inferior alveolar nerve is to the mandibular teeth and their supporting structures, there being molar and incisive branches. The mental nerve is a cutaneous branch that supplies the skin of the chin and the lower lip. It arises within the mandible in the premolar region, but soon exits onto the face via the mental foramen.

THE OTIC GANGLION (Figs 1.9, 1.10, 1.12)

This parasympathetic ganglion lies immediately below the foramen ovale on the medial surface of the main trunk of the mandibular nerve. It is concerned primarily with supplying the parotid gland (Fig. 1.10). Like other parasympathetic ganglia in the head, three types of fibres are associated with it: parasympathetic, sympathetic and sensory fibres. However, only the parasympathetic fibres synapse in the ganglion. The preganglionic parasympathetic fibres originate from the inferior salivatory nucleus in the brainstem. The fibres pass out in the glossopharyngeal nerve, appearing as the lesser (superficial) petrosal nerve from the tympanic plexus in the middle ear cavity. The lesser petrosal nerve reaches the otic ganglion by a complex course. Passing through the petrous part of the temporal bone, the lesser petrosal nerve comes to lie in the floor of the middle cranial fossa. Here, it is lateral to the

Figure 1.10 The otic parasympathetic ganglion and innervation of the parotid gland.

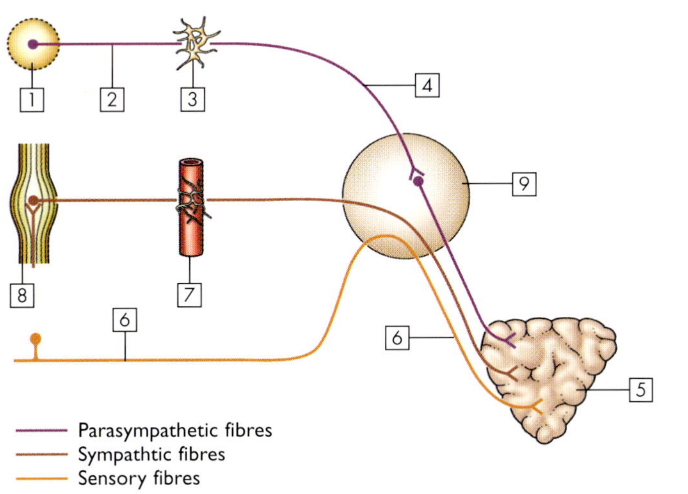

1 Inferior salivatory nucleus in brainstem
2 Tympanic branch of glossopharyngeal nerve
3 Tympanic plexus
4 Lesser petrosal nerve
5 Parotid gland
6 Auriculotemporal branch of mandibular nerve
7 Middle meningeal artery
8 Superior cervical ganglion
9 Otic ganglion

greater (superficial) petrosal branch of the facial nerve. The lesser petrosal nerve usually enters the infratemporal fossa through the foramen ovale to join the otic ganglion. On occasion, the lesser petrosal nerve passes through the sphenopetrosal fissure. The sympathetic root of the otic ganglion is derived from postganglionic fibres from the superior cervical ganglion. They are said to reach the otic ganglion from the plexus on the middle meningeal artery. Other descriptions have it that the sympathetic root arises from the deep petrosal nerve or directly from the internal carotid plexus. The sensory root is derived from the auriculotemporal nerve. The postganglionic parasympathetic fibres (with sympathetic and sensory components) reach the parotid gland by way of the auriculotemporal nerve. Parasympathetic fibres may also innervate the minor salivary glands in the cheek, passing with the buccal branch of the mandibular nerve.

The innervation of tensor veli palatini and tensor tympani is derived from the nerve to the medial pterygoid by a branch that passes through the otic ganglion.

THE MAXILLARY ARTERY

(Figs 1.4, 1.7, 1.11–1.13)

The maxillary artery is a terminal branch of the external carotid artery. It arises within the parotid gland at the level of the neck of the condyle of the mandible. It enters the infratemporal fossa between the deep surface of the condyle and the sphenomandibular ligament. At this point, it lies below the auriculotemporal nerve and above the maxillary vein. The artery can be quite firmly adherent to the capsule of the temporomandibular joint. In the infratemporal fossa, it is closely related to the lateral pterygoid muscle. Initially, it lies near the inferior border of the muscle, crossing the inferior alveolar nerve. Its subsequent course is variable, although it usually passes superficial to the lower head of the lateral pterygoid[24,30–35] before entering the pterygopalatine fossa through the pterygomaxillary fissure.

The maxillary artery has many branches. It is convenient to subdivide the artery into three parts: before the lateral pterygoid muscle (first or (retro)mandibular part), on the lateral pterygoid muscle (second or pterygoid part), and in the pterygopalatine fossa (third or pterygopalatine part) (see page 52).

The relationship of the maxillary artery to the lateral pterygoid muscle is variable, but it runs superficial to the muscle in nearly 60% of cases.[34] However, there can be asymmetry between the right and left infratemporal fossae and there appears also to be ethnic differences. In Japanese, for example, the maxillary artery runs superficial to the lateral pterygoid muscle in over 90% of persons, a much higher percentage than in Western populations.[35] In most cases, where the maxillary artery runs superficial to the lower head of the lateral pterygoid, the artery passes lateral to the inferior alveolar, lingual and buccal nerves (in 37% of persons). In 16% of cases, only the buccal nerve crosses the artery laterally

Figure 1.11 The maxillary artery.

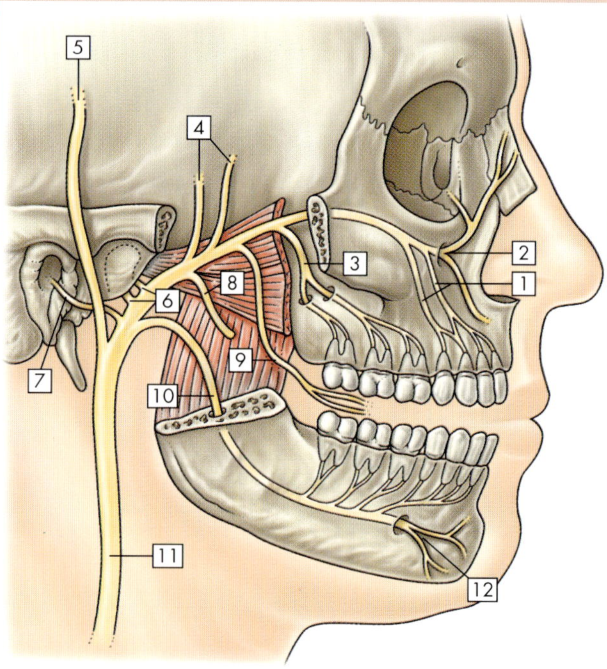

1 Anterior and middle superior alveolar arteries
2 Infra-orbital artery
3 Posterior superior alveolar artery
4 Deep temporal arteries
5 Superficial temporal artery
6 Meningeal arteries
7 Deep auricular artery
8 Lingual branch
9 Buccal artery
10 Inferior alveolar artery
11 External carotid artery
12 Mental artery

and in about 5% of cases the artery passes deep to all the branches of the mandibular nerve.[34]

The first part of the maxillary artery has five branches and all enter bone. The first branch is the deep auricular artery, supplying the skin of the external acoustic meatus and part of the tympanic membrane. A small branch contributes to the arterial supply of the temporomandibular joint. The second branch, the anterior tympanic artery, passes through the petrotympanic fissure to supply part of the lining of the middle ear. This is the companion artery to the chorda tympanic nerve. The middle meningeal artery is the main source of blood to the meninges and to the bones of the vault of the skull. The artery may arise either directly from the first part of the maxillary artery or from a common trunk with the inferior alveolar artery.[4,30] When the maxillary artery lies superficial to the lateral pterygoid muscle, the middle meningeal artery is usually the first branch of the maxillary artery. However, when the maxillary artery takes a deep course in relation to the muscle it is not usually the first branch.[24] The middle meningeal artery ascends between the two roots of the auriculotemporal nerve and leaves the infratemporal fossa through the foramen spinosum (Fig. 1.12). An accessory meningeal artery runs through the foramen ovale into the middle cranial fossa. This artery can arise directly from the maxillary artery or as a branch of the middle meningeal artery (Fig. 2.24b).[4,36] In its course in the infratemporal fossa, it is closely related to the tensor and levator veli palatini muscles and usually runs deep to the mandibular nerve. Although the accessory meningeal artery runs intracranially, its blood is mainly distributed extracranially to the pterygoid muscles, tensor veli palatini, the otic ganglion and to branches of the mandibular nerve.

The inferior alveolar artery accompanies the inferior alveolar nerve and has a similar distribution (Fig. 1.13). It very occasionally arises directly from the external carotid artery. Immediately before the inferior alveolar artery enters the mandible (at the mandibular foramen), it gives off a mylohyoid branch. There are considerable variations in the patterns of entry of the inferior alveolar artery into the mandibular foramen.[37] In the mandibular canal, the inferior alveolar artery usually runs lateral to the inferior alveolar nerve. The artery gives off branches supplying the cheek teeth before terminating in mental and incisive branches. The mental artery passes through the mental foramen onto the face to supply the lower lip, the chin and the labial mucosa related to the anterior teeth. The incisive branch continues along the incisive canal to supply the anterior teeth.

The second part of the maxillary artery also has five branches, but they differ from those of the first part in not entering bone. Muscular branches include deep temporal arteries (anterior, middle and posterior branches), pterygoid arteries and masseteric arteries. The deep temporal arteries pass between the temporalis muscle and the pericranium, producing shallow grooves in the bone. The masseteric arteries pass through the mandibular notch to enter the muscle. They can also supply the temporomandibular joint. A buccal artery accompanies the buccal nerve to supply structures in the cheek. A small lingual branch may be given off to accompany the lingual nerve and supply structures in the floor of the mouth.

Detailed knowledge of the blood supply to the human temporalis muscle is required clinically for successful flap operations. This supply is derived from three main arteries: anterior deep temporal (supplying about 20% of the muscle anteriorly), posterior deep temporal (supplying about 40% of the muscle in the posterior region) and the middle temporal arteries (supplying about just under 40% of the muscle in the middle region). A venous network accompanies the arteries, and double veins pairing one artery is a common finding.[38]

THE PTERYGOID VENOUS PLEXUS[39]

This is situated around, and within, the lateral pterygoid muscle and it surrounds the maxillary artery. Its tributaries correspond to the various branches of the maxillary artery (the plexus receives blood from the pterygoid muscles, the deep temporal veins, the middle meningeal veins and from parotid veins) (Fig. 1.14). Although it is sometimes difficult to demonstrate in the cadaver, it is very prominent in life (although the density varies considerably from individual to individual). The plexus allows for the rapid take-up of blood from regions around the infratemporal fossa.

The plexus communicates with the cavernous sinus, the facial vein, the inferior ophthalmic vein and the pharyngeal plexus. The connections with the cavernous sinus are via emissary veins passing through the foramen ovale, foramen lacerum and, where present, the emissary sphenoidal foramen. The communication with the facial vein is via the deep facial vein which accompanies the buccal nerve. The inferior ophthalmic vein communicates with the pterygoid plexus through a branch passing through the inferior orbital fissure.

The pterygoid venous plexus chiefly drains posteriorly into the maxillary vein. The maxillary vein runs with the first part of the maxillary artery, passing deep to the neck of the condyle of the mandible to enter the parotid gland. Here, it joins the superficial temporal vein to form the retromandibular vein.

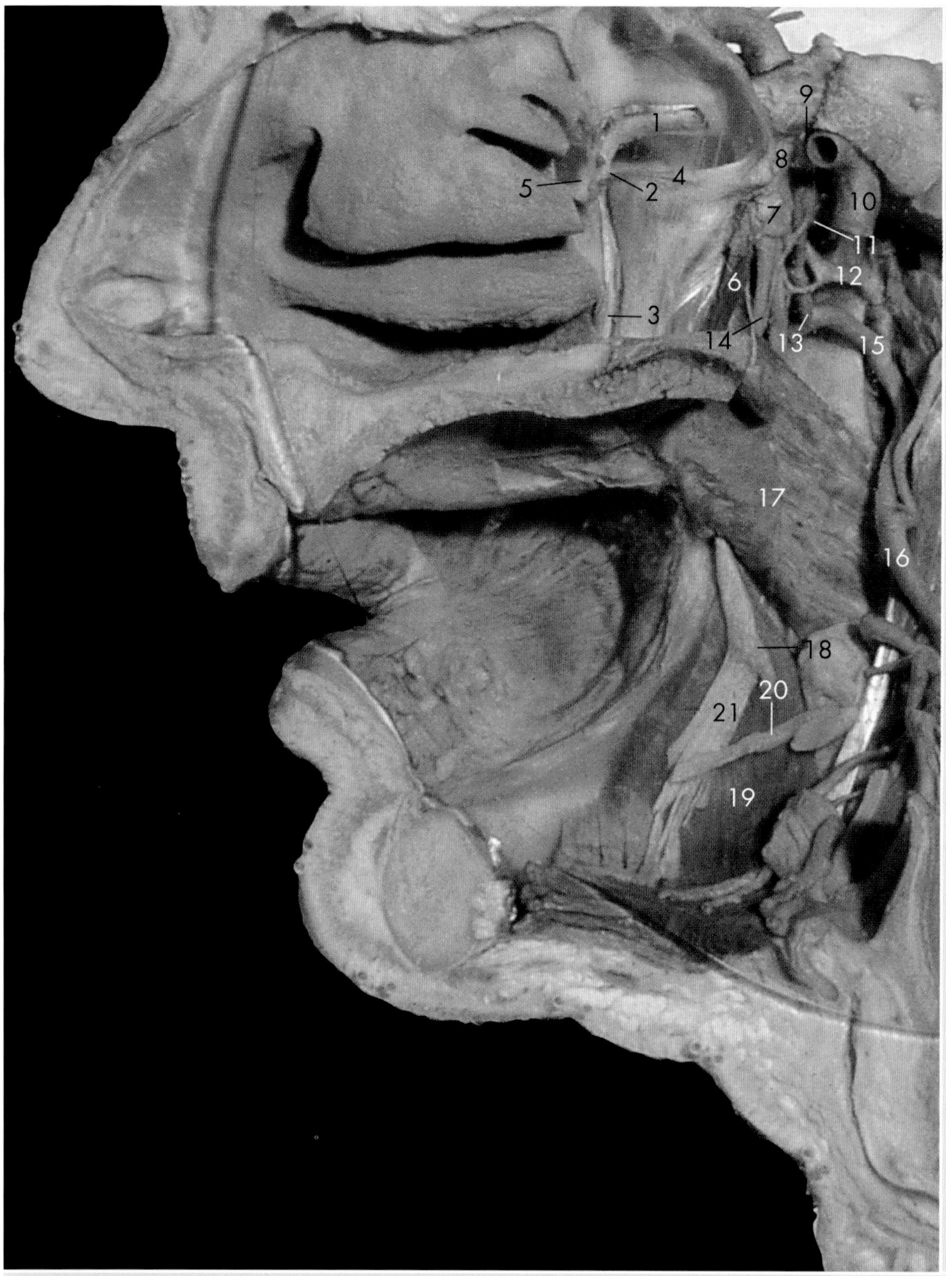

1 Maxillary nerve
2 Pterygopalatine ganglion
3 Palatine nerves
4 Nerve of pterygoid canal
5 Sphenopalatine artery at sphenopalatine foramen
6 Nerve to medial pterygoid muscle
7 Otic ganglion
8 Mandibular nerve
9 Greater petrosal nerve
10 Internal carotid artery
11 Chorda tympani nerve
12 Auriculotemporal nerve
13 Middle meningeal artery
14 Inferior alveolar nerve
15 Maxillary artery
16 External carotid artery
17 Medial pterygoid muscle
18 Submandibular ganglion
19 Mylohyoid muscle
20 Submandibular duct
21 Lingual nerve

Figure 1.12 Dissection of maxillary artery showing middle meningeal artery. (Courtesy of the Royal College of Surgeons of England.)

OTHER FEATURES OF THE INFRATEMPORAL FOSSA

In addition to the major contents described above, the infratemporal fossa also contains the sphenomandibular ligament, the tensor veli palatini muscle, the insertion of the temporalis muscle on to the coronoid process of the mandible, the maxillary nerve as it passes from the pterygopalatine fossa into the inferior orbital fissure, the posterior superior alveolar nerve(s), a loop of the facial artery (together with its ascending palatine and tonsillar branches) and the deep 'lobe' of the parotid gland (see Chapter 8).

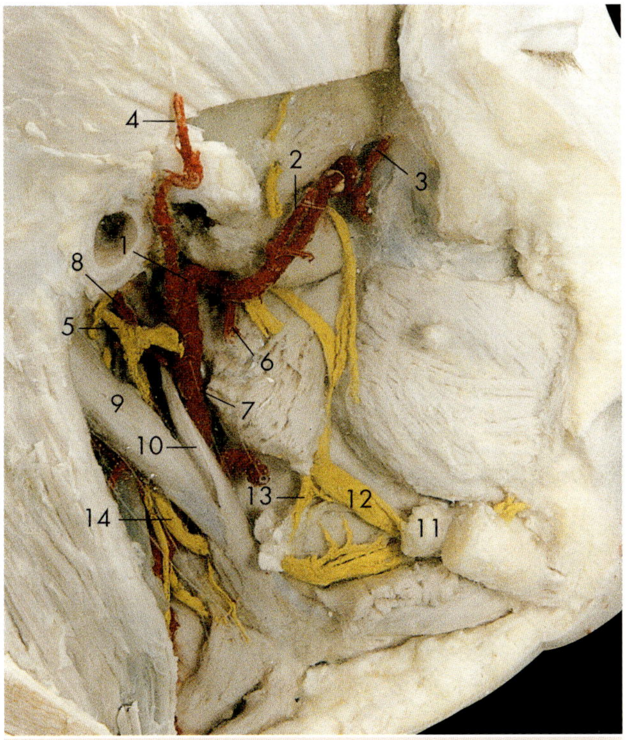

1 First (mandibular) part of maxillary artery
2 Second (pterygoid) part of maxillary artery, in this dissection passing superficial to lower head of lateral pterygoid muscle
3 Third (pterygopalatine) part of maxillary artery
4 Superficial temporal artery
5 Facial nerve
6 Inferior alveolar artery (cut end)
7 External carotid artery
8 Posterior auricular artery
9 Posterior belly of digastic muscle
10 Stylohyoid muscle
11 Sublingual gland
12 Lingual nerve
13 Submandibular ganglion
14 Hypoglossal nerve

Figure 1.13a The maxillary artery in the infratemporal fossa. (Courtesy of Professor S. Standring, GKT School of Biomedical Sciences, London.)

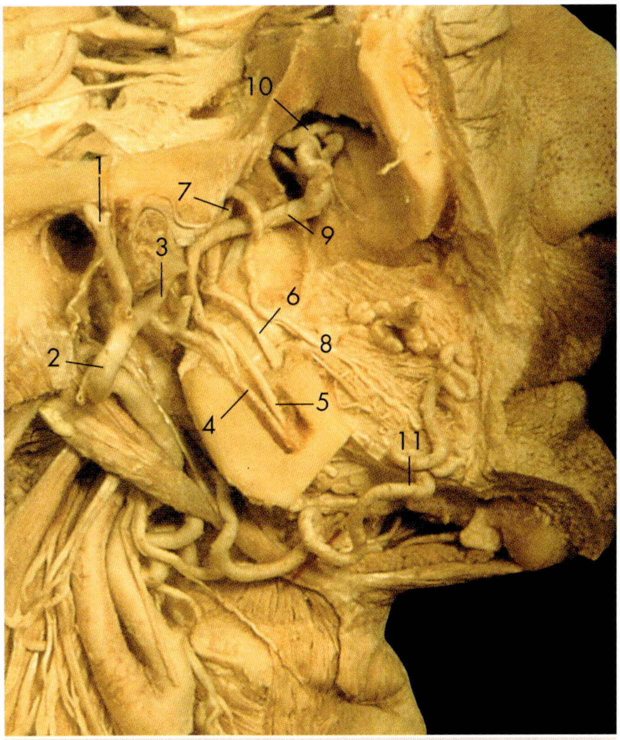

1 Superficial temporal artery
2 External carotid artery
3 First (mandibular) part of maxillary artery
4 Inferior alveolar artery
5 Inferior alveolar nerve (with mylohyoid nerve)
6 Lingual nerve
7 Middle meningeal artery
8 Buccal branch of mandibular nerve
9 Second (pterygoid) part of maxillary artery
10 Third (pterygopalatine) part of maxillary artery
11 Facial artery

Figure 1.13b Maxillary artery in the infratemporal following removal of the pterygoid muscles. (Courtesy of the Royal College of Surgeons of England.)

Figure 1.14 The veins of the neck.

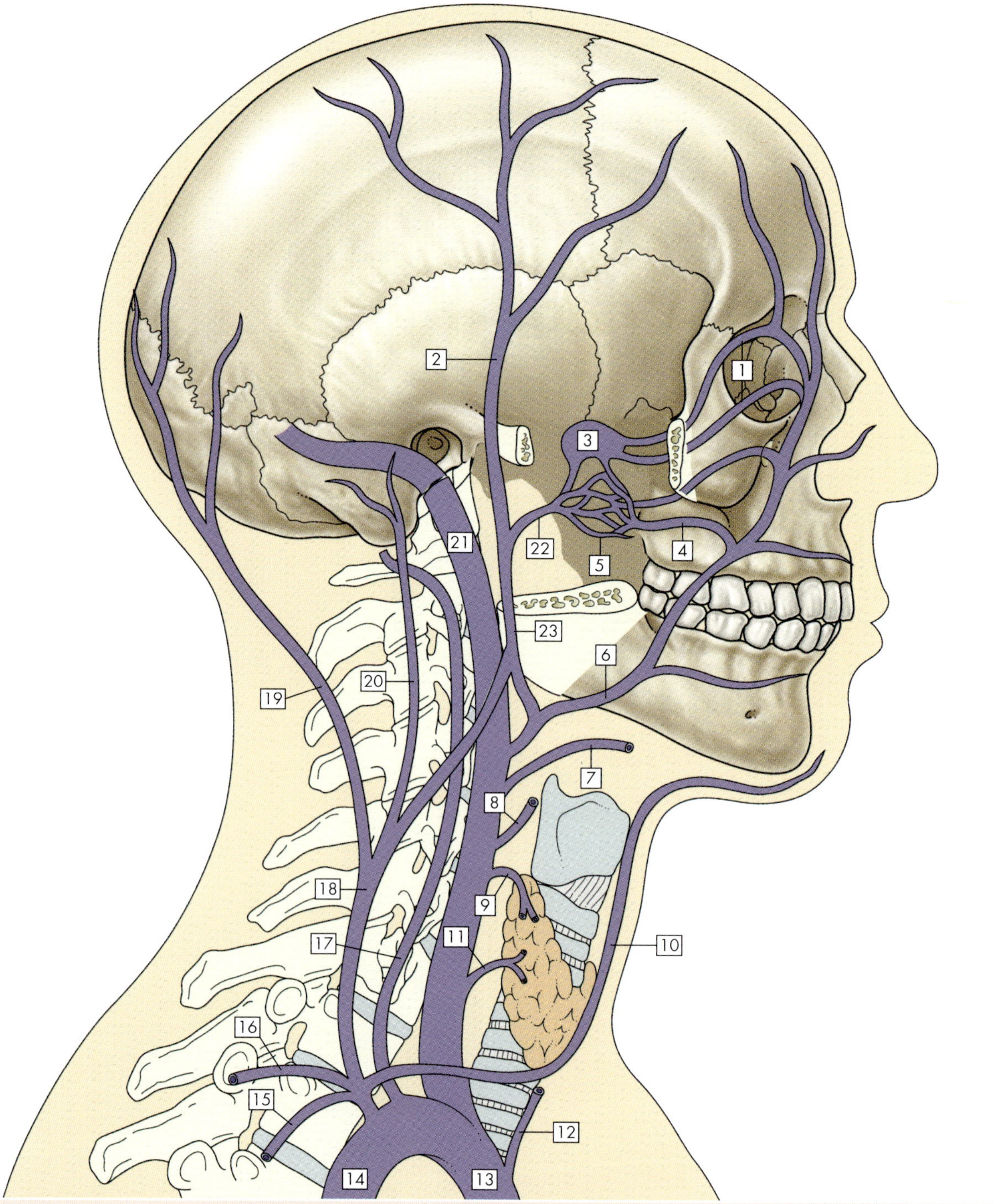

1 Superior and inferior ophthalmic veins
2 Superficial temporal vein
3 Cavernous sinus
4 Deep facial vein
5 Pterygoid plexus
6 Facial vein
7 Lingual vein
8 Pharyngeal vein
9 Superior thyroid vein
10 Anterior jugular vein
11 Middle thyroid vein
12 Inferior thyroid vein
13 Brachiocephalic vein
14 Subclavian vein
15 Suprascapular vein
16 Transverse cervical vein
17 Vertebral vein
18 External jugular vein
19 Posterior external jugular vein
20 Posterior auricular vein
21 Internal jugular vein
22 Maxillary vein
23 Retromandibular vein

THE RELATIONSHIPS OF STRUCTURES WITHIN THE INFRATEMPORAL FOSSA

The relationships of structures within the infratemporal fossa are best visualised in sectional anatomy and these are illustrated in Figs 1.15–1.22.

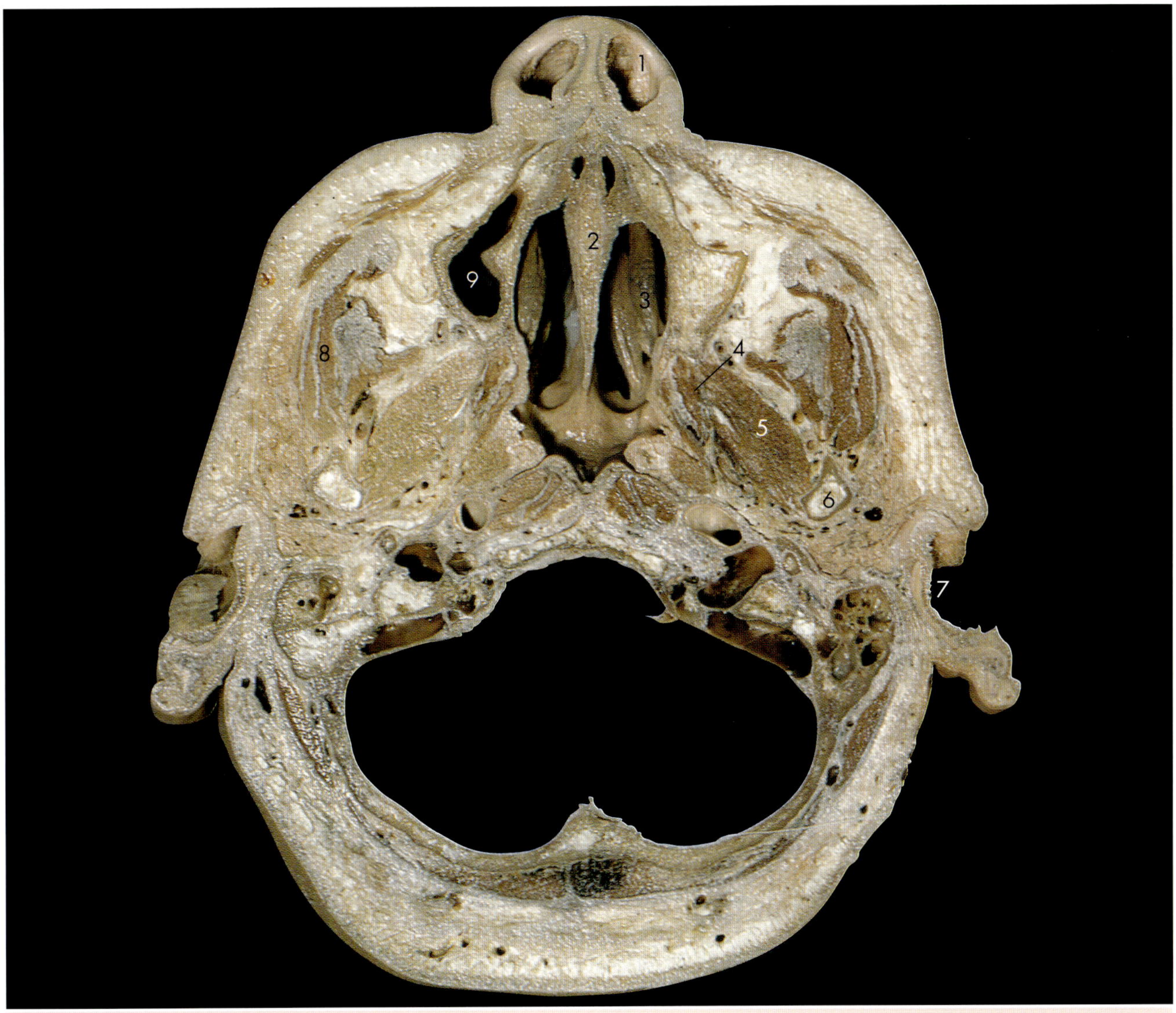

1 External nares
2 Nasal septum
3 Middle nasal concha
4 Medial pterygoid muscle
5 Lateral pterygoid muscle
6 Mandibular condyle
7 External acoustic meatus
8 Temporalis muscle
9 Maxillary air sinus

Figure 1.15 A horizontal section of the head at the level of the top of the condyle.

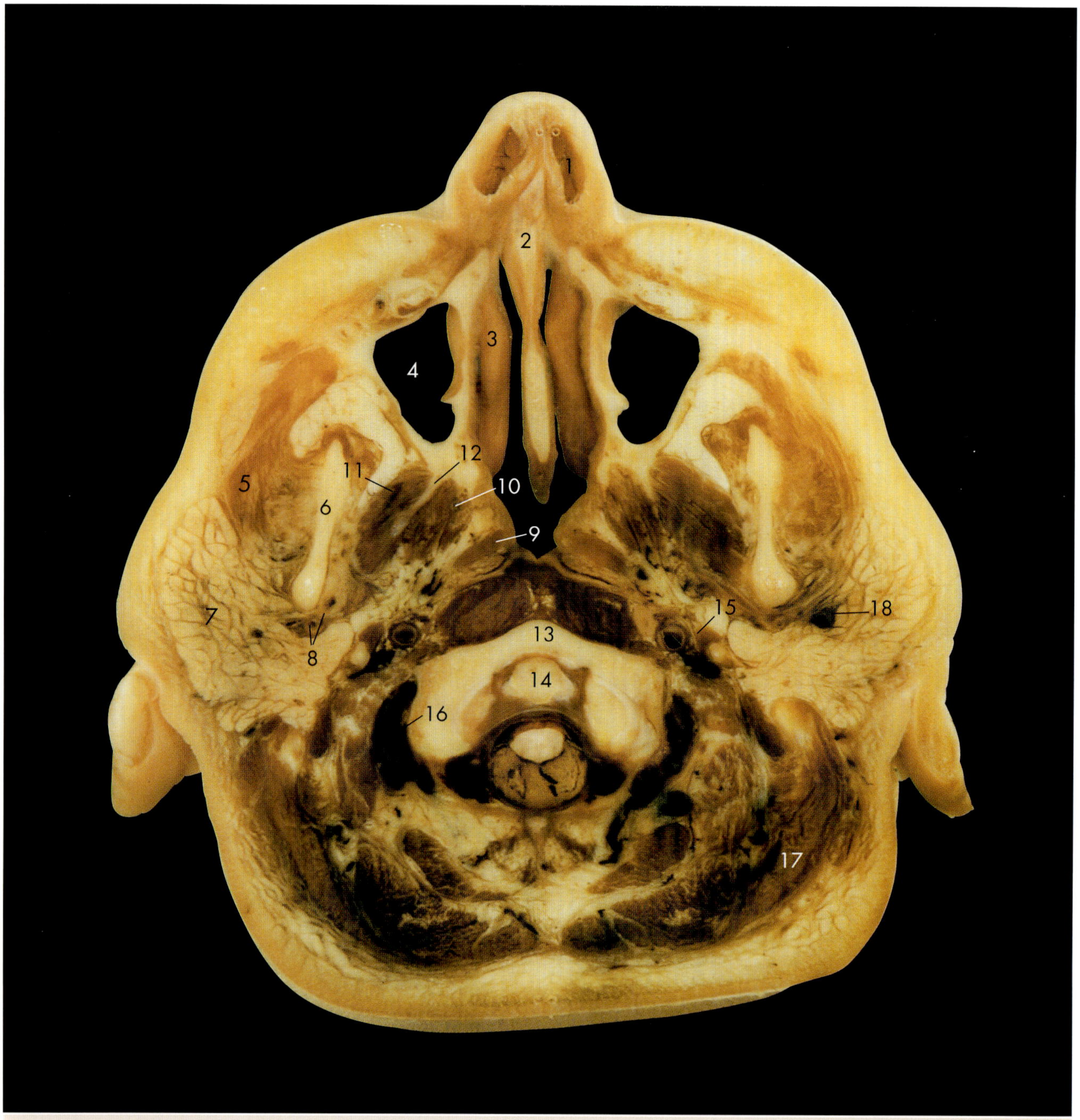

1 External naris
2 Nasal septum
3 Inferior nasal concha
4 Maxillary sinus
5 Masseter muscle
6 Ramus of mandible
7 Parotid gland
8 Superficial temporal and maxillary arteries
9 Levator veli palatini muscle
10 Medial pterygoid muscle
11 Lateral pterygoid muscle
12 Lateral pterygoid plate
13 Anterior arch of atlas
14 Dens of axis
15 Internal carotid artery, internal jugular vein and vagus nerve in carotid sheath
16 Vertebral artery
17 Sternocleidomastoid muscle
18 Retromandibular vein

Figure 1.16 A horizontal section of the head at the upper part of the ramus.

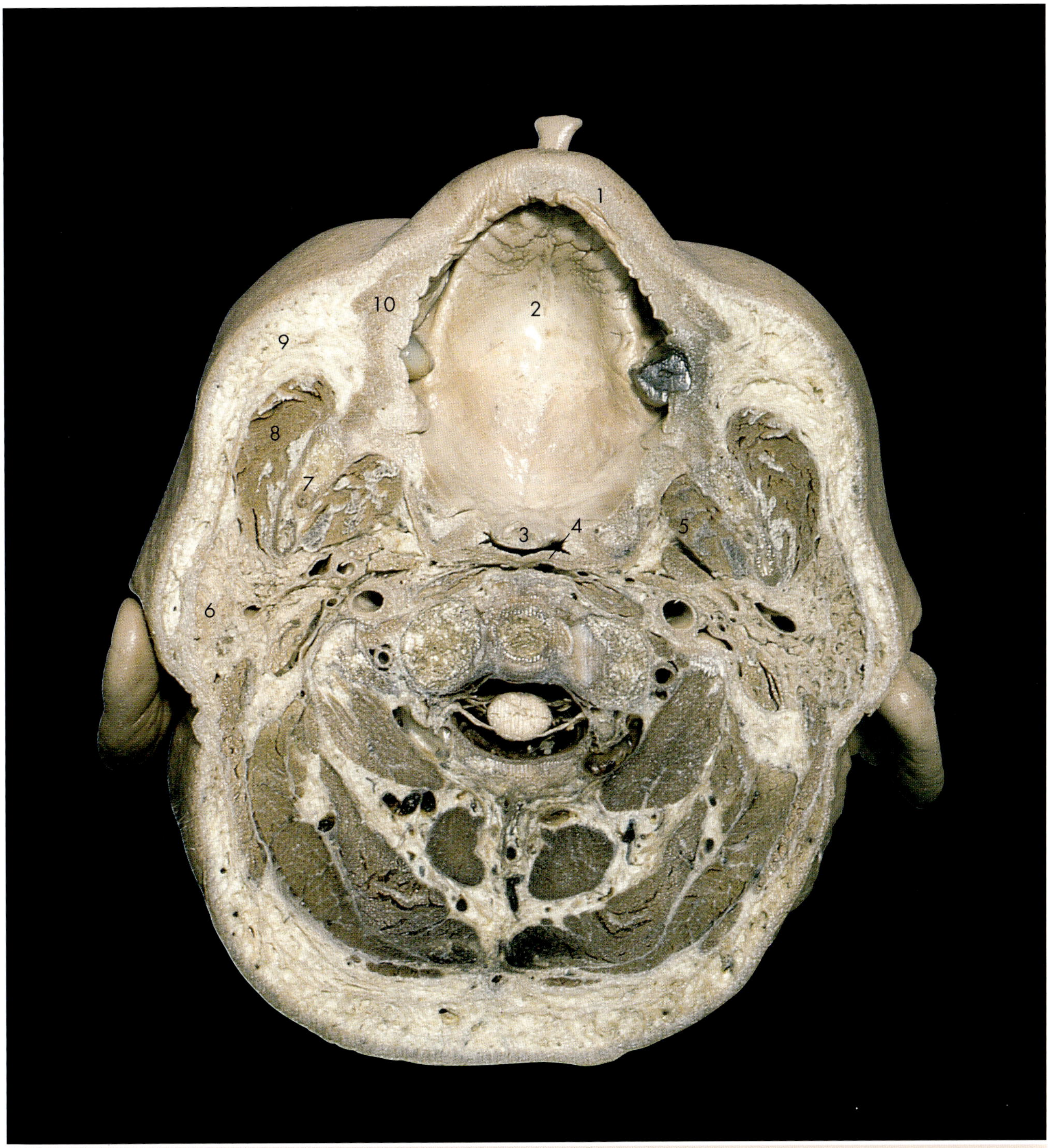

1	Upper lip	6	Parotid gland
2	Hard palate	7	Ramus of mandible
3	Uvula of soft palate	8	Masseter muscle
4	Superior constrictor of pharynx	9	Buccal pad of fat
5	Medial pterygoid muscle	10	Buccinator muscle

Figure 1.17 A horizontal section of the head at the level of the palate.

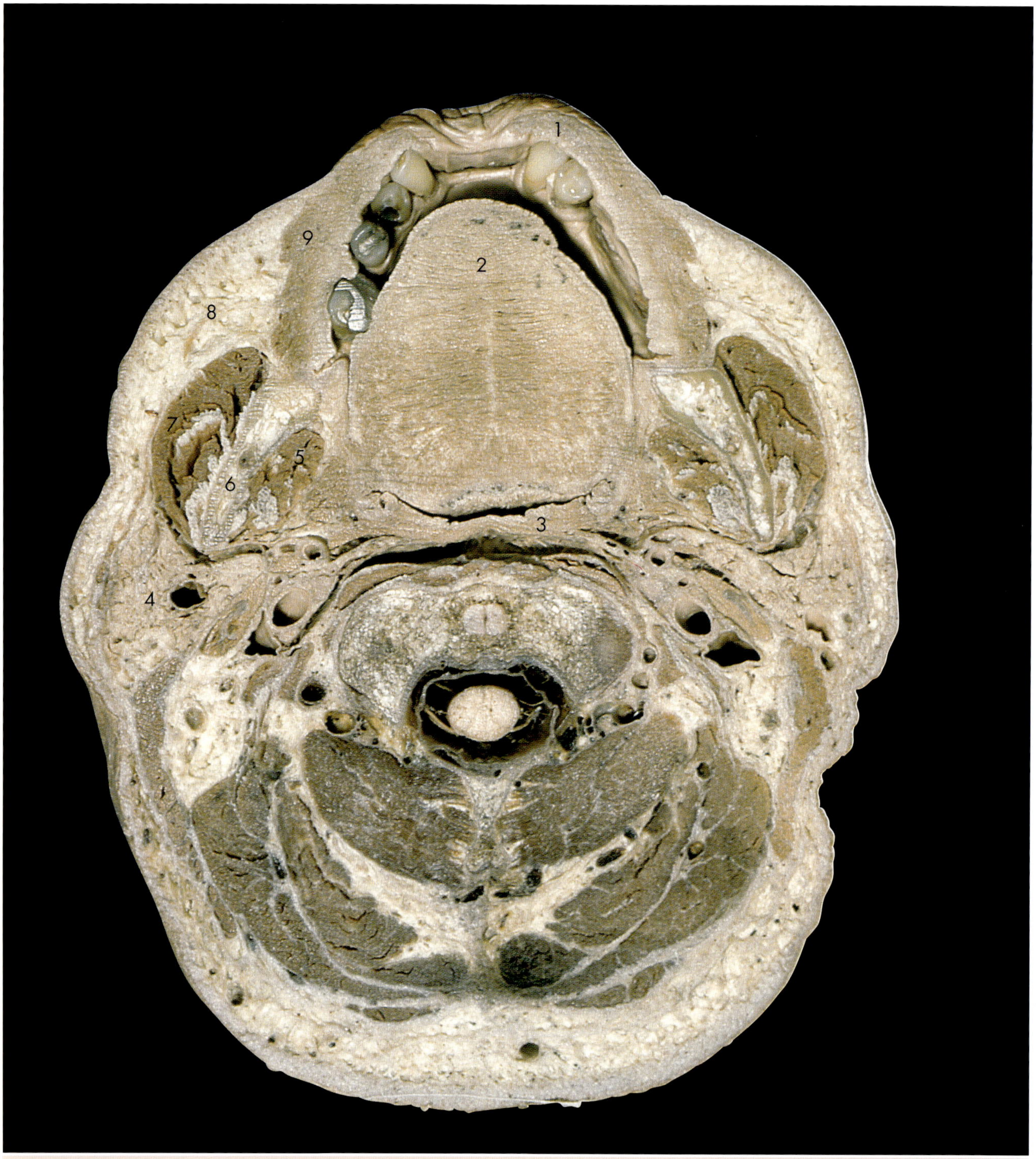

1 Lower lip
2 Tongue
3 Superior constrictor of pharynx
4 Parotid gland
5 Medial pterygoid muscle
6 Ramus of mandible
7 Masseter muscle
8 Buccal pad of fat
9 Buccinator muscle

Figure 1.18 A horizontal section of the head at the level of the tongue.

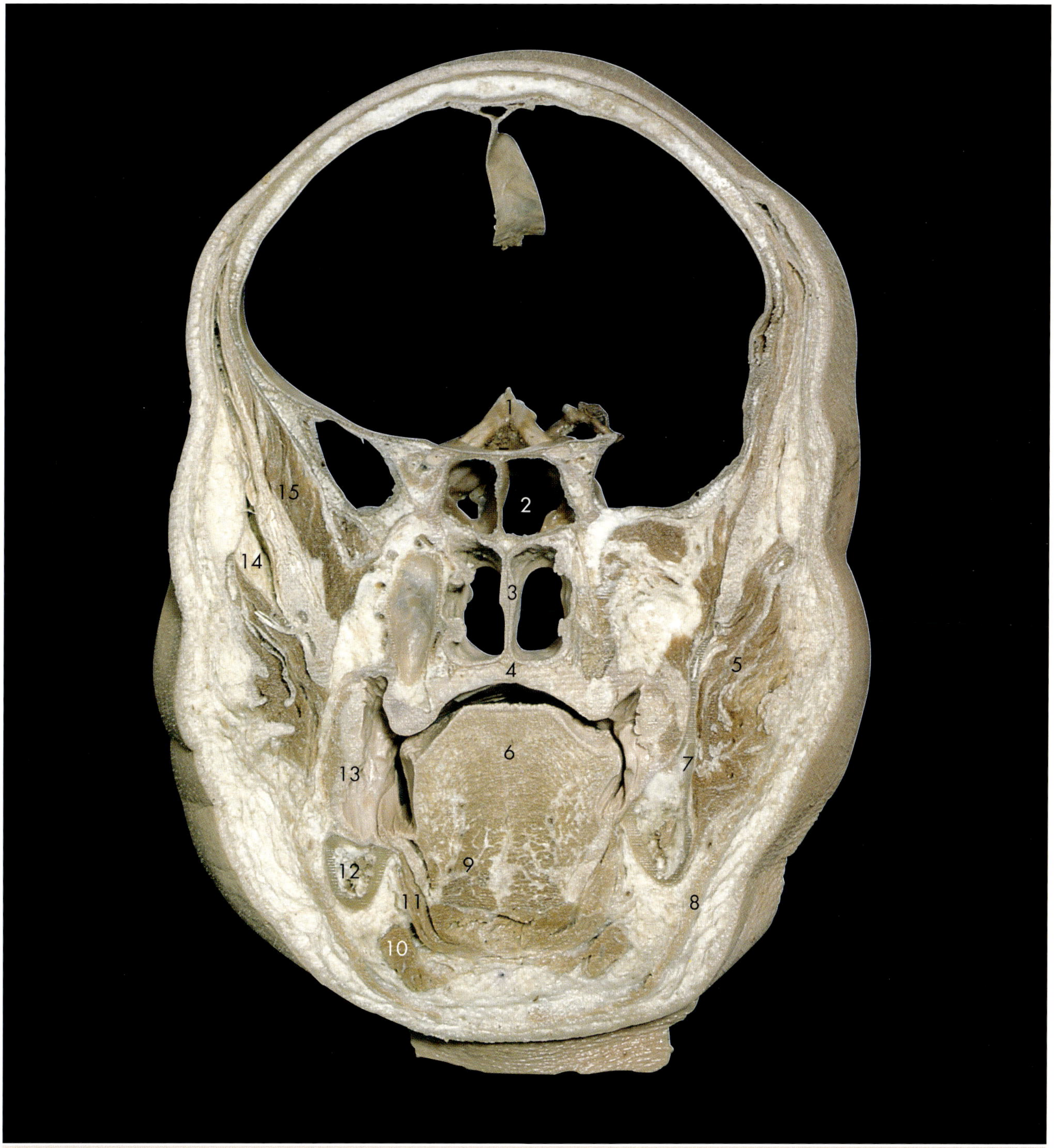

1 Optic chiasma	6 Tongue	11 Mylohyoid muscle
2 Sphenoidal air sinus	7 Ramus of mandible	12 Body of mandible
3 Nasal septum	8 Platysma muscle	13 Buccinator muscle
4 Palate	9 Genioglossus muscle	14 Zygomatic arch
5 Masseter muscle	10 Digastric muscle	15 Temporalis muscle

Figure 1.19 A coronal section of the head at the anterior border of the ramus of the mandible.

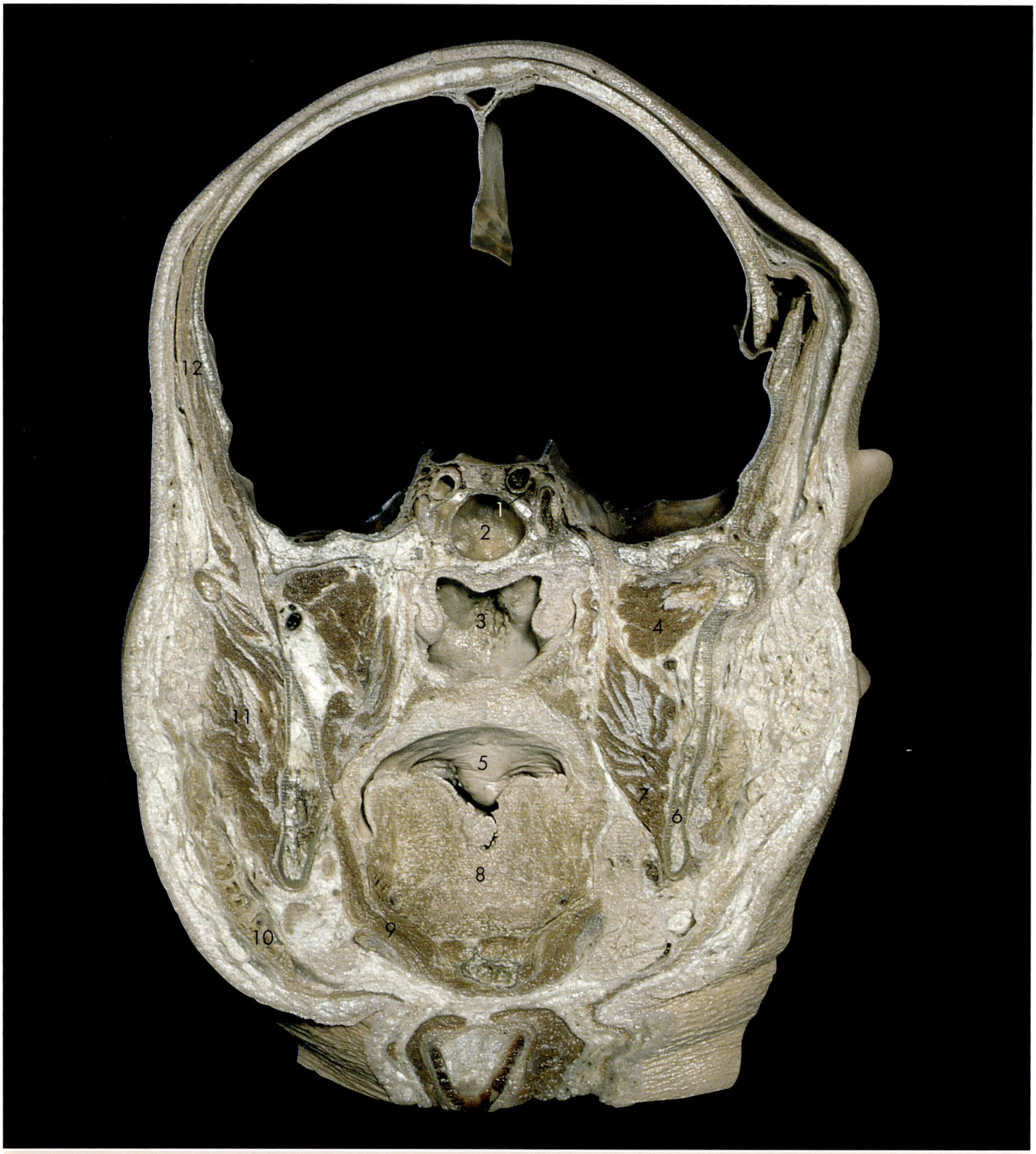

1 Region of cavernous sinus
2 Sphenoidal air sinus
3 Nasopharynx
4 Lateral pterygoid muscle
5 Soft palate and uvula
6 Ramus of mandible
7 Medial pterygoid muscle
8 Tongue
9 Mylohyoid muscle
10 Platysma
11 Masseter muscle
12 Temporalis muscle

Figure 1.20 A coronal section of the head towards the back of the ramus of the mandible.

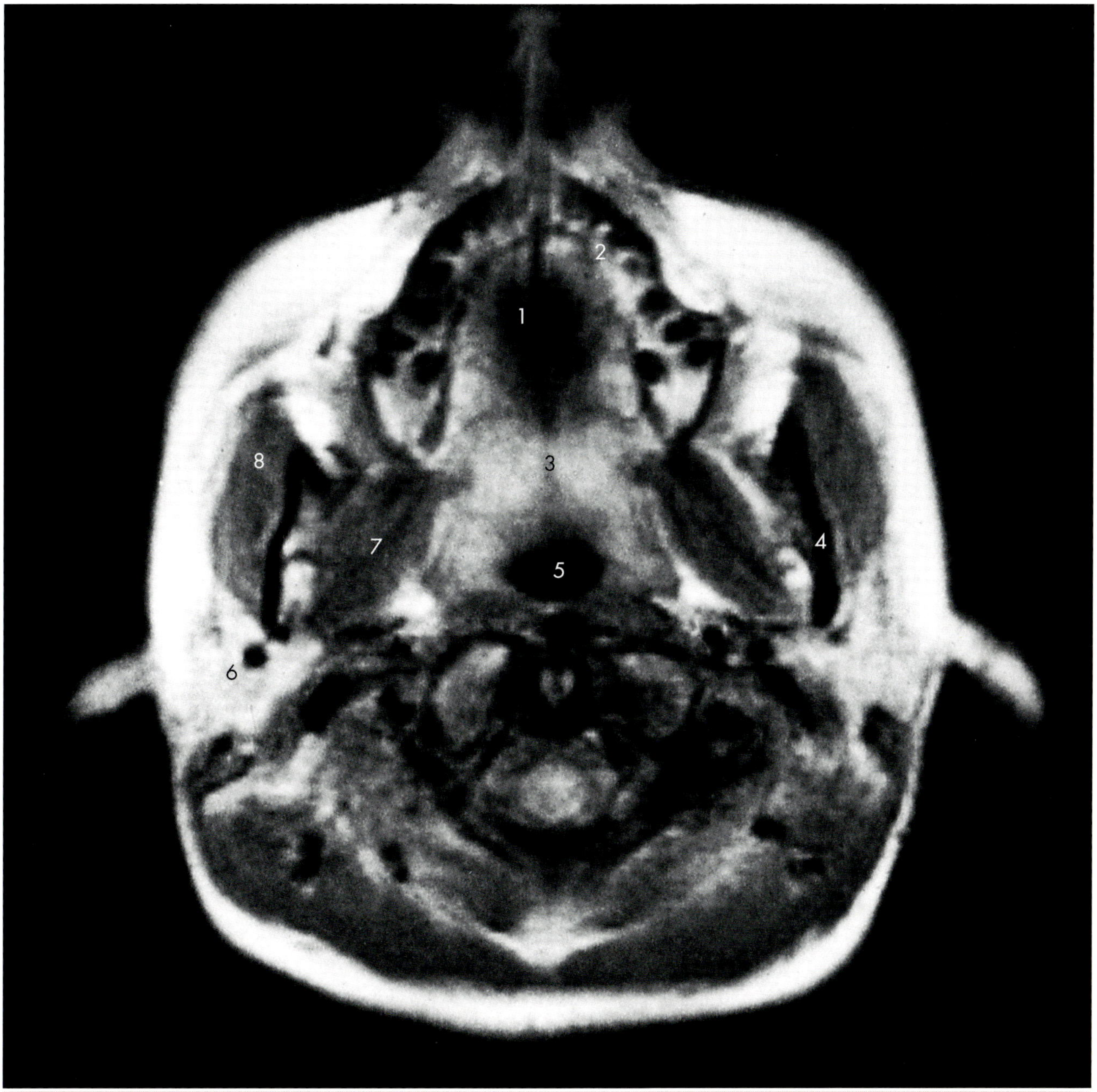

1 Hard palate
2 Maxillary alveolar bone
3 Soft palate
4 Ramus of mandible
5 Nasopharynx
6 Parotid gland
7 Lateral pterygoid muscle
8 Masseter muscle

Figure 1.21 Axial MRI scan of the head at the level of the palate.

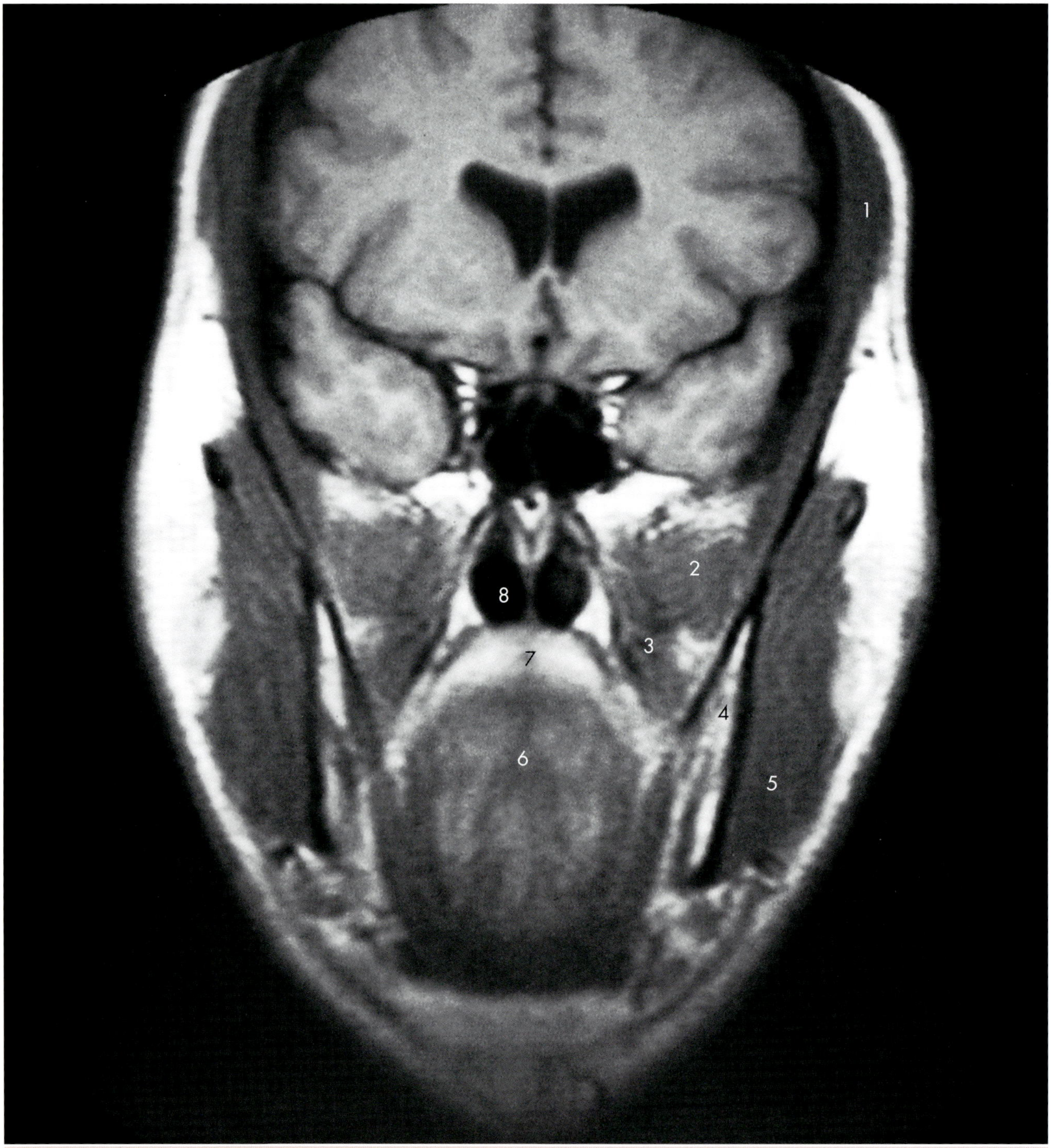

1 Temporalis muscle
2 Lateral pterygoid muscle
3 Medial pterygoid muscle
4 Ramus of mandible
5 Masseter muscle
6 Tongue
7 Palate
8 Nasal cavity

Figure 1.22 Coronal MRI scan of the head towards the back of the ramus.

REFERENCES

1. Hollingshead W H. *Anatomy for surgeons: the head and neck* (3rd edn). New York: Harper Row, 1982
2. Berkovitz B K B, Moxham B J. *A textbook of head and neck anatomy.* London: Wolfe Medical Publications, 1988
3. Williams P L (ed) *Gray's anatomy* (38th edn). Edinburgh: Churchill Livingstone, 1995
4. Lang J. *Clinical anatomy of the masticatory apparatus and peripharyngeal spaces.* New York: Thieme Medical Publishers, Inc., 1995
5. Juniper R P. The superior pterygoid muscle? *Br J Oral Surg* 1981; 19: 121–128
6. McDevitt W E. *Functional anatomy of the masticatory system.* London: Butterworth and Co., 1989
7. Bertilsson O, Strom D. A literature survey of a hundred years of anatomic and functional lateral pterygoid muscle research. *J Orofac Pain* 1995; 9: 17–23
8. Prentiss H J. Regional anatomy, emphasising mandibular movements with specific reference to full denture construction. *J Am Dent Assoc* 1923; 10: 1085–1099
9. McNamara J A Jr. The independent functions of the two heads of the lateral pterygoid muscle. *Am J Anat* 1973; 138: 197–206
10. Widmalm S E, Lillie J H, Ash M M Jr. Anatomical and electronmyographic studies of the lateral pterygoid muscle. J *Oral Rehabil* 1987; 14: 429–446
11. Honée G L J M. The anatomy of the lateral pterygoid muscle. *Acta Morphol Neerl (Scand)* 1972; 10: 331–340
12. Petersen H. *Histologie und mikroskopische anatomie.* Munchen: Bergmann, 1935
13. Porter M R. The attachment of the lateral pterygoid muscle of the meniscus. *J Prosthet Dent* 1970; 24: 555–562
14. Mosolov N N. On the anatomy of human masticatory musculature. In: Schumacher G H (ed) *Morphology of the maxillo-mandibular apparatus.* Leipzig: Thieme, 1972
15. LeToux G, Duval J M, Darnault P. The human temporomandibular joint: current anatomic and physiologic status. *Surg Radiol Anat* 1989; 11: 283–288
16. Wilkinson T M. The relationship between the disc and the lateral pterygoid muscle in the human temporomandibular joint. *J Prosthet Dent* 1988; 60: 715–724
17. Lau H. Innervation of the jaw muscles in representatives of various chewing types. In: *Morphology of the maxillo-mandibular apparatus.* Leipzig: Thieme, 1972; 133–137
18. Schumacher G H. Funktionsbedingter strukturwandel des M. masseter. *Gegenbaurs Morphol Jb* 1961; 102: 150–169
19. Xu J A, Yuasa K, Yoshiura K, Kanda S. Quantitative analysis of masticatory muscles using computed tomography. *Dentomaxillofac Radiol* 1994; 23: 154–158
20. Newton J P, Yemm R, Able R W, Menhinick S. Changes in human jaw muscles with age and dental state. *Gerodontology* 1993; 10: 16–22
21. Hannam A G, Wood W W. Relationships between the size and spatial morphology of human masseter and medial pterygoid muscles, the craniofacial skeleton, and jaw biomechanics. *Am J Phys Anthropol* 1989; 40: 429–445
22. Dunn G F, Robinson W L. Anatomical observation of a craniomandibular muscle originating from the skull base: the sphenomandibularis. *J Craniomandibular Prac* 1996; 14: 97–103
23. Zenker W. Uber einige befunde am M. temporalis des menschen. *Z Anat Entwickl Gesch* 1955; 118: 355–368
24. Lurje A. On the topographical anatomy of the internal maxillary artery. *Acta Anat* 1947; 2: 219–231
25. Krizan Z. Beitrag zur kenntnis des dritten trigeminusastes. *Pract Otol Rhinol Laryngol* 1956; 18: 144
26. Baumel J J, Vanderheiten J P, McElenney J E. The auriculotemporal nerve of man. *Am J Anat* 1971; 130: 431–440
27. Kiesselbach J E, Chamberlain J E. Clinical and anatomic observations on the relationship of the lingual nerve to the mandibular third molar. *J Oral Maxillofac Surg* 1984; 42: 565–567
28. Kiesselbach J E. Discussion in *J Oral Maxillofac Surg* 1995; 53: 1181
29. Pogrel M A, Renaut A, Schmidt B, Ammar A. The relationship of the lingual nerve to the mandibular third molar region. *J Oral Maxillofac Surg* 1995; 53: 1178–1181
30. Adachi B. *Das Arteriensystem der Japaner.* Bd I und II. Kyoto: Verlag der Kaiserlich-Japanischen Universitat, 1928. (Quoted in ref. 4)
31. Lasker G W, Opdyke D L, Miller H. The position of the internal maxillary artery and its questionable relation to the cephalic index. *Anat Rec* 1951; 109: 119–126
32. Skopakoff C. Uber die variabilitat im verlauf der A. maxillaris. *Anat Anz* 1968; 123: 534–546
33. Urban A. Topographie der pars pterygopalatina der A. maxillaris aus frontaler sicht. [thesis] Wurzburg. (Quoted in ref. 4), 1974
34. Pretterklieber M L, Skopakoff C, Mayr R. The human maxillary artery reinvestigated: 1. Topographical relations in the infratemporal fossa. *Acta Anat Basel* 1991; 142: 281–287
35. Sashi R, Tomura N, Hashimoto M, Kobayashi M, Watarai J. Angiographic anatomy of the first and second segments of the maxillary artery. *Radiat Med* 1996; 14: 133–138
36. Baumel J J, Beard D Y. The accessory meningeal artery of man. *J Anat* 1961; 95: 386–402
37. Bast R A. Alveolaris inferior, Verlauf und Verzweigungstypen. (Dissertation) Wurzburg. (Quoted in ref. 4), 1982.
38. Chung L K. The vascular anatomy of the human temporalis muscle: Implications for surgical splitting techniques. *Int J Oral Maxillofac Surg* 1996; 25: 414–421
39. Deplus S, Bremond-Gignac D, Gillot C, Lassau J P. The pterygoid venous plexuses. *Surg Radiol Anat* 1996; 18: 23–27

Chapter 2

The temporomandibular joint and pterygopalatine fossa

B.K.B. BERKOVITZ, B.J. MOXHAM AND J.D. LANGDON

For a full understanding of the surgical and clinical aspects of the infratemporal fossa, an appreciation of its relationship to adjacent structures is of importance. This chapter will consider the anatomy of the temporomandibular joint and the pterygopalatine fossa, while Chapter 8 will deal with the parotid gland.

THE TEMPOROMANDIBULAR JOINT

The temporomandibular (craniomandibular) joint (TMJ) is a synovial joint. It is formed by the condylar process of the mandible articulating in the mandibular (glenoid) fossa of the temporal bone (Fig. 2.1).

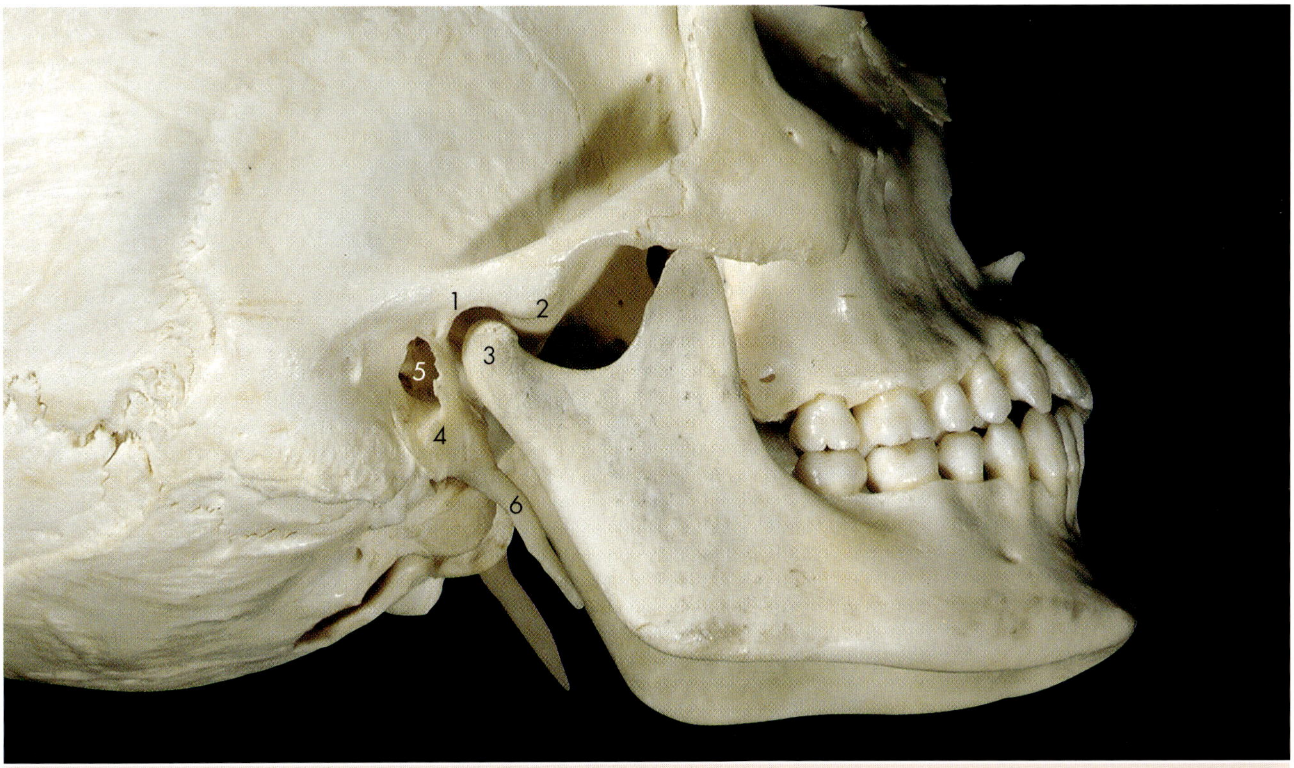

1 Mandibular fossa
2 Articular eminence
3 Mandibular condyle
4 Tympanic plate
5 External acoustic meatus
6 Styloid process

Figure 2.1 The skull from the side showing articulation of the temporomandibular joint.

Unlike most other synovial joints, the joint cavity is divided into upper and lower compartments by an articular disc. Although basically a hinge joint, the TMJ also allows for some gliding movements. Movement of the condylar head occurs within the mandibular fossa and down a bony prominence termed the articular eminence (tubercle) of the temporal bone, which is located immediately anterior to the mandibular fossa. That there are two TMJs associated with a single mandible has considerable functional significance in that movement at one joint must imply additional movement at the other. The reader is referred to additional accounts of the joint.[1–7]

THE MANDIBULAR FOSSA

The mandibular fossa is an oval depression in the temporal bone lying immediately anterior to the external acoustic meatus (Figs 1.1, 2.2, 2.15). Its mediolateral dimension is greater than its anteroposterior one in order to accommodate the mandibular condyle, and it is wider laterally than medially. The curvature of the mandibular fossa varies and may show some relationship to the nature of the occlusion. The mandibular fossa is bounded anteriorly by the articular eminence, laterally by the zygomatic process, and posteriorly by the tympanic plate. The

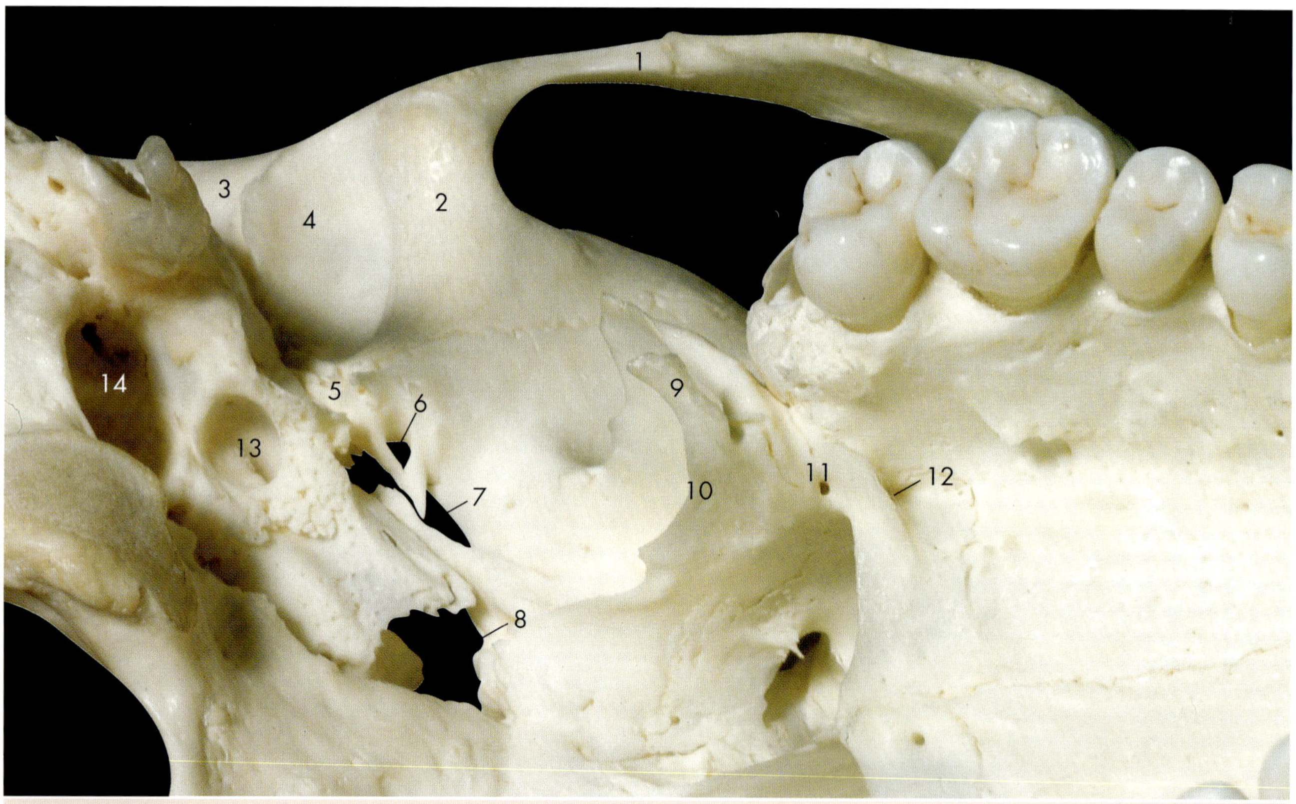

1 Zygomatic arch
2 Articular eminence
3 Squamotympanic fissure
4 Mandibular fossa
5 Spine of sphenoid bone
6 Foramen spinosum
7 Foramen ovale
8 Foramen lacerum
9 Pterygoid hamulus
10 Medial pterygoid plate
11 Lesser palatine foramen
12 Greater palatine foramen
13 Carotid canal
14 Jugular formen

Figure 2.2 The base of the skull showing the mandibular fossa and related structures.

posterior margin is elevated to form the posterior auricular ridge, which may be enlarged laterally as the postglenoid tubercle just anterior to the external acoustic meatus. Medially, the mandibular fossa may be defined by a ridge, the medial glenoid plane. The squamous and tympanic parts of the temporal bone are delineated laterally by the squamotympanic fissure. Medially, this fissure bifurcates due to the presence of a small component of the petrous portion, the tegmen tympani, giving rise to the petrosquamous fissure anteriorly and the petrotympanic fissure immediately behind (Fig. 1.1). The petrotympanic fissure is the site at which the chorda tympani nerve exits from the cranium into the infratemporal fossa.

The shape of the mandibular fossa does not exactly conform to the shape of the mandibular condyle, the articular disc moulding together the joint surfaces. The bone of the central part of the fossa is thin. This indicates that masticatory loads are not dissipated through the mandibular fossa but through the teeth and thence the facial bones and base of the cranium.

THE CONDYLAR PROCESS OF THE MANDIBLE

The mandibular condyle (Figs 2.1–2.4) varies considerably both in size and shape. When viewed from above, the condyle is roughly ovoid in outline, the anteroposterior dimension of the condyle (approximately 1 cm) being approximately half the mediolateral dimension. The medial aspect of the condyle is wider than the lateral. The long axis of the condyle is not, however, at right angles to the ramus, but diverges posteriorly from a strictly coronal plane. Thus, the lateral pole of the condyle lies slightly anterior to the medial pole and, if the long axes of the two condyles are extended, they meet at an obtuse angle (approximately 145°) at the anterior border of the foramen magnum.

The articular surfaces of the condyle are the anterior and superior surfaces. These surfaces are convex. The articular surface area of the condyle is of the order of 200 mm^2, which is about half that of the mandibular fossa.[8] The non-articular posterior surface of the condyle is broad and flat. The articular surface may be

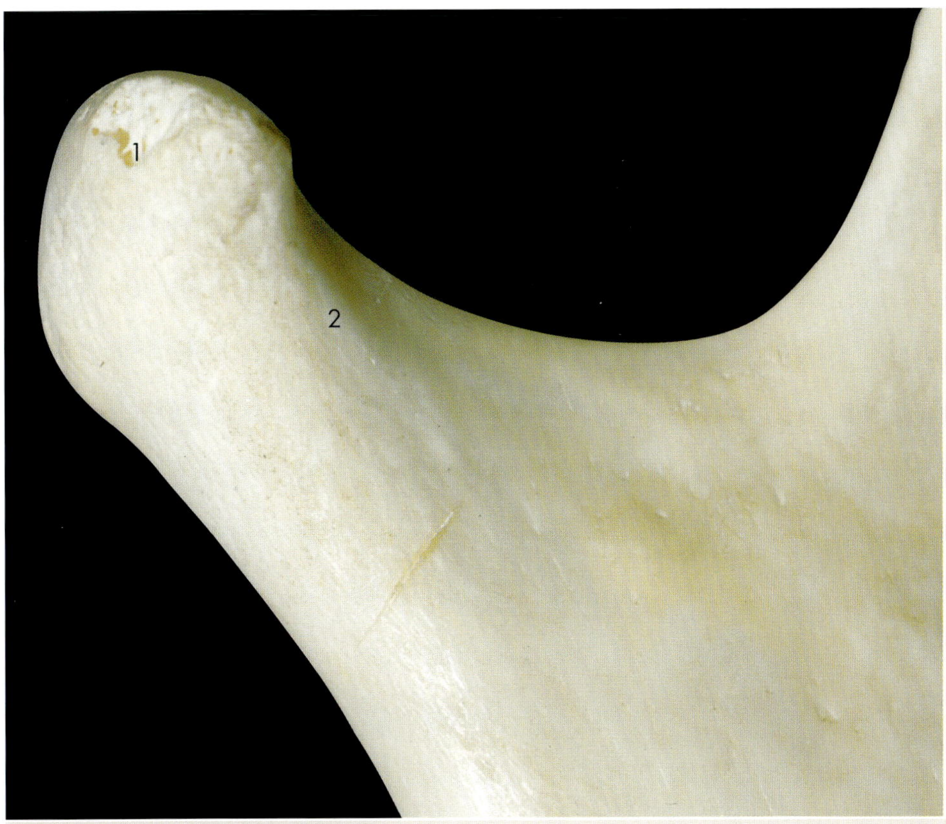

1 Head of condyle
2 Neck of condyle

Figure 2.3 The condyle viewed laterally.

separated from the nonarticular surface by a slight ridge indicating the site of attachment of the joint capsule.

The articular head of the condyle joins the ramus through a thin bony projection termed the neck of the condyle. A small depression (the pterygoid fovea) marks part of the attachment of the lateral pterygoid muscle. This fovea is situated on the anterior surface of the neck, below the articular surface.

1 Head of condyle
2 Neck of condyle
3 Pterygoid fovea
4 Coronoid process
5 Lateral pole of condyle
6 Medial pole of condyle

Figure 2.4 The condyle viewed anteriorly showing the pterygoid fovea.

The condyle is composed of a core of cancellous bone covered by a thin layer of compact bone. During the period of growth, however, a layer of hyaline cartilage lies immediately beneath the fibrous articulating surface of the condyle.

THE JOINT CAPSULE

The capsule of the TMJ is a thin slack cuff which, of itself, does not limit mandibular movements and is too weak to provide much support for the joint (Fig. 2.5).

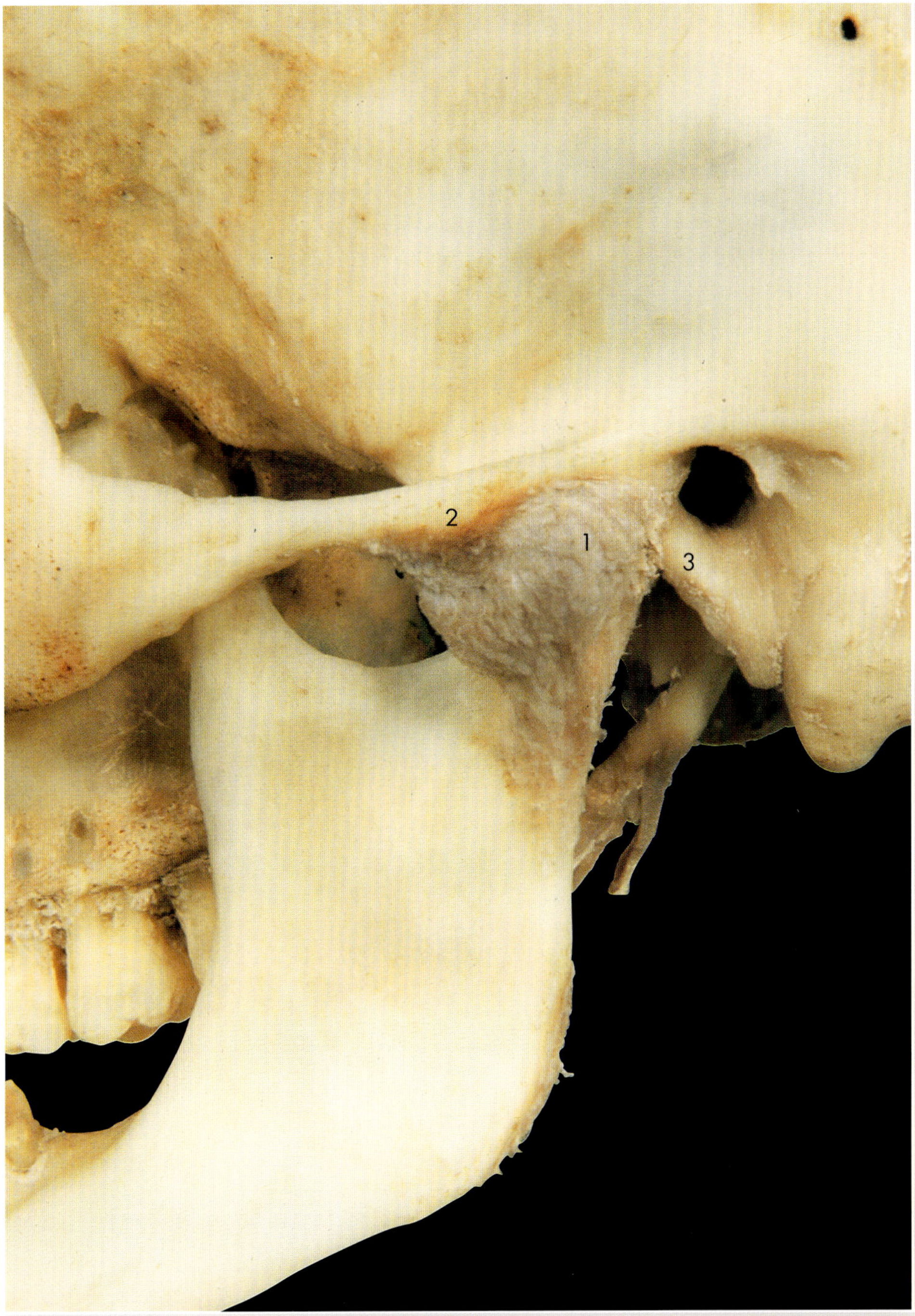

1 Capsule of temporomandibular joint
2 Articular eminence
3 Tympanic plate

Figure 2.5 The capsule viewed laterally. (Courtesy of Professor S. Standring, GKT School of Biomedical Sciences, London.)

Below, it is attached to the neck of the condyle of the mandible. Above, it is attached to the mandibular fossa, extending anteriorly to just in front of the crest of the articular eminence, posteriorly to the squamotympanic and petrotympanic fissures, medially to the medial glenoid plane and laterally between the lateral margin of the articular eminence and the postglenoid process. The capsule posteriorly is associated with the thicker, vascular, but loosely arranged connective tissue of the bilaminar zone of the articular disc. Internally, the capsule gives attachment to the articular disc.[9,10] The collagen fibres of the capsule run predominantly in a vertical direction.[5]

A synovial membrane lines the inner surface of the fibrous capsule and the margins of the articular disc but does not cover the articular surfaces of the joint. The synovial membrane secretes the synovial fluid that occupies the joint cavities. The synovial fluid lubricates the joint and may also have nutritive functions. Important components of the fluid are the proteoglycans, which aid the lubrication. At rest, the hydrostatic pressure of the synovial fluid has been reported as being subatmospheric, but this is greatly elevated during mastication.[11,12]

THE ARTICULAR DISC (Figs 2.6–2.10)

The articular disc (meniscus) is of a dense, fibrous consistency and is moulded to the bony joint surfaces above and below. Blood vessels are only evident at the periphery of the articular disc, the bulk of it being avascular. Above, it covers the slope of the articular eminence in front while below it covers the condyle. When viewed in sagittal section, the upper surface of the disc is concavoconvex from before backwards and the lower surface is concave (Figs 2.6–2.8). Viewed superiorly, the articular disc is somewhat oval in outline. The disc is of variable thickness, being thinnest centrally over the articular surface of the mandibular condyle and thickest posteriorly in the region above, and behind, the mandibular condyle. The lateral half of the disc is thinner than the medial half. The medial and lateral margins of the disc are slightly thickened (Fig. 2.9).

The articular disc has been subdivided into three portions: anterior, intermediate and posterior (Fig. 2.8). The intermediate zone is the thinnest and is the area in contact with the articular surface of the condyle. In the intermediate part, the collagen bundles have been described as running preferentially in an anteroposterior direction, while in the anterior and posterior bands, they run both anteroposteriorly and mediolaterally. The collagen fibres are crimped, perhaps evidence that the disc is subjected to tensional forces. The overall shape of the articular disc is thought to provide a self-centring mechanism, which automatically acts to maintain its correct relationship to the articular surface of the mandibular condyle during mandibular movements.

The margin of the articular disc merges peripherally with the joint capsule. Anteriorly, fibrous bands connect the disc to the anterior margin of the articular eminence above and to the anterior margin of the condyle below. Medially and laterally, the articular disc is attached to the joint capsule and, just below the medial and lateral poles of the condyle, by triangular zones of connective tissue. Posteriorly, the disc is attached to the capsule by a looser connective tissue, the retrodiscal tissue (pad) that has a bilaminar appearance. The superior lamina is loose and possesses numerous vascular elements and elastin fibres.[13] It attaches to the anterior margin of the squamotympanic fissure. The inferior lamina is relatively avascular, less extensible (as it has few elastin fibres) and is attached to the posterior margin of the condyle. The volume of the retrodiscal tissue appears to increase four to five times as a result of venous engorgement as the jaw is opened and the condyle moves downwards and forwards. This venous engorgement of the retrodiscal veins is not the result of the tissue having erectile properties,[13] but more the result of their continuity with the pterygoid venous plexus lying medial to the condyle. This activity fills the vacated space in the mandibular fossa and rapidly equilibrates any changes in intracapsular pressures which may hinder jaw movement. Changes in tissue fluid pressures during mandibular movements could also help regulate the flow of blood in the retrodiscal pad.[2] As the mandibular condyle moves backwards during jaw closure, blood leaves the retrodiscal tissues. The close relationship of elastin fibres to the walls of the blood vessels in the retrodiscal tissues has led to the view that the fibres function as a pump, facilitating blood flow during venous dilatation and compression.[14,15] It has been suggested that the retrodiscal pad may serve as an absorber of sounds produced by the temporomandibular joint.[13]

The return of the articular disc to its original position may be aided by the elastic recoil of the superior lamella. However, the finding that the superior lamina is folded on itself when the jaw is closed and only increased in length by one quarter by the time the condyle reached the crest of the articular eminence argues against this view.[16] This indicates that movement of the disc may be passive, due to the shape of the disc and its firm insertion to the lateral and medial poles of the condyle of the mandible, and also due to the finding that the superior head of the lateral pterygoid is only active on final closure.[17]

The articular disc of the TMJ divides the joint cavity into superior and inferior joint cavities. About 1 ml of synovial fluid occupies the inferior joint cavity, while a little more occupies the slightly larger superior joint cavity.

Numerous studies have been undertaken to determine the precise attachment of the lateral pterygoid muscle to the articular disc, as this may help us understand TMJ disorders such as internal derangement of

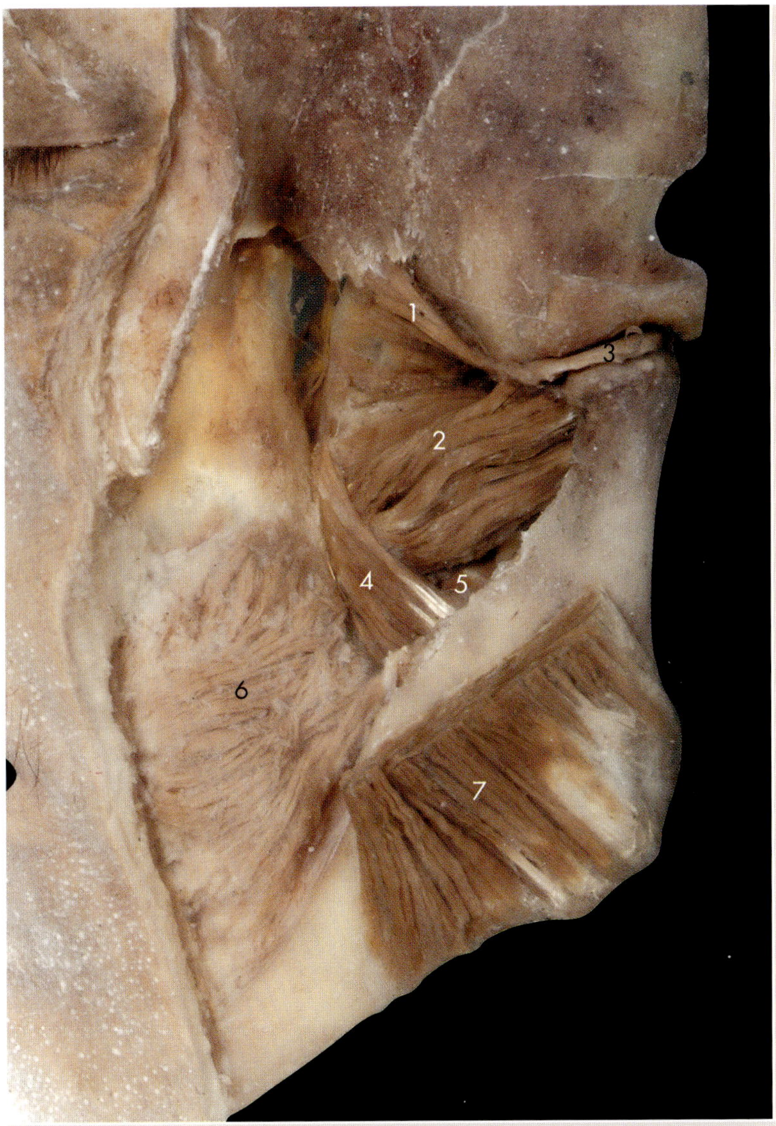

1 Fibres of upper head of lateral pterygoid muscle appearing to attach into articular disc
2 Fibres of lower head of lateral pterygoid muscle inserting into pterygoid fovea
3 Articular disc
4 Superficial head of medial pterygoid muscle
5 Deep head of medial pterygoid muscle
6 Buccinator muscle
7 Masseter muscle

Figure 2.6 Dissection showing articular disc attached to the lateral pterygoid. (Courtesy of Professor L. Garey, Anatomy Department, Imperial College Medical School, London.)

1 Articular eminence
2 Mandibular fossa
3 Articular disc
4 Mandibular condyle
5 Lateral pterygoid muscle
6 External acoustic meatus
7 Parotid gland

Figure 2.7 Sagittal section of the articular disc.

the articular disc, when the disc is displaced generally in an anteromedial direction.[18–21] These numerous studies have produced variable results which have been summarised in a major review of the literature.[22] The findings indicate that, in the majority of studies (60%), fibres from the superior head of the lateral pterygoid muscle gain a direct attachment into the capsule of the joint and to the medial aspect of the anterior border of the articular disc (as well as to the condyle). In 30% of articles, only a few muscle fibres are inserted into the disc, while in the remaining 10% of studies the superior head of the lateral pterygoid muscle is attached only to the condyle. Very rarely, some fibres from the inferior head of the lateral pterygoid may insert into the articular disc.[23] Such differences may reflect biological variation or the particular techniques used. In one recent study trying to explain the different results, the differing methodology was blamed and it was con-

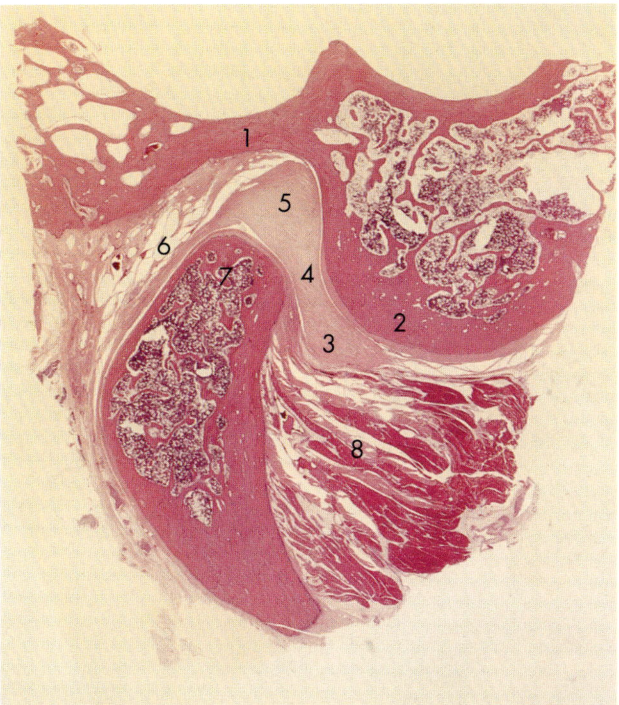

1 Mandibular fossa
2 Articular eminence
3 Anterior band of articular disc
4 Intermediate band of articular disc
5 Posterior band of articular disc
6 Retrodiscal tissue
7 Mandibular condyle
8 Lateral pterygoid muscle

Figure 2.8a Low power histological section of the articular disc. Haematoxylin and eosin.

Figure 2.8b Micrograph of articular disc viewed with differential interference contrast microscopy revealing crimped nature of collagen (x150).

cluded that part of the superior head of the lateral pterygoid muscle is attached only to the capsule and did not extend into the disc. As there is a firm adhesion between the capsule, disc and muscle, this arrangement allows them to act as a unit without the necessity of muscle fibres passing into the disc.[24]

Some reports have suggested that fibres of the masseter muscle may be attached to the anterolateral part of the articular disc of the temporomandibular joint.[25] However, there are other studies which dispute this and maintain that there is no direct muscular attachment, but rather a blending of fibrous tissue.[9,26] Rarely, fibres of the temporalis muscle may gain attachment to the disc.[25,26]

Whereas some regard the functions of the articular disc as helping to spread the joint forces and to stabilise the condyle, others see its function as primarily permitting the condyle to move more freely.[27]

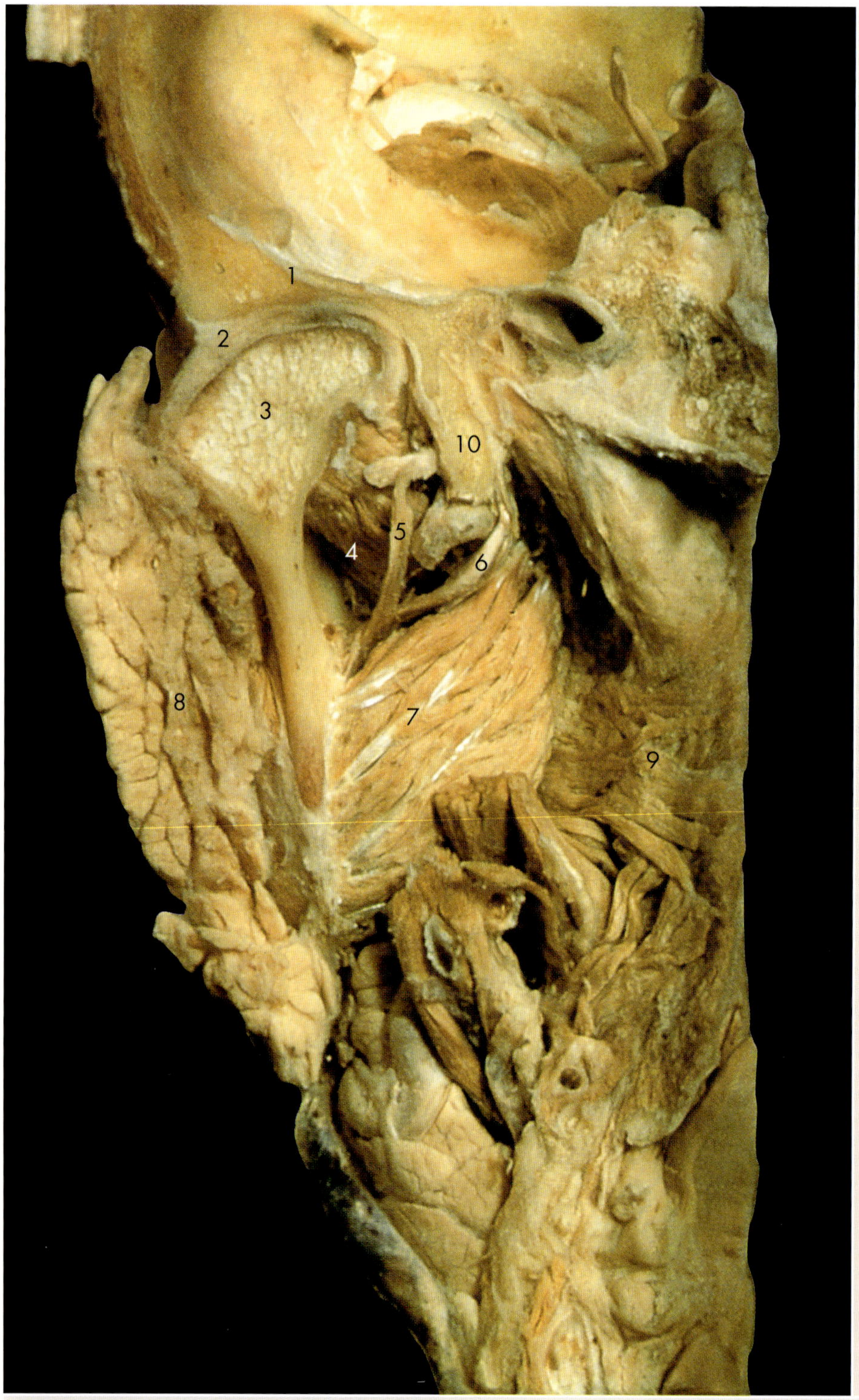

1 Mandibular fossa
2 Articular disc
3 Mandibular condyle
4 Lateral pterygoid muscle
5 Inferior alveolar nerve
6 Lingual nerve
7 Medial pterygoid muscle
8 Parotid gland
9 Superior constrictor muscle
10 Tensor veli palatini muscle

Figure 2.9 The articular disc viewed posteriorly. (Courtesy of the Royal College of Surgeons of England.)

The Ligaments of the Temporomandibular Joint

The joint capsule is attached to the neck of the condyle and to the margins of the mandibular fossa. It is strengthened by the lateral ligament (temporomandibular ligament) although this cannot readily be separated from the capsule (Fig. 2.11).[5,28] Histological studies confirm the close attachment of the lateral ligament to the capsule.[29] From the articular eminence, this ligament passes downwards and backwards to attach on to the lateral surface and posterior border of the neck of the mandibular condyle. The lateral ligament is reinforced by a horizontal band of fibres running from the articular eminence to the lateral surface of the condyle. This band restricts posterior displacement of the condylar process.[30] The lateral ligament is believed also to convert the potentially separating forces generated by the muscles opening the jaws into a force that compresses the condyle of the mandible on to the articular eminence.[20] In addition to the usual type I collagen fibres, the lateral ligament also contains a considerable amount of type III collagen.[28]

In about 15% of cases, the collagen comprising the lateral ligament may not be organised into parallel-running fibre bundles, giving it an almost non-ligamentous appearance. Irregularly arranged collagen fibre bundles are encountered as a normal feature in the posterior part of the lateral ligament.[28]

Figure 2.10 Sagittal section of the temporomandibular joint

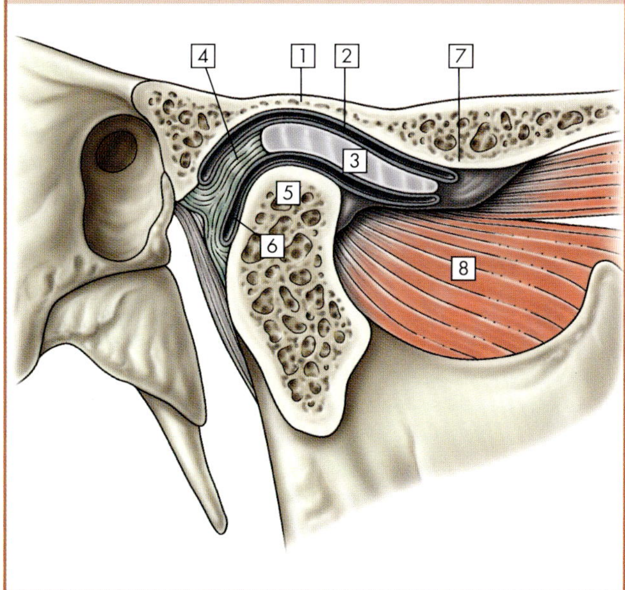

1 Mandibular fossa
2 Upper joint cavity
3 Articular disc
4 Retrodiscal pad region
5 Mandibular condyle
6 Lower joint cavity
7 Articular eminence
8 Lateral pterygoid muscle

Figure 2.11 The temporomandibular joint and its ligaments.

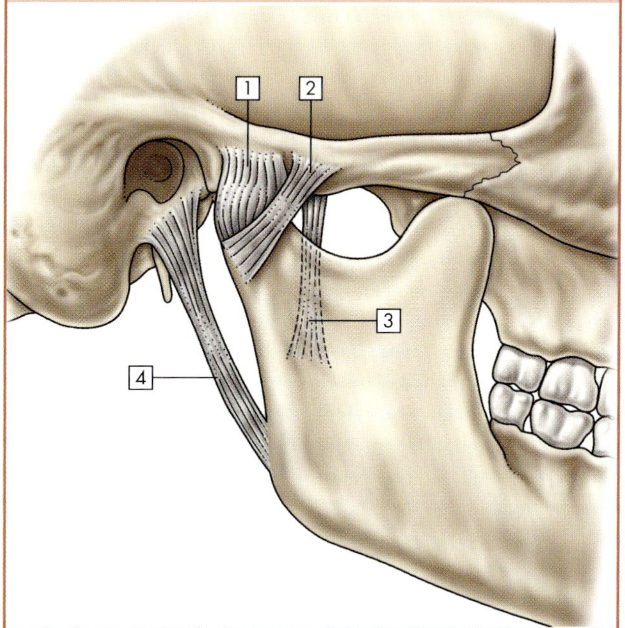

1 Capsule
2 Lateral ligament
3 Sphenomandibular ligament
4 Stylomandibular ligament

Although a medial ligament has been described, it is much less conspicuous than the lateral ligament so that the medial displacement of the TMJ is likely to be prevented by the lateral ligament of the opposite side.

The accessory ligaments of the temporomandibular joint are the stylomandibular ligament, the sphenomandibular ligament and the pterygomandibular raphe (Figs 2.11, 2.12). However, only the sphenomandibular ligament is likely to have any significant influence upon mandibular movements.

The stylomandibular ligament is a reinforced lamina of the deep cervical fascia as it passes medial to the parotid salivary gland. It extends from the tip of the styloid process and from the stylohyoid ligament to the angle of the mandible.

The sphenomandibular ligament is a remnant of the perichondrium of the embryonic first branchial arch cartilage. It runs from the spine of the sphenoid bone to the lingula near the mandibular foramen. From the spine of the sphenoid bone, it continues through the petrotympanic fissure into the middle ear to attach to the anterior process of the malleus. The sphenomandibular ligament is slack when the jaws are closed, but becomes tense at about the time when the condyle has passed in front of the lateral ligament.[20]

Recently, a new ligament has been described in association with the TMJ. This is the retinacular ligament which arises from the articular eminence, descends along the ramus of the mandible and is inserted into the fascia overlying the masseter muscle at the angle of the

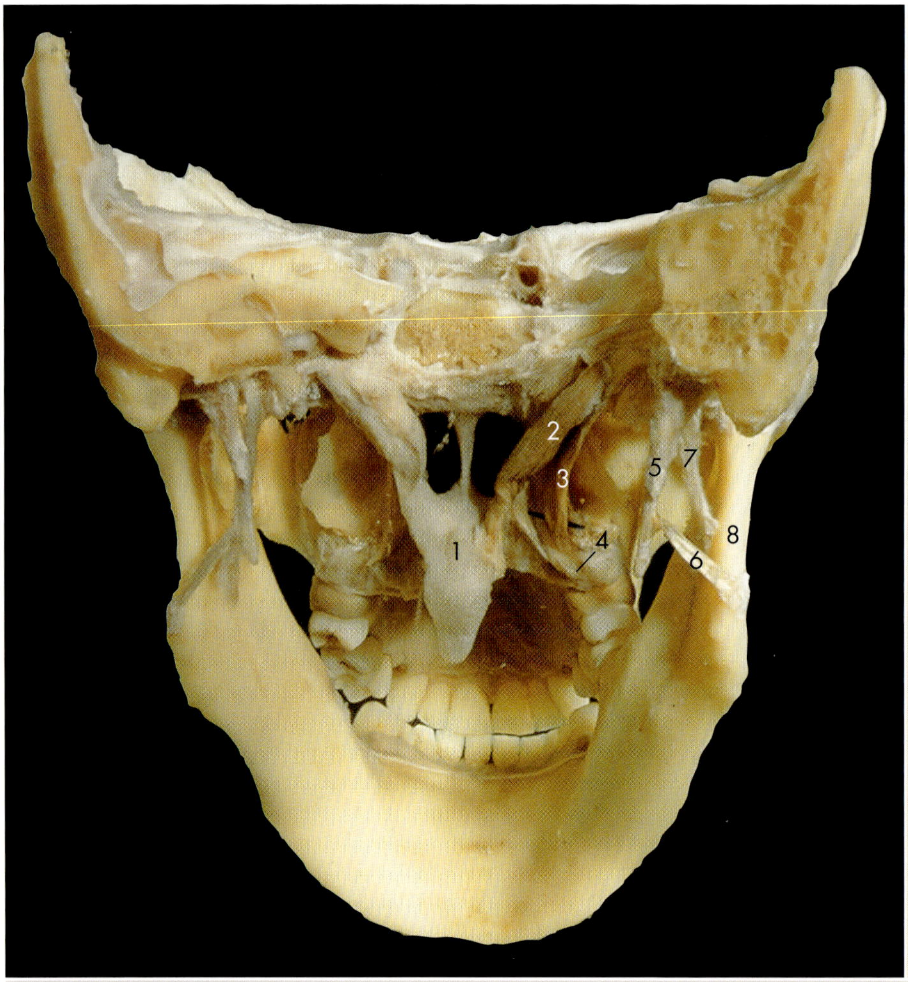

1 Soft palate
2 Levator veli palatini muscle
3 Tensor veli palatini muscle
4 Pterygoid hamulus
5 Styloid process
6 Stylomandibular ligament
7 Sphenomandibular ligament
8 Posterior border of ramus of mandibule

Figure 2.12 Posterior view of the jaws showing the stylomandibular and sphenomandibular ligaments. (Courtesy of the Royal College of Surgeons of England.)

mandible. As the ligament is connected with the posterolateral aspect of the retrodiscal tissues, and contains an accompanying vein, it may function in maintaining blood circulation during masticatory jaw movements.[31]

THE INNERVATION AND VASCULATURE OF THE TEMPOROMANDIBULAR JOINT

The TMJ is richly innervated, particularly its upper aspect. Of special significance are the encapsulated proprioceptive nerve endings important in the reflex control of mastication. Free nerve endings associated with nociception are also present.

Innervation for the joint is provided by the auriculotemporal, masseteric and deep temporal nerves.

The auriculotemporal branch of the mandibular division of the trigeminal nerve winds around the back of the TMJ, between it and the external acoustic meatus, before ascending in front of the tragus of the auricle to the temporal region (Figs 1.8, 1.9, 1.12). It provides multiple branches supplying the TMJ.

The masseteric branch of the mandibular division of the trigeminal nerve passes through the mandibular notch to enter the posterior surface of the masseter muscle and, during its course, also gives multiple branches supplying the TMJ.

The (posterior) deep temporal branch of the mandibular division of the trigeminal nerve arises in the infratemporal fossa and, passing up to supply the temporalis muscle (Fig. 1.9), provides a branch to the TMJ.

Additional sources of supply for the TMJ have been reported to be provided by the facial nerve[32] and the otic ganglion.

The vascular supply is derived from the superficial temporal artery and the maxillary artery (Fig. 1.13) (anterior tympanic and deep auricular branches). Other branches from neighbouring arteries may also contribute (e.g. deep temporal and transverse facial arteries).

HISTOLOGICAL CONSIDERATIONS

The articular surfaces of the temporomandibular joint are lined by fibrous tissue. This reflects the development of the joint. Unlike other synovial joints whose articular surfaces develop endochondrally and are therefore lined by hyaline cartilage, the temporomandibular joint develops intramembranously.

The histological appearance of the condylar cartilage is related to age, as a secondary cartilage is present until puberty. In the adult, the bony condylar head is covered superficially by the fibrous articular zone (Fig. 2.13). A layer containing an increased number of

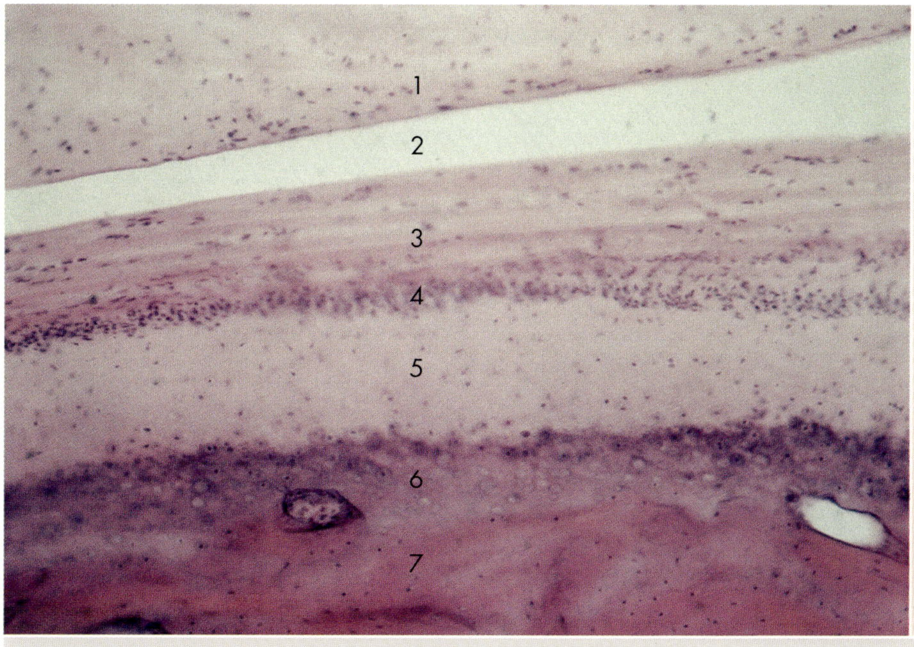

1 Articular disc
2 Lower joint cavity
3 Fibrous articular zone
4 Zone of increased cell density
5 Zone of fibrocartilage
6 Calcified cartilage
7 Bone of condyle

Figure 2.13 An histological section of part of an adult mandibular condyle. Haematoxylin and eosin, x100.

nuclei can be observed in the lower region of the covering fibrous articular layer, indicative of a proliferative zone. Two additional layers have been described below the proliferative layer: firstly, a region regarded as being composed of fibrocartilage, beneath which is a thin zone of calcified cartilage, representing the remains of the secondary condylar cartilage.

In the condyle of a child, the articular surface is again lined by a layer of fibrous tissue. Beneath the articular layer is a proliferative layer of undifferentiated cells (Fig. 2.14). Cells from this proliferative layer pass more deeply where they differentiate into chondrocytes which form the secondary condylar cartilage. The chondrocytes subsequently hypertrophy and the site undergoes endochondral ossification. Unlike a primary cartilage, the secondary condylar cartilage has less extracellular matrix, the cartilage cells themselves do not undergo cell division and do not align themselves into columns. Although once thought to be a prime causative factor in controlling mandibular growth, the secondary condylar cartilage is now not thought to have any intrinsic growth potential. The condylar cartilage disappears at about the age of 16 years.

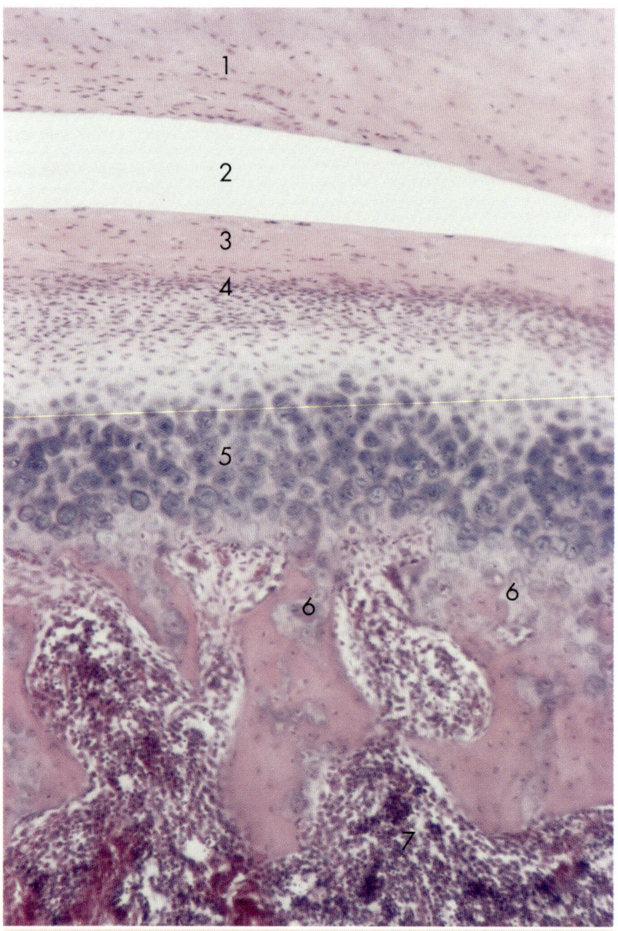

1 Articular disc
2 Lower joint cavity
3 Fibrous articular zone
4 Proliferative zone
5 Secondary condylar cartilage
6 Zone of endochondral ossification
7 Marrow space in condyle

Figure 2.14 A histological section of a condyle in a child. Haematoxylin and eosin, x90.

Regions of cartilage cells may also be seen beneath the articular fibrous covering of the mandibular fossa, including the articular eminence, but these are less conspicuous than is the case with the condyle.

The articular disc is comprised of dense fibrous tissue, the fibres being principally of type I collagen. The bulk of the cells are fibroblasts. However, more rounded, cartilage-like cells have been described within the disc (although whether their presence is age related or functionally related is not known). The disc, therefore, has been described as fibrocartilaginous.

THE PTERYGOPALATINE FOSSA

The pterygopalatine fossa lies beneath the infratemporal (posterior) surface of the maxilla and the pterygoid process of the sphenoid bone. The pterygopalatine fossa contains the maxillary nerve, the maxillary artery (third part) and the pterygopalatine parasympathetic ganglion.

The elongated cleft between the posterior surface of the maxilla and the pterygoid process of the sphenoid bone is the pterygomaxillary fissure which forms the lateral aspect of the pterygopalatine fossa (Fig. 2.15). The anterior wall of the fossa is the infratemporal surface of the maxilla. The posterior wall of the fossa is the pterygoid process below and the greater wing of the sphenoid above. The medial wall is formed by the perpendicular plate of the palatine bone. The pyramidal process of the palatine bone is situated inferiorly and articulates with the tuberosity of the maxilla. It fills the triangular gap between the lower ends of the medial and lateral pterygoid plates.

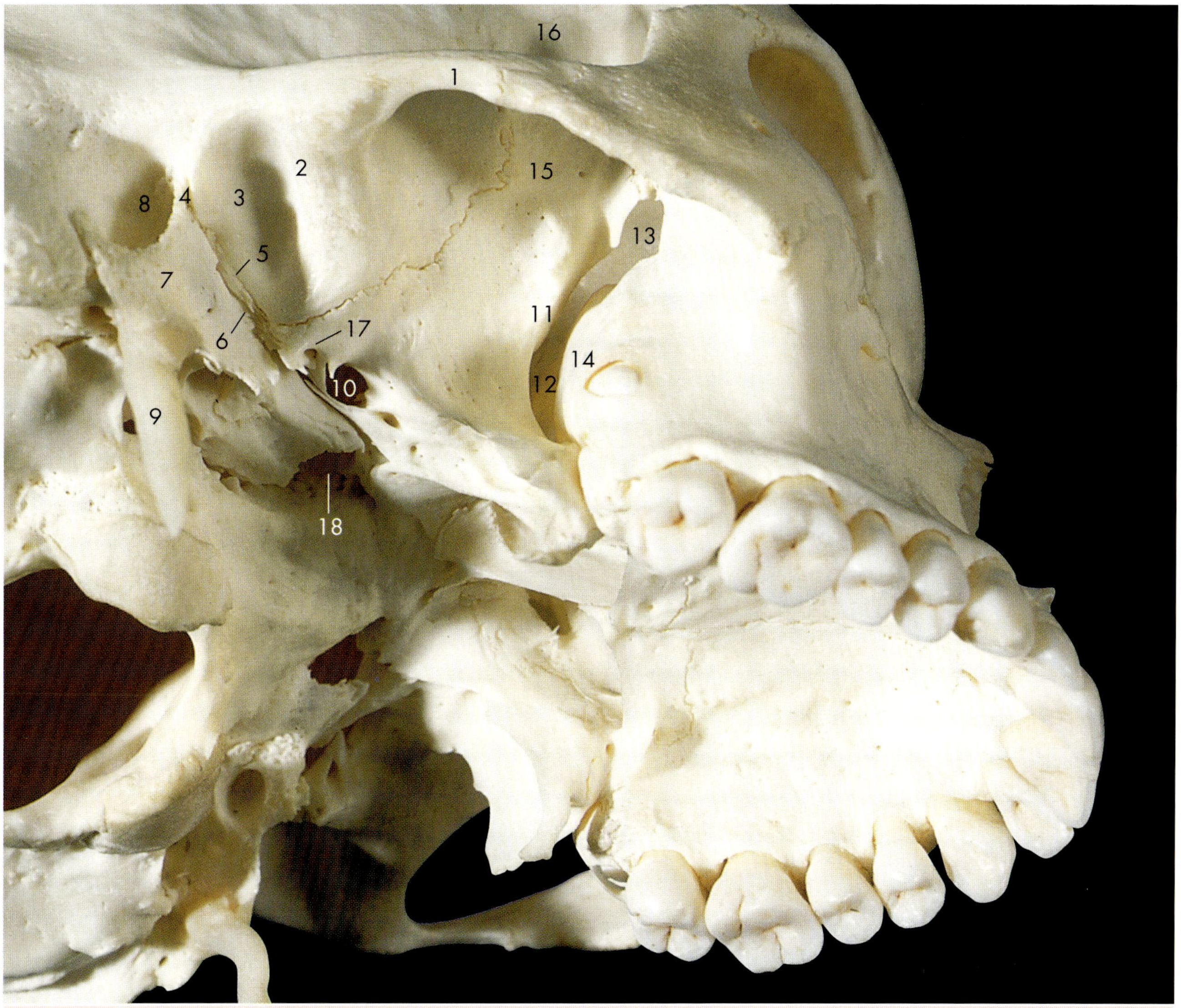

1 Zygomatic arch
2 Articular eminence
3 Mandibular fossa
4 Squamotympanic fissure
5 Petrotympanic fissure
6 Petrosquamous fissure
7 Tympanic plate
8 External acoustic meatus
9 Styloid process
10 Foramen ovale
11 Lateral pterygoid plate
12 Pterygomaxillary fissure leading into pterygopalatine fossa
13 Inferior orbital fissure
14 Posterior surface of maxilla
15 Infratemporal crest of sphenoid bone
16 Greater wing of sphenoid bone
17 Foramen spinosum
18 Foramen lacerum

Figure 2.15 The skull showing the position of the pterygopalatine fossa.

Laterally, the pterygopalatine fossa communicates with the infratemporal fossa through the pterygomaxillary fissure. The fissure continues above with the posterior end of the inferior orbital fissure in the floor of the orbit.

The pterygomaxillary fissure transmits the maxillary artery from the infratemporal fossa, the posterior superior alveolar branches of the maxillary division of the trigeminal nerve and the sphenopalatine veins. Passing through the inferior orbital fissure from the pterygopalatine fossa are the infraorbital and zygomatic branches of the maxillary nerve, the orbital branches of the pterygopalatine ganglion and the infraorbital vessels.

Entering the pterygopalatine fossa posteriorly are the foramen rotundum from the middle cranial fossa, and the pterygoid canal from the region of the foramen lacerum at the base of the skull (Fig. 2.16). The foramen rotundum (occupying the greater wing of the sphenoid bone) lies above, and lateral to, the pterygoid canal. The maxillary

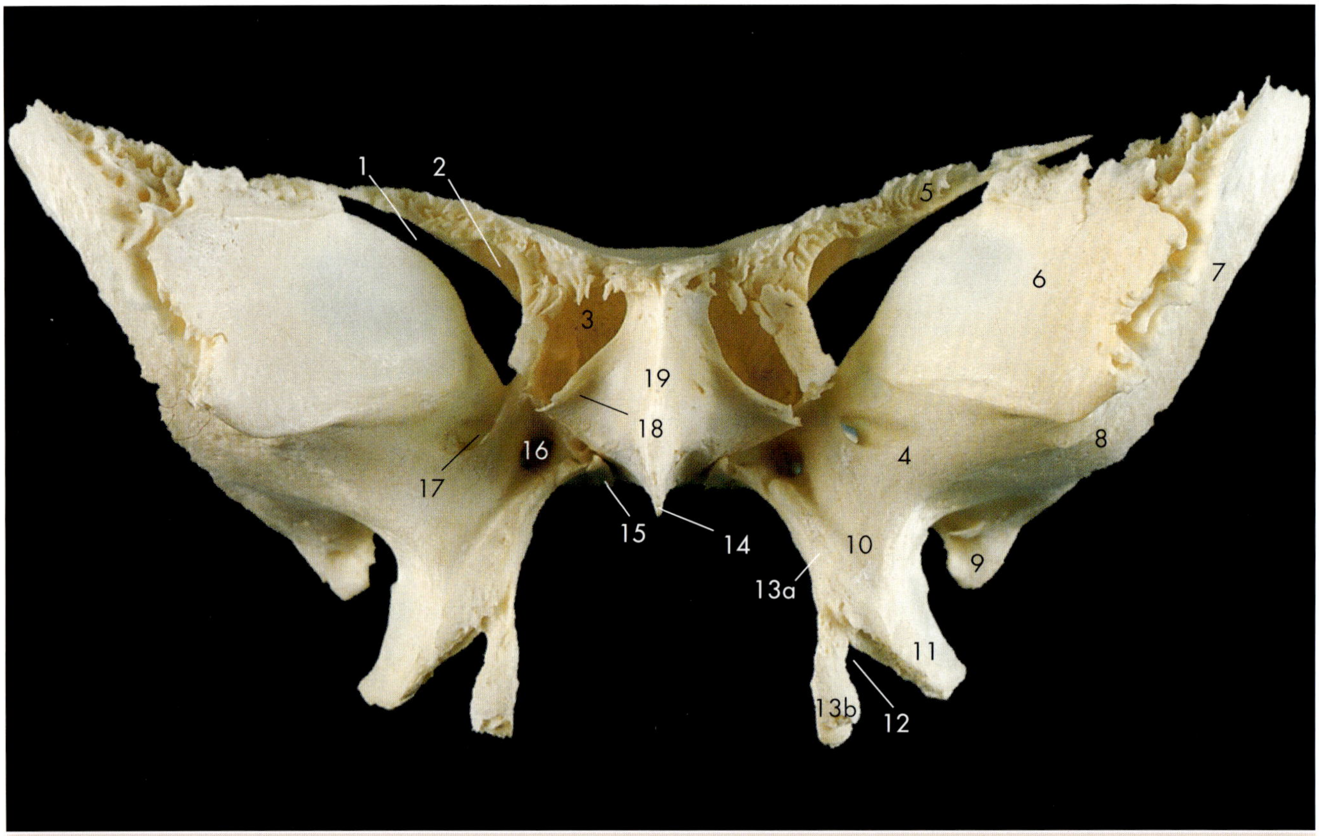

1 Superior orbital fissure
2 Optic canal
3 Sphenoidal air sinus in body of sphenoid
4 Maxillary surface of greater wing
5 Frontal margin of lesser wing
6 Orbital surface of greater wing
7 Temporal surface of greater wing
8 Infratemporal crest of greater wing
9 Spine
10 Pterygoid process
11 Lateral pterygoid plate
12 Pterygoid notch
13 Medial pterygoid plate (13a) surmounted by pterygoid hamulus (13b)
14 Rostrum
15 Vaginal process
16 Pterygoid canal
17 Foramen rotundum
18 Concha
19 Crest

Figure 2.16 The anterior surface of the sphenoid bone showing the foramen rotundum and pterygoid canal.

division of the trigeminal nerve passes through the foramen rotundum. The pterygoid canal transmits the greater petrosal and deep petrosal nerves (which combine to form the nerve of the pterygoid canal) and an accompanying artery derived from the maxillary artery.

Lying high up on the medial wall of the pterygopalatine fossa is the sphenopalatine foramen. It is formed by the notch between the orbital and sphenoid processes of the perpendicular plate of the palatine bone (Fig. 2.17), articulating with the body of the sphenoid bone. This foramen communicates with the lateral wall of the nasal cavity. It transmits the nasopalatine and posterior superior nasal nerves (from the pterygopalatine ganglion) and the sphenopalatine vessels.

At the base of the pterygopalatine fossa is found the opening of a palatine canal. This canal is formed when the greater palatine groove running down the posterior margin of the lateral surface of the perpendicular plate of

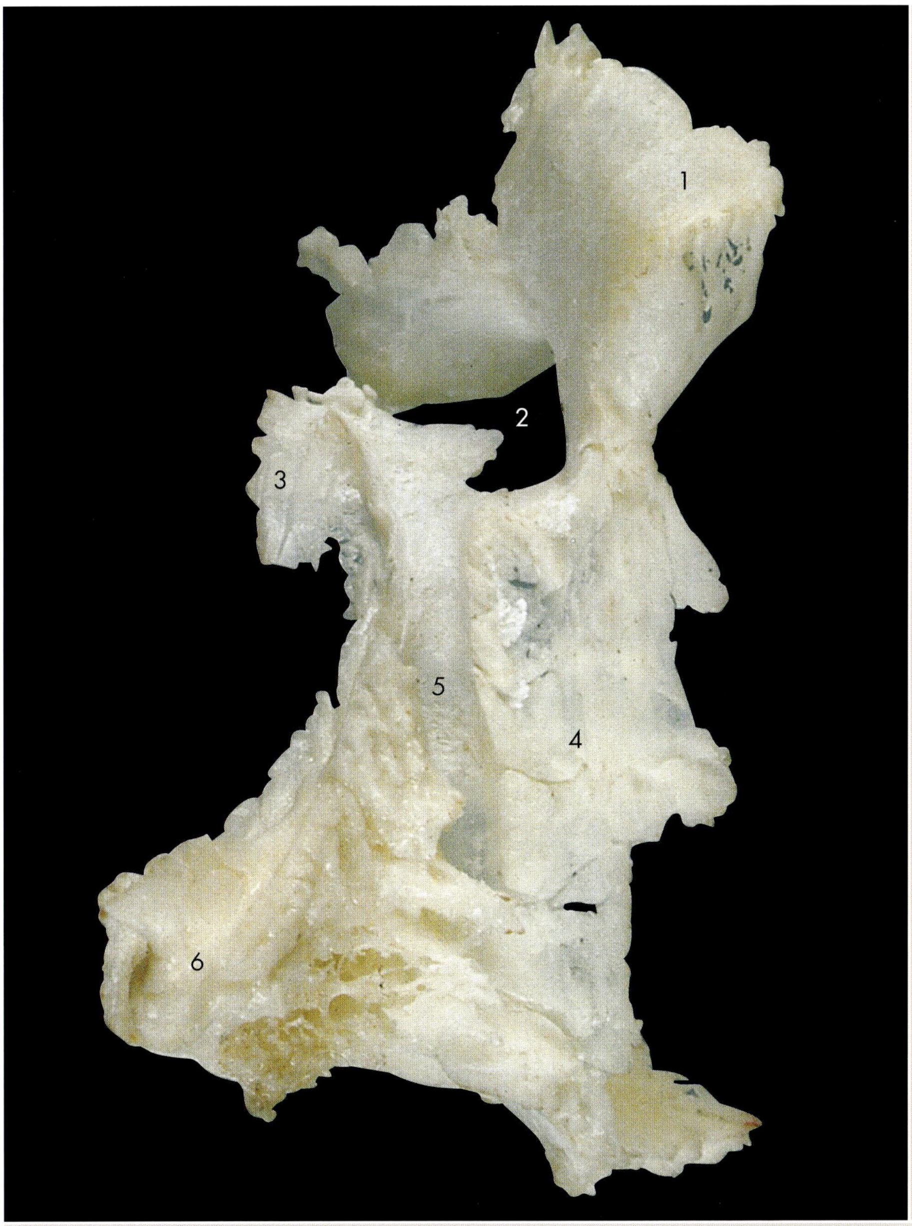

1 Orbital process
2 Sphenopalatine notch
3 Sphenoidal process
4 Perpendicular plate
5 Greater palatine groove
6 Pyramidal process

Figure 2.17 Lateral view of the palatine bone showing the sphenopalatine notch.

the palatine bone (Fig. 2.17) articulates with the posterior surface of the maxillary bone and the medial pterygoid plate. Lower down, the canal divides into greater and lesser palatine canals. The lesser palatine canal runs backwards in the pyramidal process of the palatine bone. The greater palatine canal enters the hard palate at the greater palatine foramen in the region of the transverse palatine suture. The lesser palatine canal enters the hard palate at the lesser palatine foramen (foramina). The palatine canal transmits the greater and lesser palatine nerves (and the posterior inferior nasal branches from the pterygopalatine ganglion), together with accompanying vessels, and these pass to the hard palate to emerge at the greater and lesser palatine foramina.

THE MAXILLARY NERVE

This division of the trigeminal nerve (the fifth cranial nerve) contains only sensory fibres. Functionally, it supplies the maxillary teeth and their supporting structures, the hard and soft palate, the maxillary air sinus, much of the nasal cavity, and skin overlying the middle part of the face (Figs 2.18, 2.19).

Figure 2.18 Branches of the maxillary nerve.

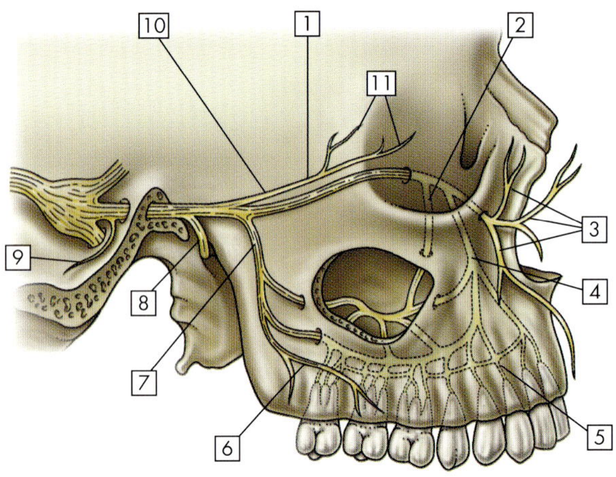

1 Infra-orbital nerve
2 Middle superior alveolar nerve
3 Terminal branches of infraorbital nerve
4 Anterior superior alveolar nerve
5 Dental plexus of nerves
6 Gingival branch of posterior superior alveolar nerve
7 Posterior superior alveolar nerve
8 Branches to pterygopalatine ganglion
9 Meningeal nerve
10 Zygomatic nerve
11 Zygomaticofacial and zygomaticotemporal branches

The maxillary nerve arises from the trigeminal ganglion on the floor of the middle cranial fossa. It passes along the lateral dural wall of the cavernous sinus to exit the cranial cavity at the foramen rotundum (Fig. 2.19). It emerges from the foramen rotundum in the upper part of the pterygopalatine fossa, where most of the branches are derived. These branches can be classified into those which come directly from the maxillary nerve, and those which are associated with the pterygopalatine parasympathetic ganglion.

- Branches from the main maxillary nerve trunk:
  - Meningeal nerve
  - Ganglionic branches
  - Zygomatic nerve
    - zygomaticotemporal nerve
    - zygomaticofacial nerve
  - Posterior superior alveolar nerve
  - Infraorbital nerve
    - middle superior alveolar nerve
    - anterior superior alveolar nerve
- Branches from the pterygopalatine ganglion:
  - Orbital nerve
  - Nasopalatine nerve
  - Posterior superior nasal nerve
  - Posterior inferior nasal nerve
  - Greater (anterior) palatine nerve
  - Lesser (posterior) palatine nerve
  - Pharyngeal branch

The meningeal nerve

This is the only branch from the main trunk of the maxillary nerve that does not originate in the pterygopalatine fossa; it arises within the middle cranial fossa, before the foramen rotundum. It runs with the middle meningeal artery and innervates the dura mater lining the middle cranial fossa.

The ganglionic branches

These are usually two in number and connect the maxillary nerve to the pterygopalatine ganglion.

The zygomatic nerve

This leaves the pterygopalatine fossa through the inferior orbital fissure. It passes along the lateral wall of the

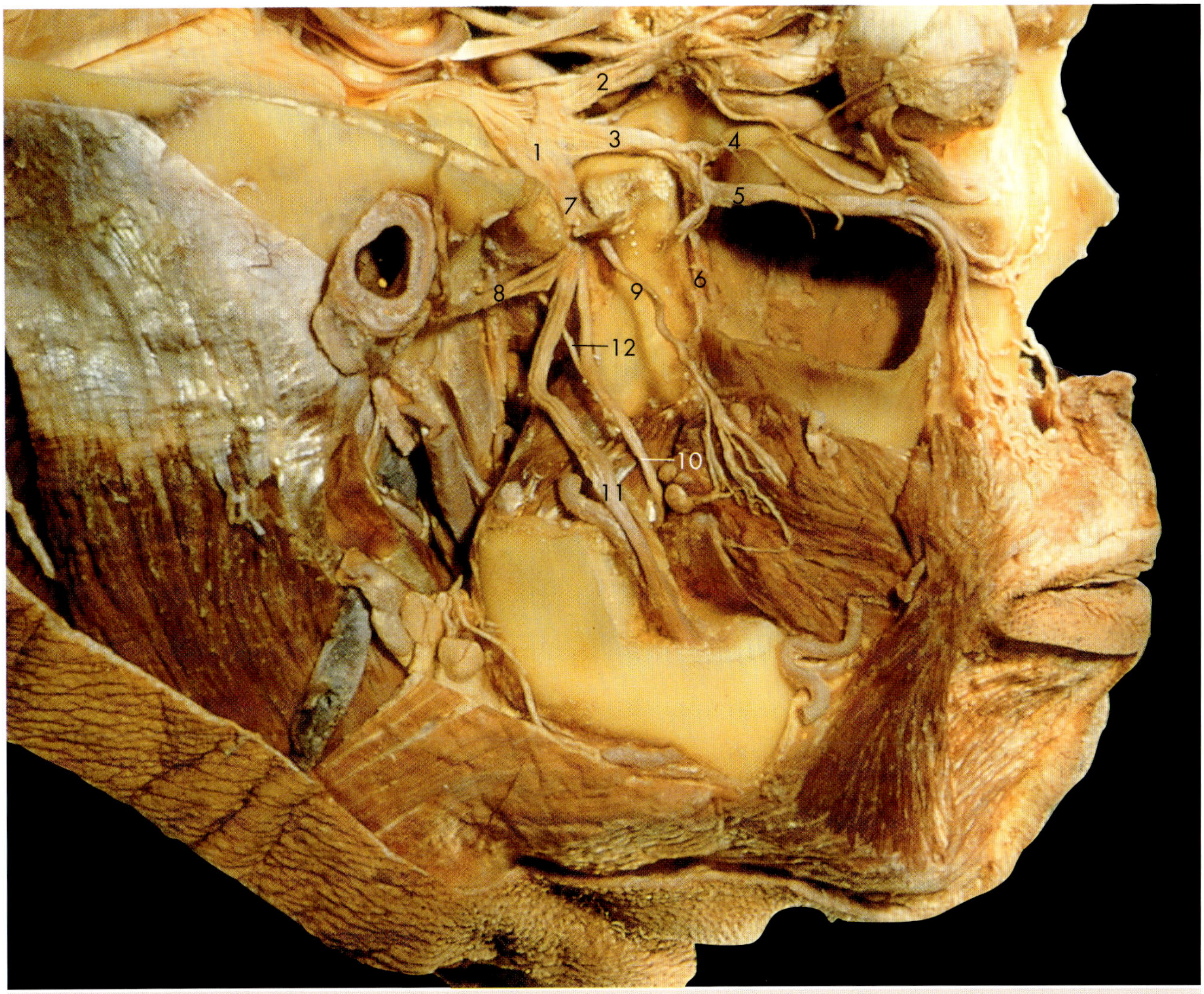

1 Trigeminal ganglion
2 Ophthalmic nerve
3 Maxillary nerve
4 Zygomatic nerve dividing into zygomaticofacial and zygomaticotemporal branches
5 Infraorbital nerve
6 Posterior superior alveolar nerve
7 Mandibular nerve
8 Auriculotemporal nerve
9 Buccal nerve
10 Lingual nerve
11 Inferior alveolar nerve
12 Chorda tympani nerve

Figure 2.19 Dissection of the maxillary and mandibular nerves. (Courtesy of the Royal College of Surgeons of England.)

orbit before dividing into zygomaticotemporal and zygomaticofacial branches. These pass through the zygomatic bone to supply overlying skin. The zygomaticotemporal nerve also gives a branch to the lacrimal nerve, which carries autonomic fibres to the lacrimal gland.

The posterior superior alveolar nerve(s)

(Figs 2.19, 2.20)

This is one of three superior alveolar nerves that supply the maxillary teeth. The middle and anterior superior alveolar nerves are branches of the infraorbital nerve (see below). The posterior superior alveolar nerve(s) leaves the pterygopalatine fossa through the pterygomaxillary fissure. Thence, it runs onto the tuberosity of the maxilla and eventually pierces the bone to supply the maxillary molar teeth and the maxillary sinus (Fig. 2.20). Before entering the maxilla, the nerve provides a gingival branch which innervates the buccal gingivae around the maxillary molars. The extra-bony course of the posterior superior alveolar nerve is variable. The nerve can subdivide into several branches just before, or just after, it enters the maxilla. Alternatively, it may arise as several distinct branches at the main trunk of the maxillary nerve.

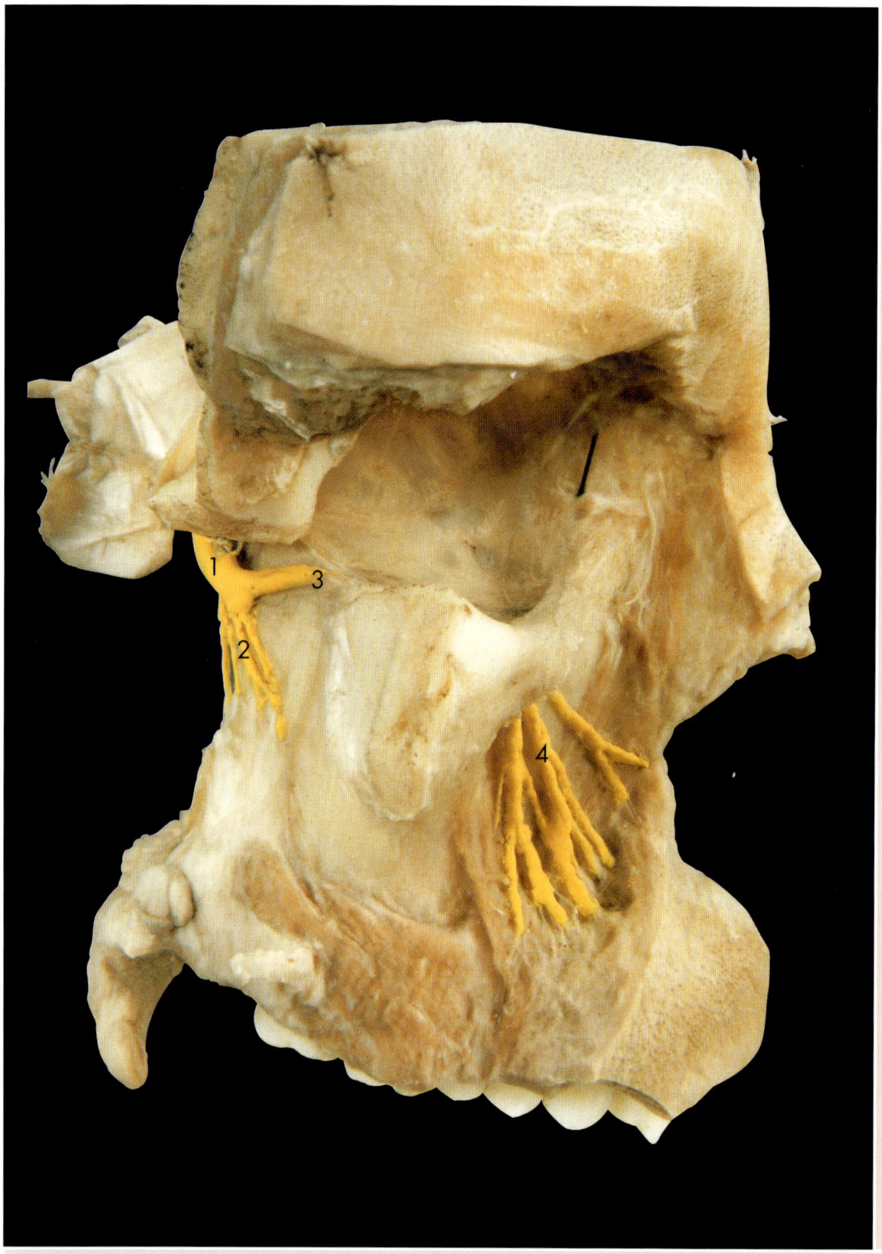

1 Maxillary nerve
2 Posterior superior alveolar nerves
3 Infraorbital nerve in floor of orbit
4 Infraorbital nerve entering face at infraorbital foramen

Figure 2.20 Frontal aspect of the face showing the posterior superior alveolar and infraorbital nerves. (Courtesy of Professor C. Dean, Department of Anatomy and Developmental Biology, University College London.)

The infraorbital nerve (Figs 2.19, 2.20)

This can be regarded as the terminal branch of the maxillary nerve proper. It leaves the pterygopalatine fossa to enter the orbit at the inferior orbital fissure. Initially lying in a groove in the floor of the orbit (the infraorbital groove), the infraorbital nerve runs into a canal (the infraorbital canal) and passes onto the face at the infraorbital foramen. The middle and anterior superior alveolar nerves arise from the infraorbital nerve in the orbit.

The branches of the maxillary nerve that arise with the pterygopalatine ganglion contain not only sensory fibres from the maxillary nerve, but also autonomic fibres from the ganglion, which are mainly distributed to glands and blood vessels.

The orbital nerve

This passes from the pterygopalatine ganglion into the orbit through the inferior orbital fissure. It supplies periosteum and, via sympathetic fibres, the orbitalis muscle. The orbital nerve can also supply part of the maxillary sinus and may pass through the posterior ethmoidal foramen to innervate posterior ethmoidal air cells and the sphenoid air sinus.

The nasopalatine nerve (Fig. 2.21)

This nerve runs medially from the pterygopalatine ganglion into the nasal cavity through the sphenopalatine foramen. It passes across the roof of the nasal cavity to reach the back of the nasal septum. The nasopalatine nerve then passes downwards and forwards within a groove on the vomer to supply the posteroinferior part of the nasal septum. It passes through the incisive canal, where it usually forms a single nerve with its fellow of the opposite side, and emerges on the hard palate at the incisive fossa to supply the oral mucosa around the incisive papilla and palatal gingiva of the anterior teeth.

The posterior superior nasal nerve (Figs 2.19, 2.20)

This nerve enters the back of the nasal cavity through the sphenopalatine foramen. It divides into lateral and medial branches. The lateral branches supply the posterosuperior part of the lateral wall of the nasal fossa. The medial branches cross the roof of the nasal cavity to supply the nasal septum overlying the posterior part of the perpendicular plate of the ethmoid.

The posterior inferior nasal nerve

This supplies the inferior part of the lateral wall of the nose in the region of the inferior nasal concha. It may arise directly from the pterygopalatine ganglion or appear as a branch from the anterior palatine nerve.

The greater (anterior) palatine nerve (Figs 2.21, 2.23)

This nerves passes downwards from the pterygopalatine ganglion, through the palatine canal, and onto the hard palate at the palatine foramen (Fig. 2.21). Within the greater palatine canal, it can give off nasal branches that innervate the posteroinferior part of the lateral wall of the nasal fossa. On the palate, it runs forwards at the interface between the palatine process and the alveolar process of the maxilla to supply much of the mucosa of the hard palate and palatal gingivae (except around the incisive papilla).

The lesser (posterior) palatine nerve(s) (Figs 2.21, 2.23)

This passes downwards from the pterygopalatine ganglion initially through the palatine canal. It then passes through the lesser palatine canal in the pyramidal process of the palatine bone and onto the palate at the lesser palatine foramen (or foramina). It runs backwards to supply the soft palate.

The pharyngeal branch

This originates from the pterygopalatine ganglion and passes through the palatovaginal canal to supply the mucosa of the nasopharynx. The palatovaginal canal is formed when the groove on the undersurface of the vaginal process of the sphenoid bone articulates with the upper surface of the sphenoid process of the palatine bone. The pharyngeal branch has also been reported to pass through the vomerovaginal canal, which generally transmits the pharyngeal branch of the sphenopalatine artery. The vomerovaginal canal lies between the upper surface of the vaginal process of the sphenoid bone and the ala of the vomer and is often continuous with the pterygoid canal.

The Pterygopalatine Ganglion (Figs 2.21–2.23; see also Fig. 1.5)

This parasympathetic ganglion is situated below the maxillary nerve in the pterygopalatine fossa, connected by

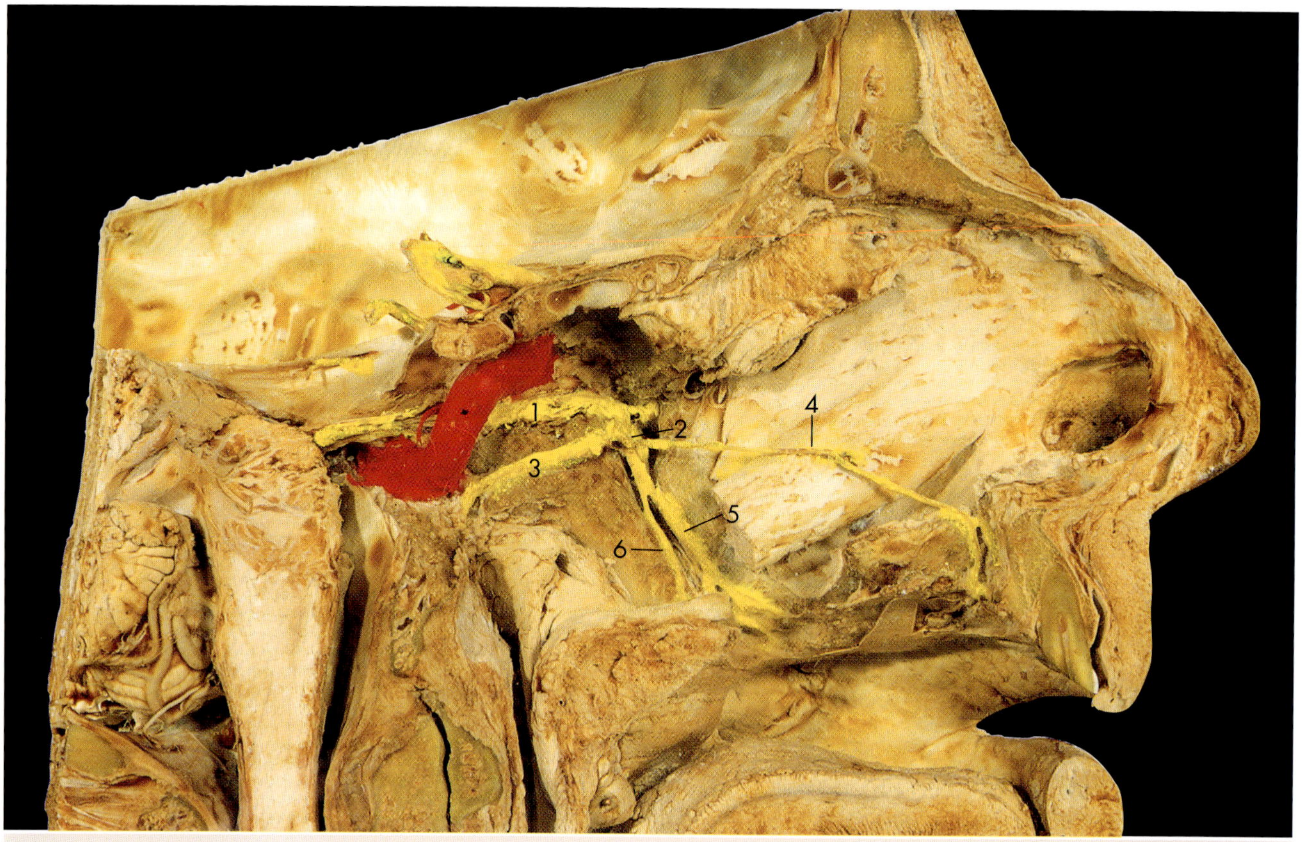

1 Maxillary nerve
2 Pterygopalatine ganglion
3 Nerve of pterygoid canal
4 Nasopalatine nerve dissected from nasal septum
5 Greater palatine nerve
6 Lesser palatine nerve

Figure 2.21 Dissection showing the pterygopalatine ganglion and the palatine nerves. (Courtesy of Professor C. Dean, Department of Anatomy and Developmental Biology, University College London.)

two ganglionic branches. It is concerned primarily with supplying the nose, palate, and lacrimal gland (Fig. 2.22).

As with other parasympathetic ganglia in the head, three types of fibres enter the pterygopalatine ganglion: parasympathetic, sympathetic, and sensory fibres. However, only the parasympathetic fibres synapse in the ganglion. The preganglionic parasympathetic fibres originate from the superior salivatory nucleus in the brainstem. The fibres pass with the nervus intermedius of the facial nerve. They subsequently emerge as the greater (superficial) petrosal nerve. This occurs within the facial canal of the temporal bone, close to the geniculate ganglion of the facial nerve. The greater petrosal nerve then passes through the bone to appear on the floor of the middle cranial fossa. It then runs medially in a shallow groove to the foramen lacerum. Passing within the foramen lacerum, the greater petrosal nerve enters the pterygoid canal which lies at the base of the pterygoid process (Fig. 2.16). After passing along the pterygoid canal, the nerve emerges into the pterygopalatine fossa and joins the pterygopalatine ganglion (Fig. 2.23). Postganglionic sympathetic fibres run to the pterygopalatine ganglion by a complex course. From the superior cervical ganglion, sympathetic fibres run to the internal carotid plexus surrounding the internal carotid artery (Fig. 2.23). From this plexus, a branch called the deep petrosal nerve is given off that enters the pterygoid canal to reach the

Figure 2.22 The pterygopalatine parasympathetic ganglion.

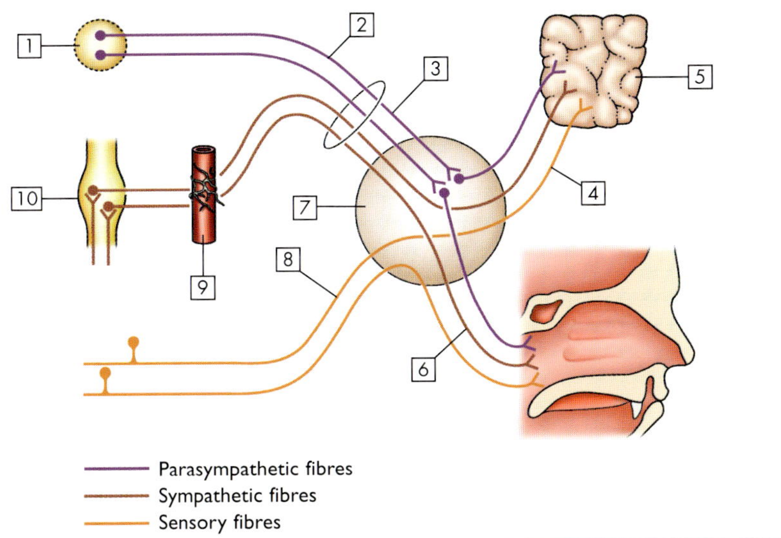

1 Superior salivatory nucleus in brainstem
2 Greater petrosal nerve
3 Nerve of pterygoid canal
4 Zygomatic and lacrimal nerves
5 Lacrimal gland
6 Pharyngeal, nasal and palatine branches of pterygopalatine ganglion
7 Pterygopalatine ganglion
8 Maxillary nerve (ganglionic branch)
9 Internal carotid artery
10 Superior cervical ganglion

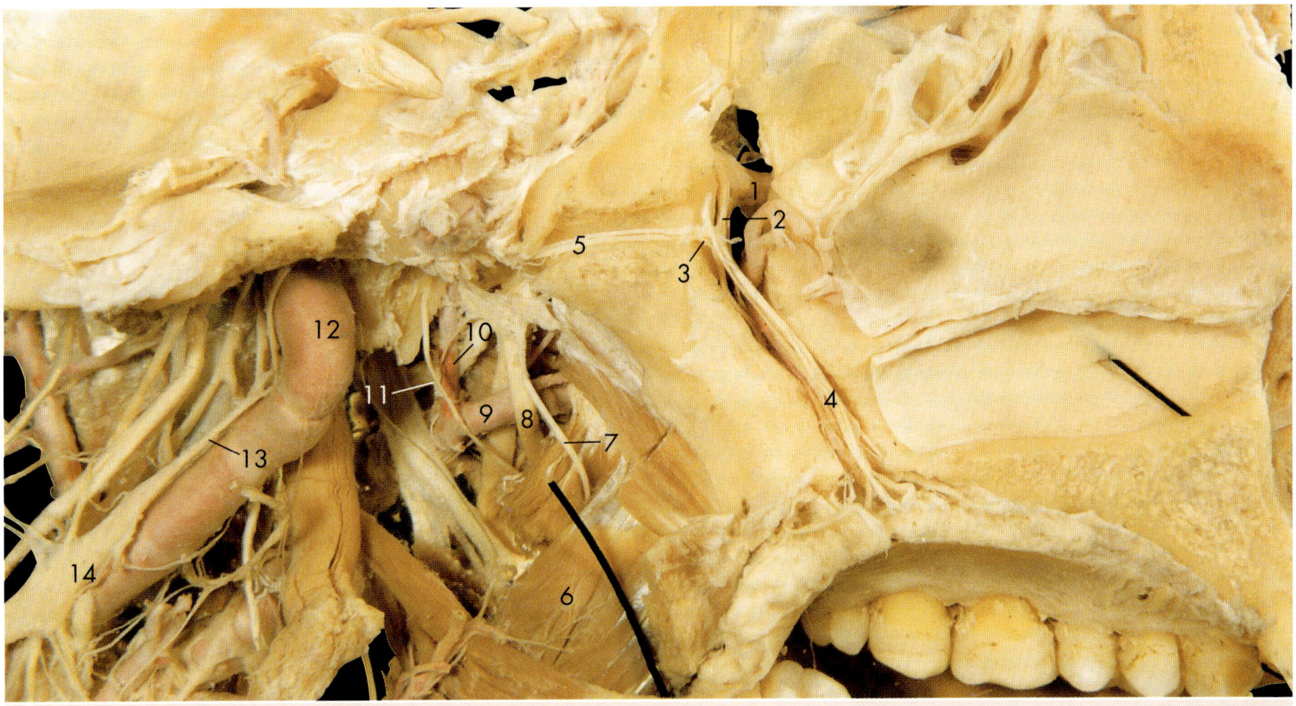

1 Maxillary nerve
2 Ganglionic branches of maxillary nerve
3 Pterygopalatine ganglion
4 Palatine nerves from pterygopalatine ganglion
5 Nerve of pterygoid canal
6 Medial pterygoid muscle
7 Mylohyoid nerve
8 Inferior alveolar nerve
9 Maxillary artery
10 Middle meningeal artery
11 Auriculotemporal nerve
12 Internal carotid artery
13 Internal carotid nerve from superior cervical sympathetic ganglion
14 Superior cervical sympathetic ganglion

Figure 2.23 A dissection of the pterygopalatine ganglion. (Courtesy of Professor S. Standing, GKT School of Biomedical Sciences, London.)

pterygopalatine ganglion. The greater petrosal nerve and the deep petrosal nerve join within the pterygoid canal to become the nerve of the pterygoid canal. The sensory fibres to the ganglion run in the ganglionic branches of the maxillary nerve.

The nerves leaving the pterygopalatine ganglion are the orbital nerve, the nasopalatine nerve, the greater and lesser palatine nerves, the posterior superior and inferior nasal nerves, and the pharyngeal nerve. These nerves are described above with the maxillary nerve. The parasympathetic component will be distributed within these nerves to supply the minor salivary glands. The parasympathetic component of the pterygopalatine ganglion is also responsible for supplying the lacrimal gland. The fibres first pass from the ganglion in one of the ganglionic branches to the maxillary nerve. They then travel with the zygomatic and zygomaticotemporal branches. Within the orbit, they pass from the zygomaticotemporal nerve to the lacrimal nerve (of the ophthalmic nerve) to reach the lacrimal gland.

THE MAXILLARY ARTERY (Figs. 1.13, 2.24, 5.8)

The maxillary artery continues from the infratemporal fossa into the pterygopalatine fossa through the pterygomaxillary fissure. It terminates within the pterygopalatine fossa, where it is called the third part of the maxillary artery. The third part of the maxillary artery gives branches that accompany the branches of the maxillary nerve (including those associated with the pterygopalatine ganglion).

The posterior superior alveolar artery arises from the maxillary artery within the pterygopalatine fossa (or occasionally from the infraorbital artery) and runs through the pterygomaxillary fissure onto the maxillary tuberosity. It supplies the maxillary molar and premolar teeth, their buccal gingivae, and the maxillary air sinus (Fig. 2.24).

The infraorbital artery enters the orbit through the inferior orbital fissure. It runs along the floor of the orbit in the infraorbital groove and the infraorbital canal to emerge onto the face at the infraorbital foramen. The infraorbital artery gives off the anterior superior alveolar artery within the infraorbital canal. This branch runs downwards to supply the anterior teeth and the anterior part of the maxillary sinus. The infraorbital artery on the face supplies the lower eyelid, part of the cheek, the side of the external nose, and the upper lip.

The artery of the pterygoid canal passes through the canal to provide branches to part of the auditory tube and the tympanic cavity of the ear, and the upper part of the pharynx. The maxillary artery also provides a pharyngeal branch which passes through the vomerovaginal canal to the nasopharynx. The pharyngeal branch of the maxillary nerve passes through the palatovaginal canal, accompanying its respective nerve from the pterygopalatine ganglion, to be distributed to the region of the nasopharynx.

The descending palatine artery leaves the pterygopalatine fossa through the palatine canal. Within this canal, it divides into the greater and lesser palatine arteries. The greater palatine artery supplies the inferior meatus of the lateral wall of the nose before passing onto the roof of the palate at the greater palatine foramen. It runs forwards to supply the hard palate and the palatal gingivae of the maxillary teeth. It also provides a branch which runs up into the incisive canal to anastomose with the sphenopalatine artery, thereby contributing to the supply of nasal septum. The lesser palatine artery (or arteries) emerges on to the palate at the greater palatine foramen (or foramina). It supplies the soft palate.

The sphenopalatine artery is the last branch of the maxillary artery to be considered, arising beyond the origin of the descending palatine artery. It enters the lateral wall of the nose through the sphenopalatine foramen (Fig.1.12). The artery initially accompanies the posterior superior nasal nerve and gives off branches to supply much of the posterior part of the lateral wall of the nose. The sphenopalatine artery then crosses the roof of the nose to accompany the nasopalatine nerve and to supply the posteroinferior part of the nasal septum.

THE VEINS OF THE PTERYGOPALATINE FOSSA

The veins of the pterygopalatine fossa are small and variable. The most consistent is the sphenopalatine vein. This vein drains the posterior aspect of the nose and passes into the pterygopalatine fossa through the sphenopalatine foramen. It drains into the pterygoid venous plexus via the pterygomaxillary fissure.

The inferior ophthalmic vein in the floor of the orbit provides a connecting branch to the pterygoid venous plexus (Fig. 1.14). This vein passes through the inferior orbital fissure in the region of the pterygopalatine fossa.

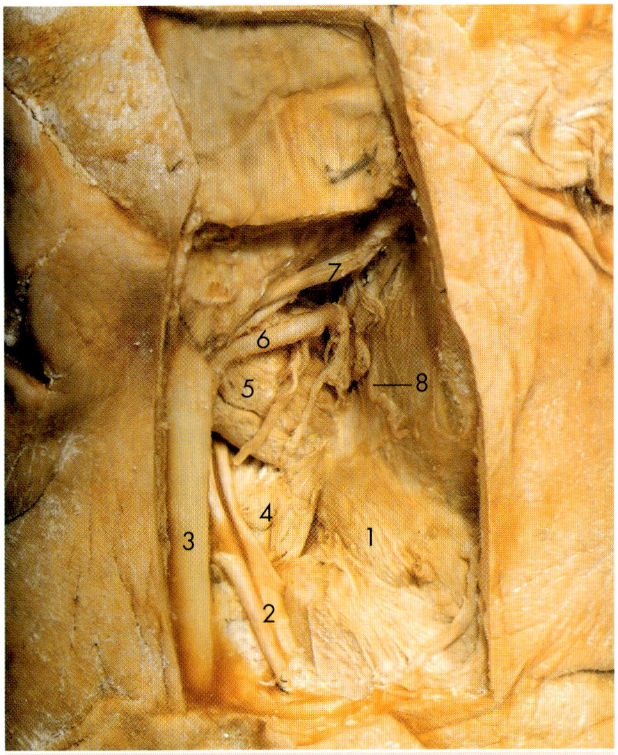

1 Buccinator muscle
2 Inferior alveolar nerve and artery
3 External carotid artery
4 Medial pterygoid muscle
5 Lateral pterygoid muscle (lower head)
6 Maxillary artery
7 Lateral pterygoid muscle (upper head)
8 Posterior superior alveolar artery

Figure 2.24a The maxillary artery after removal of the lateral pterygoid muscle. (Courtesy of Professor L. Garey, Anatomy Department, Imperial College Medical School, London.)

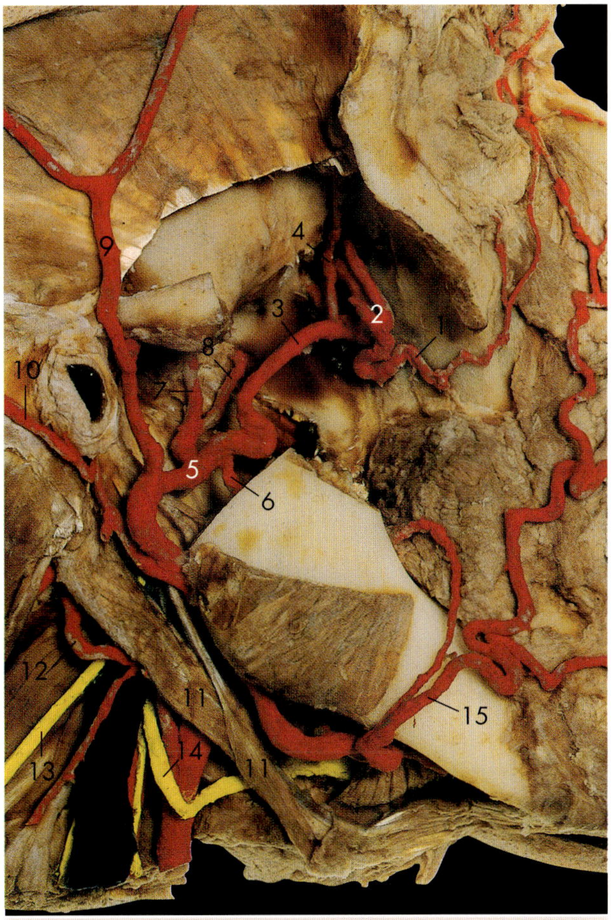

1 Posterior superior alveolar artery
2 Third part (pterygopalatine part) of maxillary artery
3 Second part (pterygoid part) of maxillary artery
4 Deep temporal artery
5 First part (mandibular part) of maxillary artery
6 Inferior alveolar artery
7 Accessory meningeal artery
8 Middle meningeal artery
9 Superficial temporal artery
10 Posterior auricular artery
11 Posterior belly of digastric and stylohyoid muscles
12 Occipital artery
13 Spinal accessory nerve
14 Hypoglossal nerve
15 Facial artery

Figure 2.24b Lateral view of infratemporal fossa with pterygoid muscles removed to show course of maxillary artery. (Courtesy of Professor L. Garey, Anatomy Department, Imperial College Medical School, London.)

SOME SURGICAL ASPECTS OF THE TEMPOROMANDIBULAR JOINT

An understanding of the surgical anatomy of the TMJ is essential when undertaking either arthroscopy or open surgical procedures on the joint.

TMJ arthroscopy was first described by Ohnishi in 1980.[33] It is used both as a diagnostic procedure and also to treat some TMJ disorders. Arthroscopic diagnosis of TMJ disorders is useful for abnormalities of the synovium, capsule, articular disc and condylar head. Biopsies can be obtained and synovial fluid aspirated for diagnostic purposes.

The breakdown of fibrous adhesions and the lavage of the joint to eliminate debris often results in rapid relief of pain and can sometimes help with the 'closed lock' where the disc is anteriorly displaced. The disc can be released, returned to its rest position and, in expert hands, it can be sutured in place. In hypermobile joints with recurrent subluxation or dislocation sclerosants can be injected. Synovitis accompanying inflammatory changes responds to joint lavage and the instillation of steroids. Irregular and rough surfaces of fibrocartilage or subchondral bone can be smoothed and reshaped using powered rotary cutting instruments introduced through the endoscope.[34–38]

Figure 2.25 Trago-canthal line measurements; posterolateral, middle lateral and anterolateral puncture points. (Redrawn after Phillips, 1998.[38])

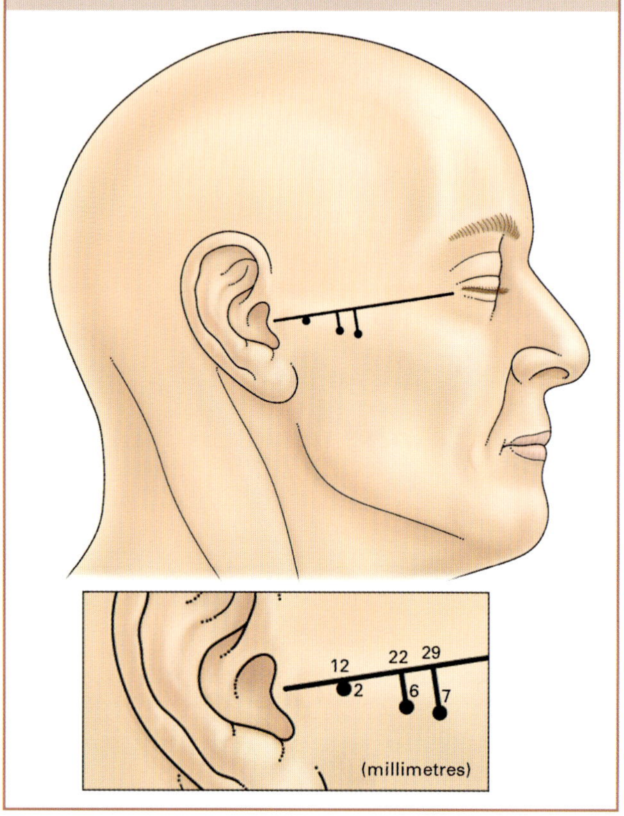

SURGICAL ANATOMY OF TMJ ARTHROSCOPY

Following routine skin preparation, a line is drawn from the outer canthus of the eye to the midpoint of the tragus of the ear. Measuring from the most distal point of the tragus, points are marked at 12 mm, 22 mm and 29 mm along this line. Perpendicular lines are now drawn from these points 2 mm, 6 mm, and 7 mm in length respectively (Fig. 2.25).[39] These lines mark the entry points of the arthroscope.

After the injection of approximately 2 ml of 0.25% bupivacaine into the upper joint space, a small stab incision is made at the 12/2 mm point (Figs 2.26 and 2.27). This avoids the auriculotemporal nerve, facial nerve branches and the superficial temporal blood

Figure 2.26 Distension of upper joint space with bupivacaine. (Redrawn after Phillips, 1998.[38])

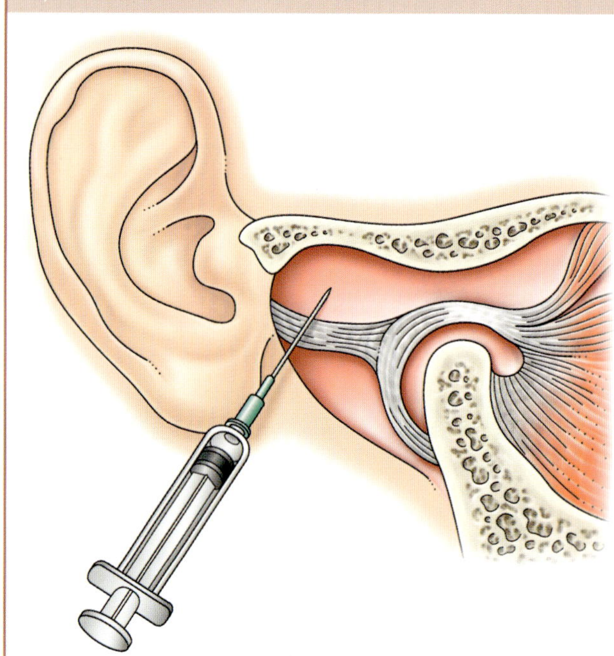

Figure 2.27 Optional stab incisions. (Redrawn after Phillips, 1998.[38])

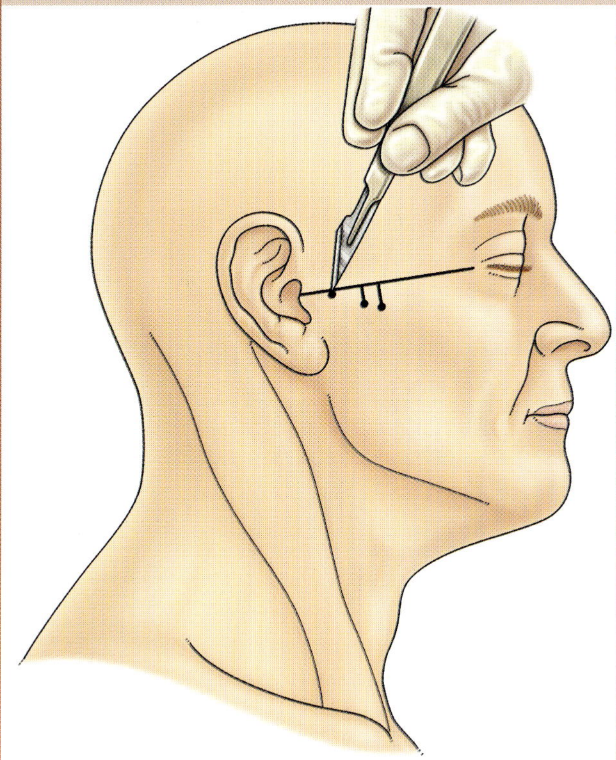

vessels. Using a sharp trocar, the cannula is introduced through the skin pointing forwards, upwards and inwards at approximately 45° (Fig. 2.28). The lateral surface of the capsule of the joint is about 2 cm deep and offers some resistance to the trocar. With a gentle twisting movement, the trocar enters the upper joint cavity. The trocar is then removed and, if the cannula is in the joint space, some clear fluid (bupivacaine) will run out. The endoscope can now be passed through the cannula and into the joint (Fig. 2.29). The assistant now maintains pressure with the irrigating solution to maintain distension of the joint. A 21G hypodermic needle is next inserted into the joint at the 22/6 mm point (Fig. 2.30). Initially, bloodstained fluid drains from the needle but, with continuous irrigation, the fluid becomes clear.

The upper joint compartment can be examined systematically looking for hypervascularity of the synovium, villonodular folds, adhesions or fibrosis, as well as anatomical changes such as disc displacement (Fig.2.30). Considerable experience is required to identify the arthroscopic anatomy of the joint. The anterior slope of the articular eminence (tubercle) cannot be visualised with the arthroscope entering the joint space from

Figure 2.28 Posterolateral puncture with rotatory movements of sharp trocar and cannula. (Redrawn after Phillips, 1998.[38])

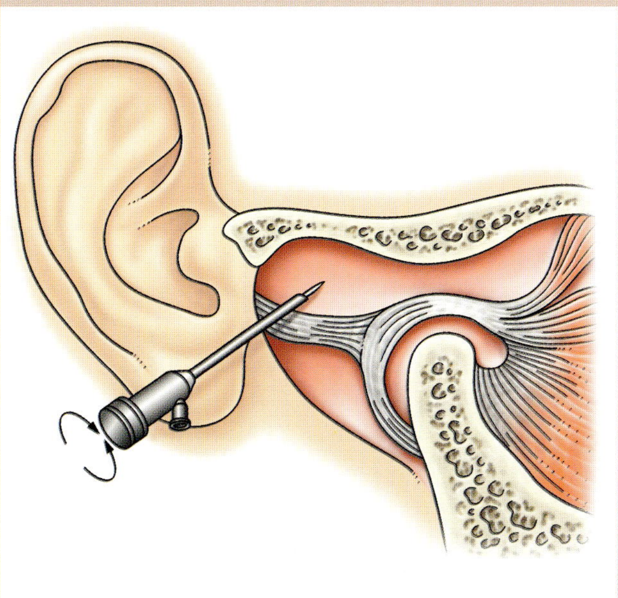

Figure 2.29 Passing the arthroscope from the back of the superior joint space into the anterior region. (Redrawn after Phillips, 1998.[38])

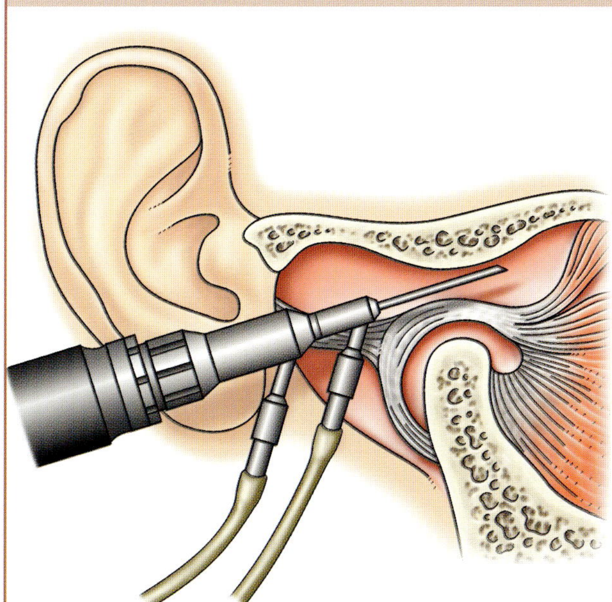

behind via the 12/2 mm point. Instead, the arthroscope is introduced from in front via an additional stab incision at the 29/7 mm point (Fig. 2.31). A second cannula should be used for this entry, leaving the original cannula in place.

Surgical complications are relatively few and transient in nature.[40] The tympanic membrane should always be examined both before and following arthroscopy. Should a perforation occur, the patient should be prescribed antibiotic–steroid eardrops for 2 weeks. Most small perforations will heal spontaneously. If they do not, the patient should be referred to an otologist surgeon. Perforation of the roof of the mandibular fossa may occur with resultant leakage of cerebrospinal fluid. The roof of the fossa is on average only 0.9 mm thick. Should such a leak occur, formal neurosurgical closure of the dura will be required.

Figure 2.30 Endaural triangulation of the back of the upper joint space. (Redrawn after Phillips, 1998.[38])

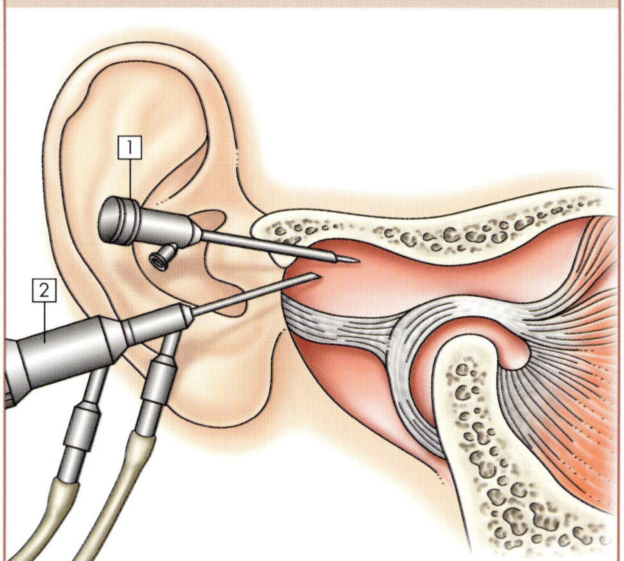

1 Endaural cannula and trocar
2 Posterolateral arthroscope

Figure 2.31 Triangulation using two separate portals of entry. (Redrawn after Phillips, 1998.[38])

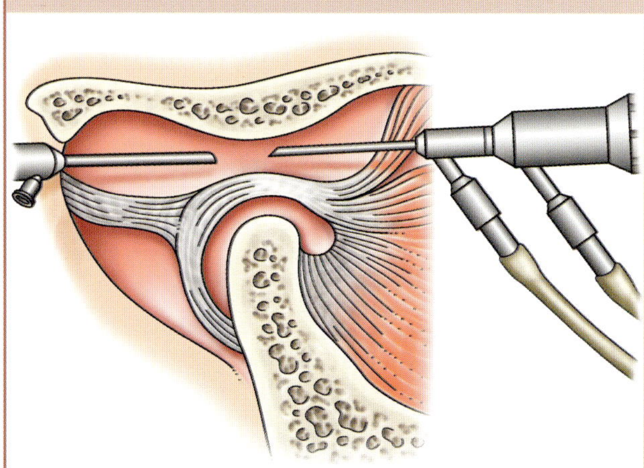

Nerves (auriculotemporal branch of the mandibular nerve or temporal and zygomatic branches of the facial nerve) may be damaged during TMJ arthroscopy. Spontaneous recovery is usual, although the patient may require a temporary lateral tarsorrhaphy to protect the cornea. Bleeding from the joint is prevented by adding adrenaline (1:200,000) to the irrigating fluid. Bleeding from the overlying soft tissues is controlled with a pressure dressing.

OPEN ACCESS TO THE TEMPOROMANDIBULAR JOINT

Open surgical access to the TMJ is often compromised by efforts to protect the facial nerve and its branches. Various approaches have been advocated, preauricular,[41] submandibular,[42] postauricular,[43] endaural,[44] intraoral,[45] temporal,[46] or by direct incision along the zygomatic arch.[47] None of these have gained universal acceptance due to either restricted access to the joint or unacceptable complications. In 1979, Al-Kayat and Bramley[48] described their modified preauricular approach to the TMJ which now forms the basis of all open surgical procedures on this joint. Their approach is based upon experiences gained by dissecting 56 cadaveric facial halves noting the detailed anatomical relations of the external auditory canal and the trunk and branches of the facial nerve. One of us (JL) uses a modification of the technique described by them.[49]

The skin incision is in two parts which meet at the insertion of the upper border of the auricle onto the temporal skin (Fig. 2.32). The preauricular incision starts at the lower border of the insertion of the ear lobe onto the skin and runs upwards around the edge of the tragus and up to the insertion of the auricle. The temporal incision runs from this point in a straight line at 45° to the zygomatic arch. This incision is deepened to the temporal fascia and root of the zygomatic arch (Fig.

2.33). It may be necessary to clamp and divide branches of the superficial temporal artery or vein. The preauricular incision is deepened by sharp and blunt dissection in front of the tragal cartilage in the avascular plane just anterior to the cartilaginous external acoustic meatus (Figs 2.34).

The superficial layer of the superficial temporal fascia is incised along the length of the temporal skin incision (to reveal the underlying fat) and through the periosteum at the root of the zygomatic arch. The incised layer of superficial temporal fascia is separated from the fat by blunt dissection down to the upper border of the zygomatic arch to which the fascia is attached (Fig. 2.35). The periosteum overlying the arch

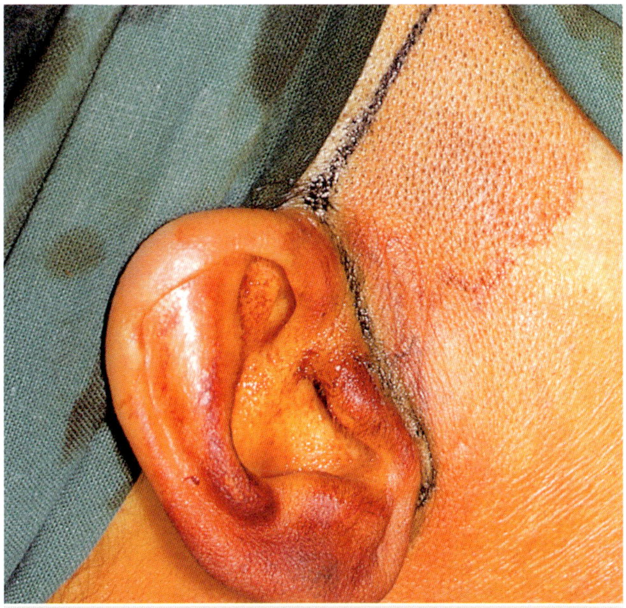

Figure 2.32 Open access to the TMJ: skin incision.

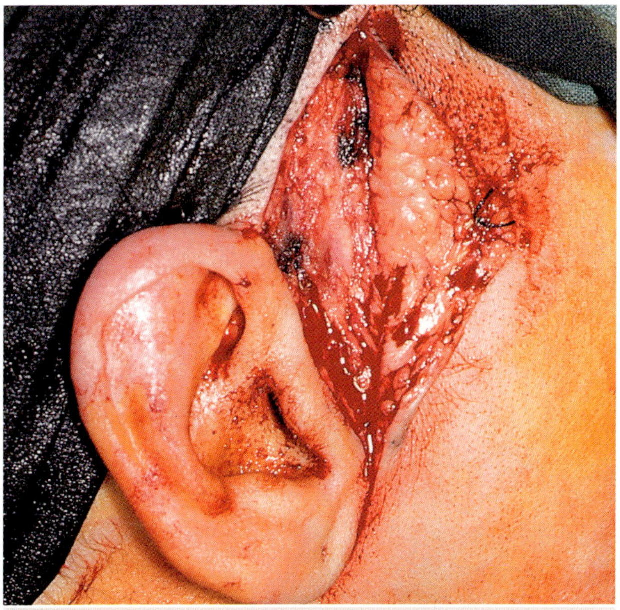

Figure 2.33 Open access to the TMJ: temporalis fascia.

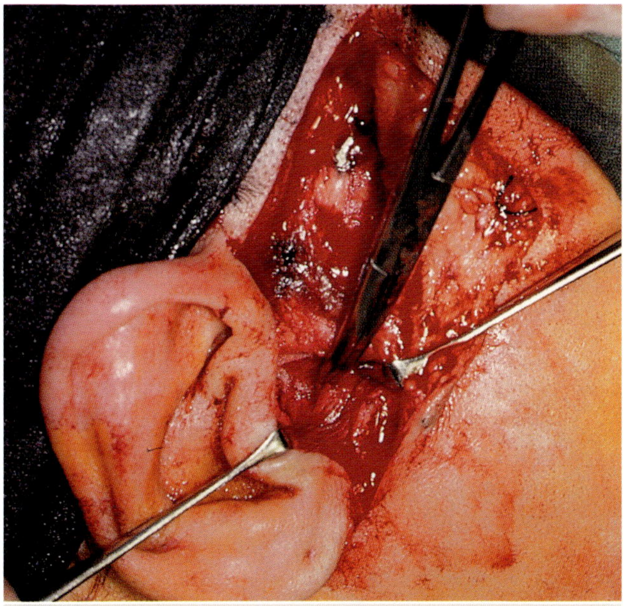

Figure 2.34 Open access to the TMJ: avascular plane.

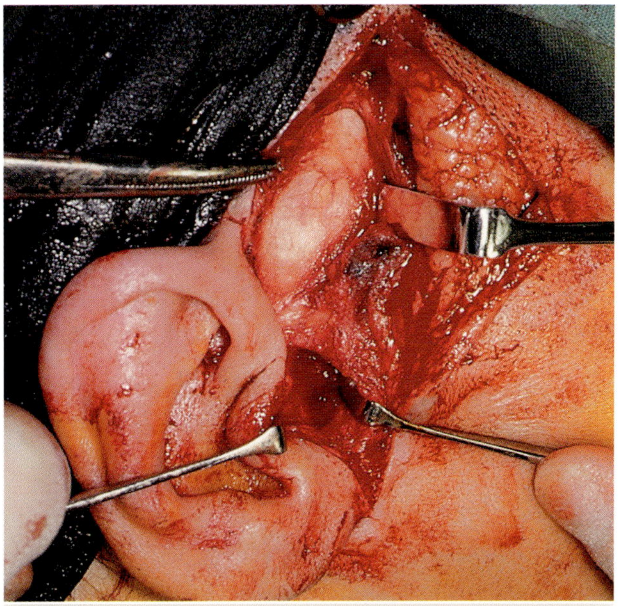

Figure 2.35 Open access to the TMJ: zygomatic arch.

is then elevated anteriorly for about 15 mm. The capsule of the joint is then identified below the arch by sharp and blunt dissection. The flap is then developed in this plane to reveal the articular tubercle and condylar neck (Fig. 2.36). At this point, the bifurcation of the facial nerve is no nearer than 2.4 cm in an inferoposterior direction from the postglenoid tubercle. All branches of the facial nerve run in a more superficial plane within the soft tissues raised in the flap.

On completion of TMJ surgery, a suction drain is laid behind the condylar neck and the wound closed in layers, repairing the outer layer of the temporal fascia with resorbable sutures.

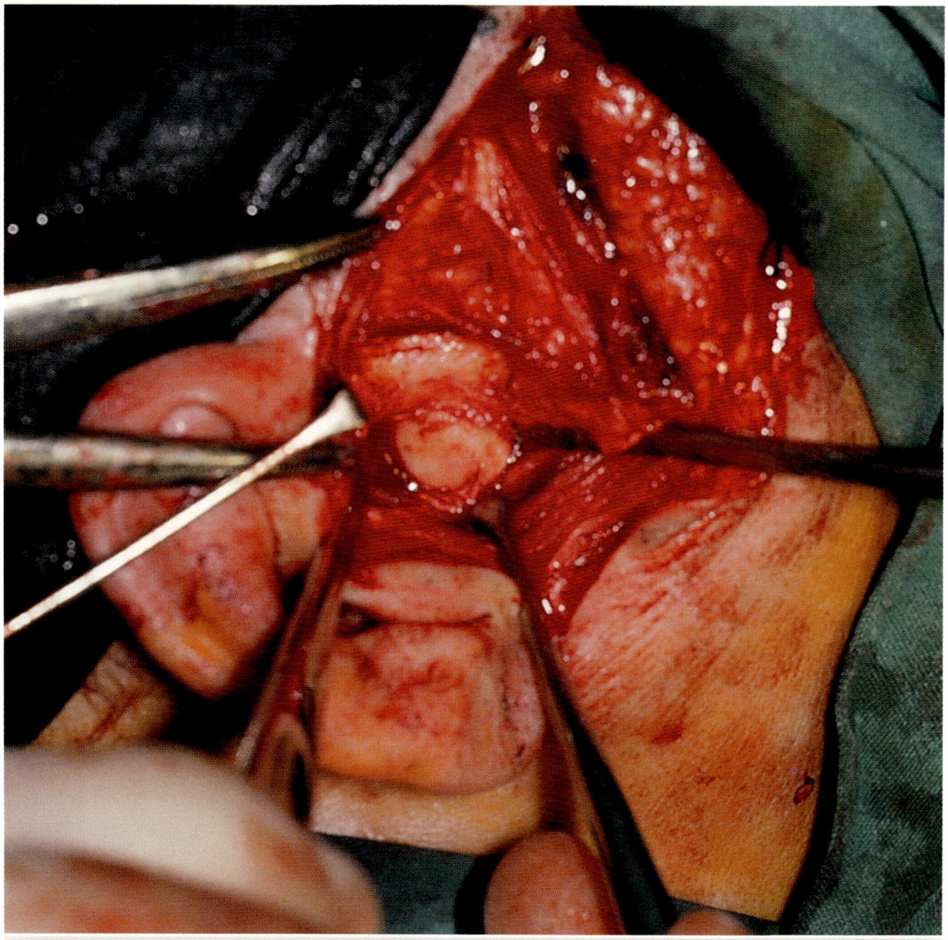

Figure 2.36 Open access to the TMJ: joint exposure.

REFERENCES

1. Oberg T, Carlsson G E. *Temporomandibular joint function and dysfunction.* St Louis: Mosby, 1979
2. McDevitt W E. *Functional anatomy of the masticatory system.* Wright (ed). London: Butterworth, 1989
3. Hawthorn R, Flatau A. Temporomandibular joint anatomy. In: Norman J E de B and Bramley P A (eds). *A textbook and colour atlas of the temporomandibular joint.* London: Wolfe, 1990; 1–51
4. Assael L A. Functional anatomy. In: Kaplan A S, Assael L A. *Temporomandibular disorders.* Philadelphia: Saunders, 1991; 2–10
5. McKay G S, Yemm R. The structure and function of the temporomandibular joint. *Br Dent J* 1992; 173: 127–132
6. Lang J. *Clinical anatomy of the masticatory apparatus and peripharyngeal spaces.* Stuttgart: Thieme, 1995
7. Dijkgraaf L C, De Bont L G M, Boering G, Liem R S B. Structure of the normal synovial membrane of the temporomandibular joint. *J Oral Maxillofac Surg* 1996; 54: 332–338
8. Hoffman H. Der supra- und intradiscoidale Gelenkspalt des menschlichen Kiefergelenks. [Thesis.] Wurzburg; 1973 (quoted by Lang in ref. 6)
9. Yung J P, Carpentier P, Marguelles-Bonnet R, Meunissier M. Anatomy of the temporomandibular joint and related structures in the frontal plane. *J Craniomandib Pract* 1990; 8: 101–107
10. Schmolke C. The relationship between the temporomandibular joint capsule, articular disc and jaw muscles. *J Anat* 1994; 184: 335–345
11. Nitzan D W, Mahler Y, Simkin A. Intra-articular pressure measurements in patients with suddenly developing severely limited mouth opening. *J Oral Maxillofac Surg* 1992; 50: 1038–1042
12. Dijkgraaf L C, De Bont L G M, Boering G, Liem R S B. Function, biochemistry, and metabolism of the normal synovial membrane of the temporomandibular joint: a review of the literature. *J Oral Maxillofac Surg* 1996; 54: 95–100
13. Wish-Baratz S, Ring G D, Hiss J *et al.* The microscopic structure and function of the vascular retrodiscal pad of the human temporomandibular joint. *Arch Oral Biol* 1993; 38: 265–268
14. Scalpino R P. The posterior attachment: its structure, function and appearance in TMJ imaging studies. Part 1. *J Craniomandib Diord* 1991; 5: 83–95
15. Scalpino R P. The posterior attachment: its structure, function and appearance in TMJ imaging studies. Part 2. *J Craniomandib Diord* 1991; 5: 155–166
16. Wilkinson T M, Crowley C M. A histologic study of retrodiscal tissues of the human temporomandibular joint in the open and closed position. *J Orofac Pain* 1994; 8: 7–17
17. Wilkinson T M. The relationship between the disk and lateral pterygoid muscle in the human temporomandibular joint. *J Prosthet Dent* 1988; 60: 715–724
18. McCarty W L. Internal derangements of the temporomandibular joint. In: Keith D A (ed) *Surgery of the temporomandibular joint.* Oxford: Blackwell, 1988
19. Wongwatana S, Kronman J H, Clark R E *et al.* Anatomic basis for disc displacement in temporomandibular joint (TMJ) dysfunction. *Am J Orthod Dentofac Orthop* 1994; 105: 257–264
20. Osborn J W. Internal derangement and the accessory ligaments around the temporomandibular joint. *J Oral Rehabil* 1995; 22: 731–740
21. Dolwick M F. A re-evaluation of the importance of disc position in temporomandibular disorders. *Am Dent J* 1996; 41: 184–187
22. Bertilsson O, Ström D. A literature survey of a hundred years of anatomic and functional lateral pterygoid muscle research. *J Orofac Pain* 1995; 9: 17–23
23. Naidoo L C D. Lateral pterygoid muscle and its relationship to the meniscus of the temporomandibular joint. *Oral Surg Oral Med Oral Pathol* 1996; 82: 4–9
24. Heylings D J A, Nielsen I L, McNeil C. Lateral pterygoid muscle and the temporomandibular disc. *J Orofac Pain* 1995; 9: 915
25. Velasco M J R, Vezquez R J F, Collado J J. The relationships between temporomandibular joint disc and related masticatory muscles in humans. *J Oral Maxillofac Surg* 1993; 51: 390–395
26. Loughner B A, Gremillion H A, Larkin L H *et al.* Muscle attachment to the lateral aspect of the articular disk of the human temporomandibular joint. *Oral Surg Oral Med Oral Pathol* 1996; 82: 139–144
27. Osborn J W. The disc of the human temporomandibular joint: design, function and failure. *J Oral Rehabil* 1985; 12: 279–293
28. Sato I, Shindo K, Ezure H, Shimada K. Morphology of the lateral ligament in the human temporomandibular joint. *Oral Surg Oral Med, Oral Pathol Oral Radiol Endod* 1996; 81: 151–156
29. Savalle W P M. Some aspects of the morphology of the human temporomandibular joint capsule. *Acta Anat* 1988; 131: 292–296
30. DuBrul E L. *Sicher and DuBrul's oral anatomy,* 8th edn. St Louis: Ishiyaku EuroAmerica, 1988
31. Shiraishi Y, Hayakawa M, Tanaka S, Hoshino T. A new retinacular ligament and vein of the human temporomandibular joint. *Clin Anat* 1995; 8: 208–213
32. Guerrier Y, Bolonyi F. L'innervation de l'articulation temporo-maxillaire. *Ann Otol Laryngol* 1948; 65: 109–111
33. Ohnishi M. Clinical application of arthroscopy in temporomandibular joint diseases. *Bull Tokyo Med Dent Univ* 1980; 27: 141–150
34. Israel H A, Technique for placement of a distal traction suture during temporomandibular joint arthroscopy. *J Oral Maxillofac Surg* 1989; 47: 311–313
35. Moses J P, Poker I D. TMJ arthroscopic surgery: an analysis of 237 patients. *J Oral Maxillofac Surg* 1989; 47: 790–794

36. Ohnishi M. Arthroscopic surgery for hypermobility and recurrent mandibular dislocation. *Oral Maxillofac Surg Clin N Am* 1989; 1: 153–164
37. Tarro A W. Arthroscopic treatment of anterior disc displacement; a preliminary report. *J Oral Maxillofac Surg* 1989; 47: 353–358
38. Tarro A W. A fully visualised arthroscopic disc suturing technique. *J Oral Maxillofac Surg* 1994; 52: 362–369
39. Phillips J G. Temporomandibular joint arthroscopy and endoscopy. In: Langdon J D, Patel M F (eds) *Operative maxillofacial surgery*. London: Chapman and Hall, 1998
40. Carls F R, Engleke W, Locher M C, Sailer H F. Complications following arthroscopy of the temporomandibular joint: analysis covering a 10 year period. *J Cranio-maxillofac Surg* 1996; 24: 12–15
41. Rowe N L. Surgery of the temporomandibular joint. *Proc R Soc Med* 1972; 65: 383
42. Ward T G. Surgery of the mandibular joint. *Ann R Coll Surg Eng* 1961; 28: 139
43. Alexander RW. Post-auricular approach for surgery of the temporomandibular articulation. *J Oral Surg* 1975; 33: 346
44. Hosxe G. Une voie d'abord de l'articulation temporo-mandibulaire. *Rev Stomatol Chir Maxillofac* 1972; 73: 375
45. Quinn J H. Lateral coronoid approach for intra-oral reduction of fractures of the zygomatic arch. *J Oral Surg* 1977; 35: 321
46. Gillies H D, Kilner TP, Stone D. Fractures of the malar–zygomatic compound with a description of a new X-ray position. *Br J Oral Surg* 1927; 14: 651
47. Hueston J T. The surgical exposure of the dislocated mandibular condyle. *Br J Plast Surg* 1959; 12: 275
48. Al-Kayat A, Bromley P. A modified preauricular approach to the temporomandibular joint and malar arch. *Br J Oral Surg* 1979; 17: 91–103
49. Juniper R P. Temporomandibular joint surgery. In: Langdon J D, Patel M F (eds) *Operative maxillofacial surgery*. London: Chapman and Hall, 1998

Chapter **3**

Local anaesthesia and the infratemporal fossa

J.D. LANGDON AND J.P. ROOD

BACKGROUND

A detailed knowledge of the infratemporal fossa and its contents is fundamental to successful regional anaesthesia in both the upper and lower jaws. The trigeminal nerve is predominantly sensory and the cell bodies of the sensory fibres are in the trigeminal ganglion which lies in Meckel's cave in the middle cranial fossa. Three large trunks originate in the ganglion: the ophthalmic, maxillary and the mandibular divisions.

The *ophthalmic* division is the smallest, which just before it enters the orbit via the superior orbital fissure, divides into three branches: the lacrimal nerve, the nasociliary nerve and the frontal nerve. The *lacrimal nerve* courses in a superoanterolateral direction to reach the lacrimal gland. It also innervates the conjunctiva and the skin of the lateral canthus of the eye. Postganglionic secretory fibres from the pterygopalatine ganglion reach the lacrimal nerve via the communicating branch of the zygomatic nerve. The *nasociliary nerve* crosses the orbital cavity in an anteromedial direction towards the medial orbital wall. The terminal branches (anterior ethmoidal, infratrochlear and external nasal) innervate the mucous membrane of the superoanterior part of the nasal cavity and also the skin between the nose and medial canthus of the eye. The *frontal nerve* continues in the direction of the ophthalmic nerve trunk dividing in the orbital cavity. The largest branch is the supraorbital nerve which leaves the orbit at the supraorbital notch to supply the skin of the upper eyelid, the forehead and the anterior of the scalp. The smaller supratrochlear nerve leaves the frontal nerve deep within the orbit and approaches the superomedial aspect of the orbit to innervate the upper lid and forehead.

The *maxillary* division is also exclusively sensory. It emerges through the base of the skull into the pterygopalatine fossa via the foramen rotundum (Fig. 3.1; see also page 46). Within the pterygopalatine fossa it gives off a number of branches: two large ganglionic branches (which enter the pterygopalatine ganglion), the zygomatic nerve and the *posterior superior alveolar nerves*. From the ganglion, the *greater and lesser palatine nerves* descend through the greater palatine canal. In the canal the greater palatine nerve gives branches to the nasal mucosa, and then emerges through the greater palatine foramen to supply the hard palate. The lesser palatine nerves emerge through the smaller lesser palatine foramina to supply the soft palate (see also page 49).

The *nasopalatine nerve* leaves the pterygopalatine ganglion via the sphenopalatine foramen to pass forward and downward on the nasal septum to reach the incisal canal where it gives off its terminal branches (Fig. 2.21). They innervate the mucous membrane and gingivae of the anterior of the hard palate.

The maxillary nerve then runs forward through the infraorbital canal where the *middle* (when present) *and anterior superior alveolar nerves* arise.

The maxillary (superior) dental plexus (Fig. 3.2) is formed by the intercommunication of the posterior, middle and anterior superior alveolar nerves, supplying the maxillary teeth and associated gingivae and buccal soft tissues. The palatal soft tissues are supplied by the palatine nerves, which may contribute to the dental plexus.

The *infraorbital nerve* emerges from the infraorbital foramen to supply the skin between the eye and the nostril, and the upper lip (Fig.2.20). Its inferior palpebral branches innervate the lower eyelid, the external nasal branches supply the skin on the side of the nose, the internal nasal branches innervate the mucous membrane of the vestibule of the nose

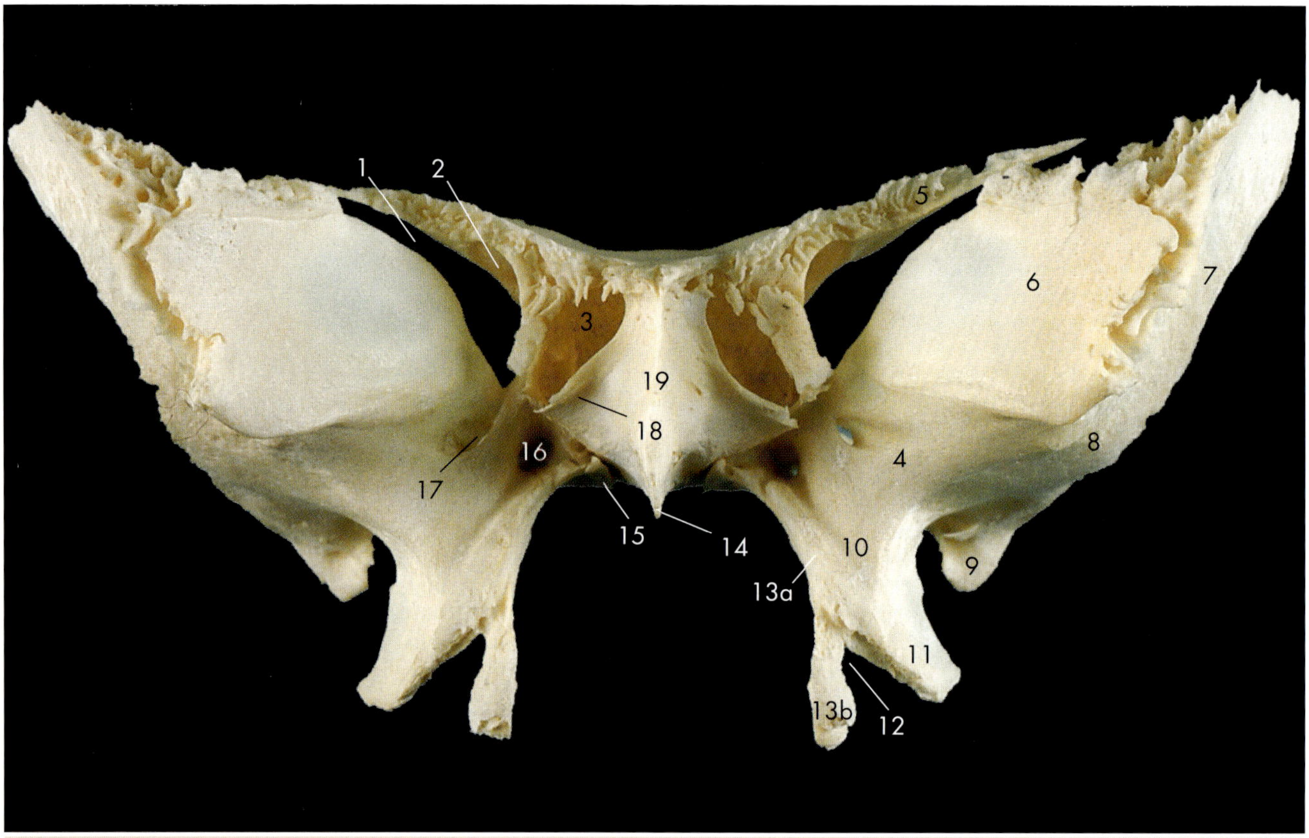

1 Superior orbital fissure
2 Optic canal
3 Sphenoidal air sinus in body of sphenoid
4 Maxillary surface of greater wing
5 Frontal margin of lesser wing
6 Orbital surface of greater wing
7 Temporal surface of greater wing
8 Infratemporal crest of greater wing
9 Spine
10 Pterygoid process
11 Lateral pterygoid plate
12 Pterygoid notch
13 Medial pterygoid plate (13a) surmounted by the pterygoid hamulus (13b)
14 Rostrum
15 Vaginal process
16 Pterygoid canal
17 Foramen rotundum
18 Concha
19 Crest

Figure 3.1 The anterior surface of the sphenoid bone showing the foramen rotundum through which the maxillary division of the trigeminal nerve emerges into the infratemporal fossa.

and the superior labial branches supply the skin and mucous membrane of the upper lip.

The *mandibular* division (also described on pages 8–13), though mainly sensory, also carries the motor branches to the muscles of mastication. The mandibular nerve enters the infratemporal fossa via the foramen ovale. The motor branches leave the trunk within the fossa. The nerve then gives off several branches. The *auriculotemporal nerve* leaves the main trunk medial to the neck of the mandibular condyle. It passes behind the condyle to innervate the external auditory canal and the skin of the temple. The *buccal nerve* is entirely sensory

Figure 3.2 Branches of the pterygopalatine ganglion.

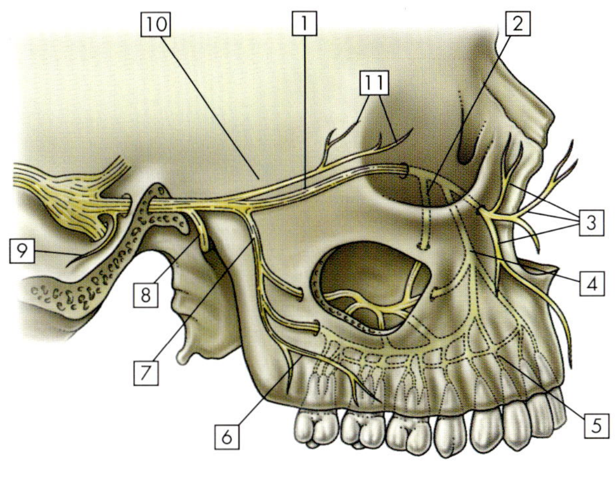

1 Infra-orbital nerve
2 Middle superior alveolar nerve
3 Terminal branches of infraorbital nerve
4 Anterior superior alveolar nerve
5 Dental plexus of nerves
6 Gingival branch of posterior superior alveolar nerve
7 Posterior superior alveolar nerve
8 Branches to pterygopalatine ganglion
9 Meningeal nerve
10 Zygomatic nerve
11 Zygomaticofacial and zygomaticotemporal branches

and passes along the medial aspect of the mandibular ramus anterior to the inferior alveolar nerve; its distribution is variable. It crosses the anterior border of the ramus inferiorly and usually innervates the cheek mucosa and buccal gingivae in the molar and sometimes premolar regions, occasionally supplying sensation in the maxillary sulcus. Sometimes it has a very small area of distribution, and maxillary nerves supply some mandibular tissues. Uncommonly, the area supplied by the buccal nerve is more extensive, and rarely it can innervate mucosa up to the corner of the mouth and onto the vermilion border.[1]

The *inferior alveolar nerve* also passes downwards along the medial aspect of the mandibular ramus within the pterygomandibular space to enter the mandibular foramen (Fig. 3.3).

Before entering the mandibular foramen, it gives off the *mylohyoid nerve*, which penetrates the sphenomandibular ligament and runs forward in a groove on the medial aspect of the mandible. This mixed motor and sensory branch innervates the posterior belly of the digastric muscle and frequently gives sensation to the skin over the point of the chin. Occasionally the mylohyoid nerve will communicate with the inferior dental plexus, usually in the molar region.[2]

The inferior alveolar nerve is usually a single trunk running just below the apices of the mandibular teeth (Fig. 3.4); however, its position is variable and it can be situated much nearer the lower border of the mandible.[3] Sometimes the nerve can divide and run through the mandible in several canals. Branches from the inferior alveolar nerve provide the main sensory supply to the teeth and gingivae. The inferior alveolar nerve runs forward to divide into its terminal incisive and mental branches, the latter emerging via the mental foramen at the level of the premolar teeth.

The mandibular (inferior) dental plexus provides the innervation to the teeth and associated structures.[4] This plexus is composed primarily of intercommunicating branches of the inferior alveolar nerve: infrequently the lingual, mylohyoid and buccal nerves (and possibly cervical nerves) communicate with the plexus.[5]

In the pterygomandibular space, the *lingual nerve* passes downwards together with the inferior alveolar nerve, lying anterior to it and sometimes receiving branches from it, and communicates with the chorda tympani of the facial nerve before reaching the mandibular foramen. This connection carries secretomotor branches to the submandibular and sublingual salivary glands via the submandibular ganglion and taste fibres to the tongue. The trunk of the lingual nerve gives off small branches to the lingual gingivae in the molar region. The lingual gingivae of the anterior part of the lower jaw and the mucosa of the floor of the mouth are also supplied by branches of the lingual nerve. The terminal branches of the lingual nerve supply sensation to the tongue on that side. Sometimes this nerve contributes to the mandibular (inferior) dental plexus in the molar region.[5]

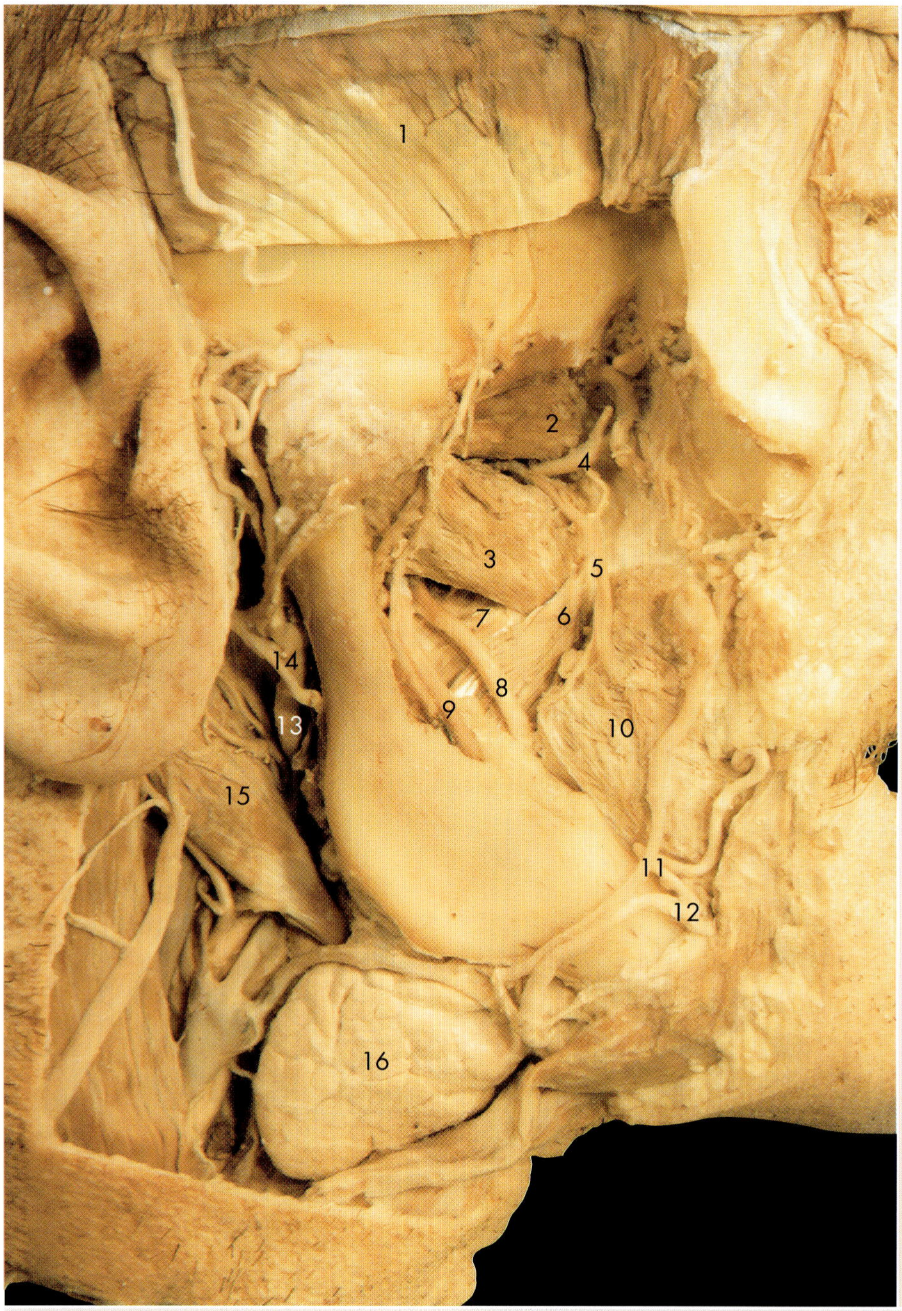

1 Temporalis muscle
2 Lateral pterygoid muscle (upper head)
3 Lateral pterygoid muscle (lower head)
4 Maxillary artery
5 Buccal branch of mandibular nerve
6 Medial pterygoid muscle (superficial head)
7 Medial pterygoid muscle (deep head)
8 Lingual nerve
9 Inferior alveolar nerve and vessel
10 Buccinator muscle
11 Facial vein
12 Facial artery
13 External carotid artery
14 Facial nerve
15 Digastric muscle (posterior belly)
16 Submandibular gland

Figure 3.3 The inferior alveolar nerve passing downwards along the medial aspect of the mandibular ramus within the pterygomasseteric space. (Courtesy of Professor S. Standring, GKT School of Biomedical Sciences, London.)

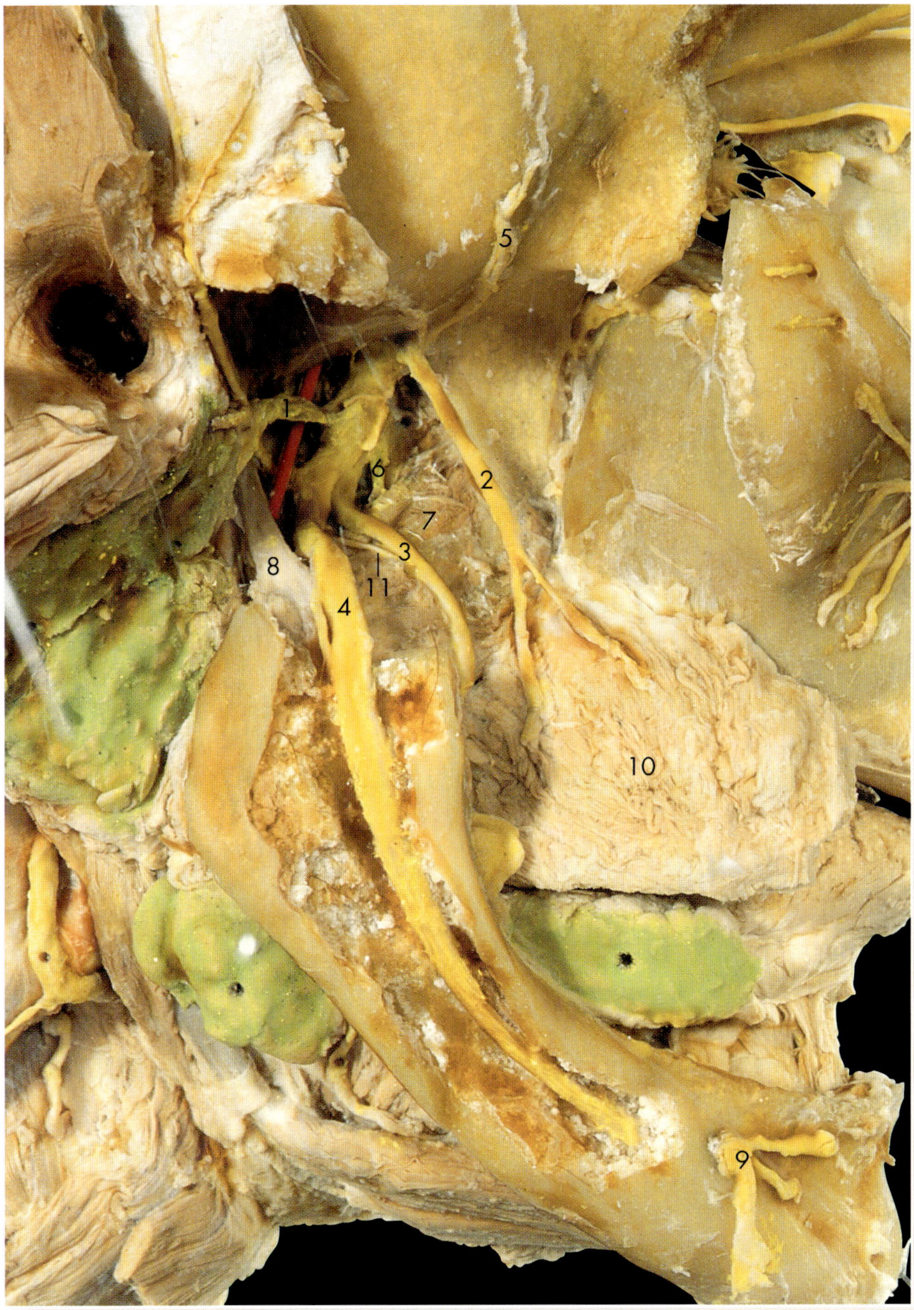

1 Roots of auriculotemporal nerve surrounding middle meningeal artery
2 Buccal branch of mandibular nerve
3 Lingual nerve
4 Inferior alveolar nerve with mylohyoid branch
5 Deep temporal nerve
6 Nerve to medial pterygoid muscle
7 Medial pterygoid muscle
8 Sphenomandibular ligament
9 Mental nerve
10 Buccinator muscle
11 Chorda tympani

Figure 3.4 The mandibular nerve with the lateral pterygoid muscle removed. (Courtesy of Professor C. Dean, Department of Anatomy and Developmental Biology, University College London.)

MAXILLARY ANAESTHESIA

Anaesthesia for procedures within the mouth can usually be achieved using buccal supraperiosteal infiltration injections and injections near the incisive and greater palatine foramina. Injections at the infraorbital foramen may be employed in the anterior aspects of the upper jaw.

Nerve blocks administered in the infratemporal region are:

1. Posterior superior alveolar;
2. Maxillary: for which there are two intraoral techniques and one extraoral approach.

POSTERIOR SUPERIOR ALVEOLAR NERVE BLOCK (FIG. 3.5)

The posterior superior alveolar nerve runs within the pterygopalatine fossa and enters the posterior aspect of the maxilla (and therefore the infratemporal fossa) as a single nerve, or as several branches.[6] The nerves descend within the maxilla, in the posterior aspect of the maxillary antrum, supplying sensation to the antrum and then contributing largely to the posterior aspects of the superior dental plexus.

The site of the injection is about 1 cm above and behind the apices of the maxillary third molar. The distal aspect of the zygomaticoalveolar crest is palpated with the forefinger high in the buccal sulcus, and the mucosa is held under tension. The needle enters at the height of the vestibule, and will normally be adjacent to the distal root of the second molar. The syringe is held laterally, so that the needle is directed medially as well as posteriorly and superiorly. Usually the needle is about 45° to the patient's occlusal and sagittal planes. This is achieved more easily if the patient's mouth is partially closed; and if there is still restriction in positioning the syringe correctly to achieve the optimum angulation, the patient can be asked to move the mandible towards the side which is being treated. The needle is advanced for about 1.5 cm (no more than 2 cm) close to the posterior wall of the maxilla, and then aspiration is carried out because of the proximity of the small arteries accompanying the nerve and the pterygoid venous plexus. At least 1.5 ml of local anaesthetic solution should be injected slowly to achieve satisfactory anaesthesia.

This injection is useful for procedures involving the posterior aspects of the maxilla and the maxillary antrum. It is not routinely used for simple intraoral or dental procedures. Some dentists are cautious about adopting this injection because of the possibility of haematomas arising from laceration of the posterior superior alveolar artery or from damage to the pterygoid venous plexus; this is uncommon.

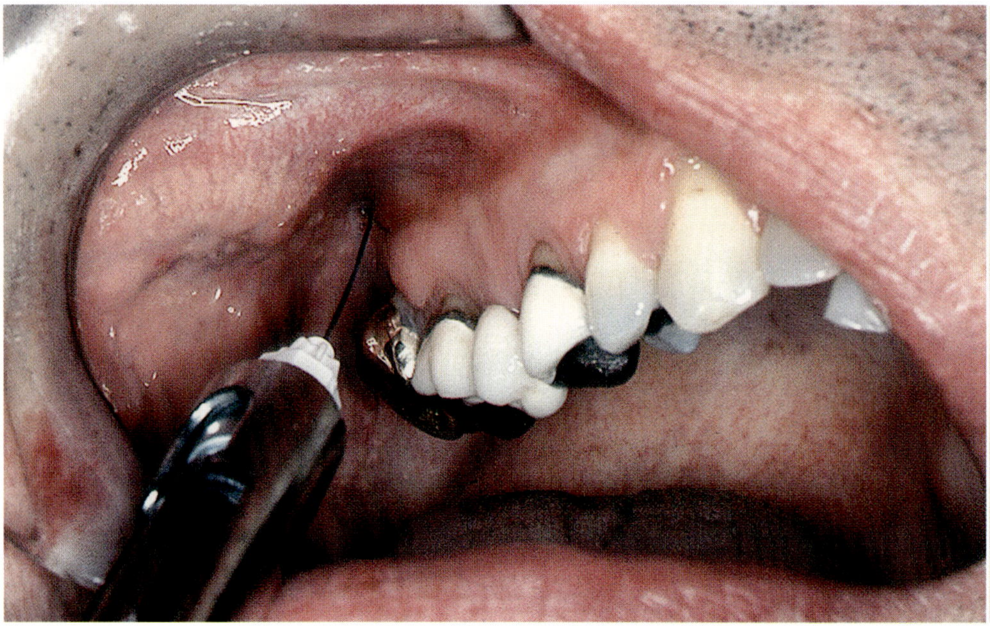

Figure 3.5 The posterior superior alveolar nerve block.

Maxillary nerve block

Blockade of the complete maxillary nerve is rarely indicated, but is useful for extensive surgery of the maxilla, or in the treatment of acute trigeminal neuralgia of the maxillary division. These nerve blocks can be utilised in the differential diagnosis of facial pain. These injections must be undertaken in hospital, where facilities exist to deal with the rare (but reported) complications, such as interference with the contents of the orbit, or anaesthesia of the brainstem.

Intraoral techniques

Two methods of approaching the maxillary nerve have been described – one being the buccal or tuberosity approach at the posterior aspect of the maxilla, and the other being via the greater palatine canal.

Tuberosity method (Fig. 3.6)

This is similar to the posterior superior alveolar nerve block, but the needle is advanced further to enter the pterygomaxillary fissure. This requires the syringe to be held more laterally to ensure that the needle stays close to the infratemporal aspect of the maxilla and it is often performed more successfully when the patient has the mouth almost completely closed. Improved access can be achieved with a needle mounted on an angled hub. Some surgeons bend the needle to achieve improved access, but this must be done with extreme caution, and the needle kept constantly in view in case it fractures.

Once the needle has been inserted high in the buccal sulcus just distal to the second molar, it is slowly advanced to a maximum depth of about 3 cm and then, after aspiration, 2.0 ml of anaesthetic solution will be required: this must then diffuse to block the maxillary nerve within the pterygopalatine fossa. Even at this depth, the tip of the needle is likely to be 1 cm below the main trunk of the maxillary nerve.[7] If the needle does not enter the pterygomaxillary fissure, there is sometimes inadequate anaesthesia and the injection needs to be repeated with another cartridge of solution to improve diffusion.

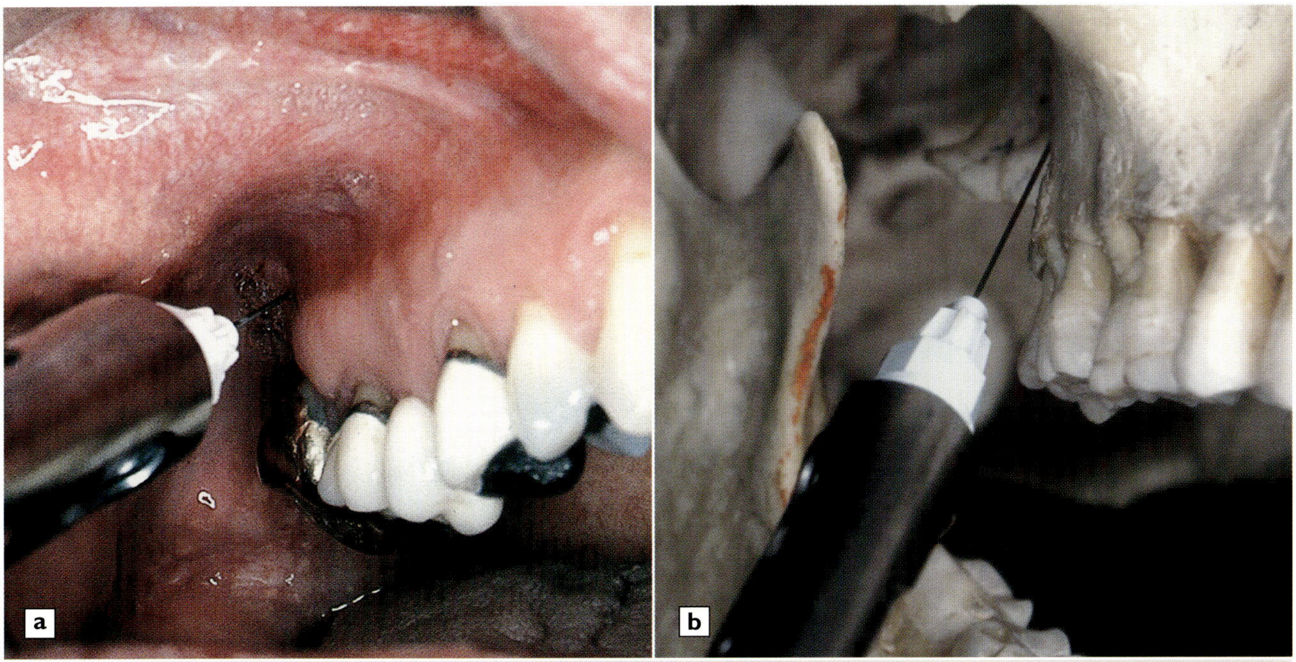

Figure 3.6a, b Maxillary nerve block using the tuberosity method.

Greater palatine canal method (Fig. 3.7)
For this technique, the greater palatine foramen must be located. It is sometimes palpable and is usually located close to the third molar tooth, although it can sometimes be slightly further forward opposite the distal aspects of the second molar.

The patient's head must be extended to get a clear view of the posterior of the palate, and a long needle is inserted into the mucosa at the appropriate site and a small amount of local anaesthetic solution infiltrated. The position of the foramen can now be confirmed by probing with the needle which must be inclined so that it passes upwards and backwards into the canal. The needle is advanced slowly into the canal, and it is sometimes necessary to change the direction slightly to ensure that the needle continues to advance along the canal which is sometimes inclined laterally. Some surgeons advocate bending the needle at the hub to ensure ease of access, but this should be avoided if possible because of the risk of needle fracture. The depth of injection is usually 2.5–3 cm, but it is often impossible to achieve this. Again aspiration must be performed and 2.0 ml of solution slowly injected.

It has been shown that the greater palatine canal is sometimes obstructed.[6] If resistance is met before an adequate depth of penetration has been achieved, it is

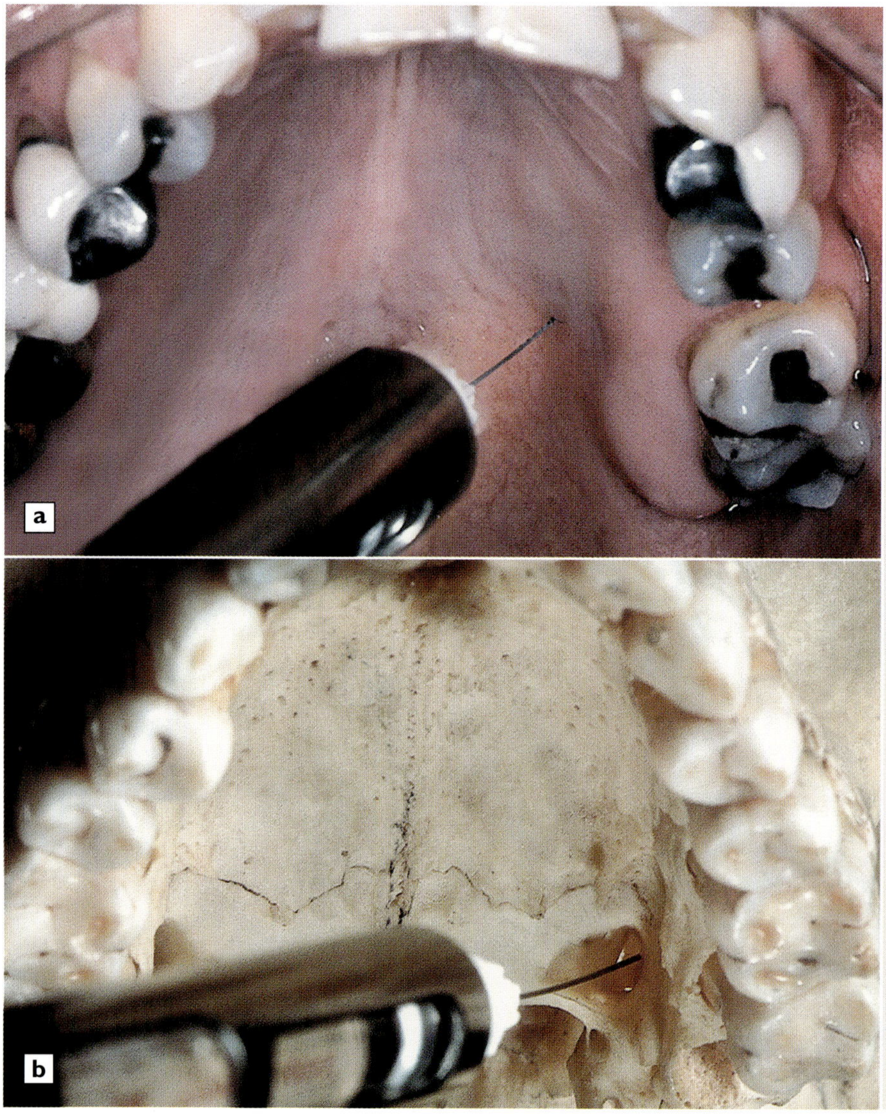

Figure 3.7a, b Maxillary nerve block using the greater palatine canal method.

unwise to attempt to force the needle further as it might jam and fracture. An alternative approach should be adopted.

Extraoral approach (Fig. 3.8)
This injection is rarely indicated. It may be employed for diagnostic blocks when the patient has trismus.

A heavy gauge needle (20 or 18G) is employed and it is often necessary to inject 3.0 ml of solution to achieve satisfactory anaesthesia; the solution should be drawn up into an appropriate syringe.

The skin is prepared with a suitable antiseptic, and infiltrated with local anaesthetic at the site of penetration. This is selected by identifying, by palpation, the zygomatic arch and the coronoid process of the mandible. The needle will pass behind the coronoid, beneath the zygomatic arch.

The needle is inserted through the skin surface until the lateral pterygoid plate is contacted. The needle is now redirected to be inclined upwards and forwards, which requires it to be withdrawn partially before reinserting in the new direction. The total depth of insertion would be no more than 5 cm, at which point the tip of the needle should be at the pterygopalatine fossa. After aspiration, the anaesthetic solution is slowly injected.

An alternative approach has been suggested[8] in which the entry site is above the zygomatic arch, at its junction with the frontal process. The needle is angled 60° to the sagittal plane and 10° to the horizontal plane, and advanced in small steps, each time aspirating and injecting increments of anaesthetic to reduce discomfort. The depth of injection is about 5 cm when the tip of the needle should be in the pterygopalatine fossa – and bone may have been contacted. After aspirating, 3.0 ml of anaesthetic solution is slowly deposited.

These are not easy injections to carry out, and are rarely undertaken. For those additional reasons, they should be administered in hospital surroundings, so that any complications can be managed.

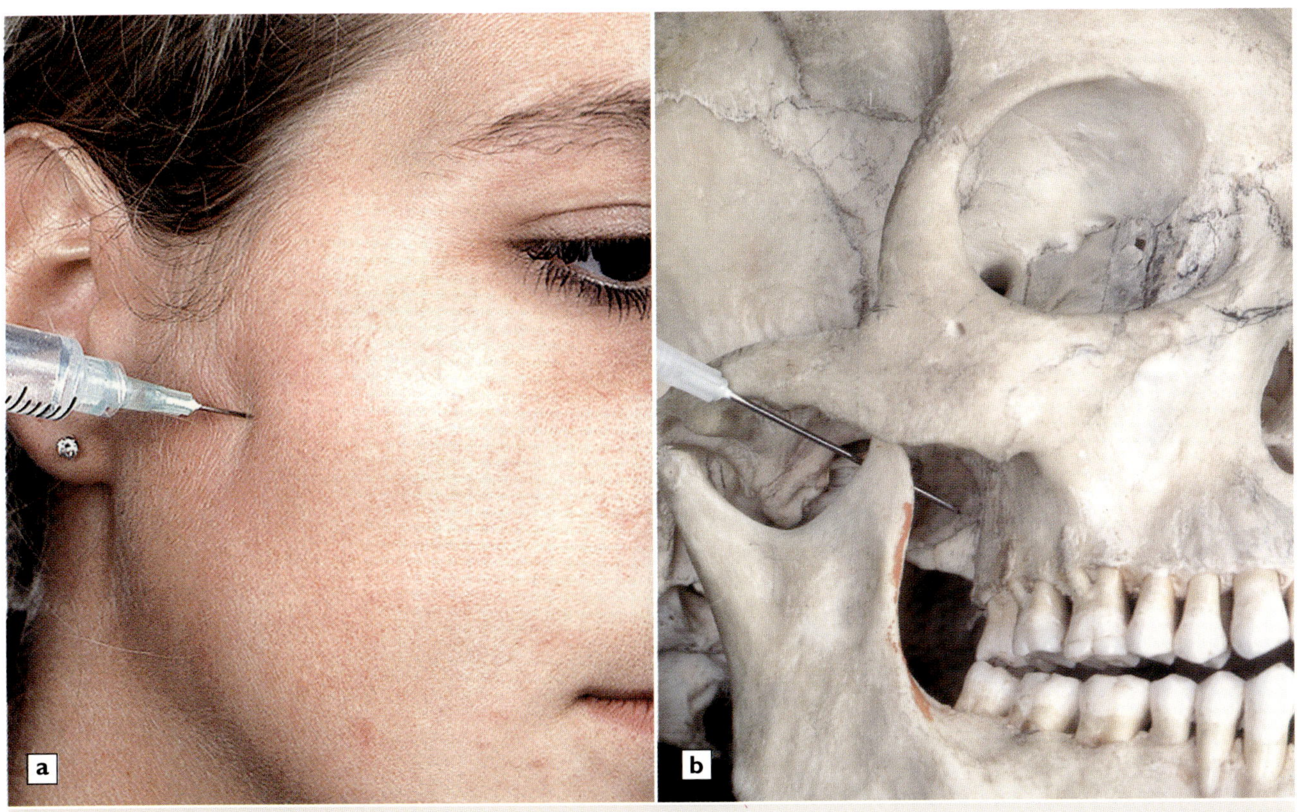

Figure 3.8a, b Maxillary nerve block using the external approach.

COMPLICATIONS OF INFRATEMPORAL NERVE BLOCKS FOR THE MAXILLA

In addition to the common minor problems associated with injections, specific hazards relate to these infratemporal nerve blocks.

Posterior superior alveolar injection

This is remarkably safe. The posterior superior alveolar artery runs with the nerve(s), but is no more likely to be damaged than arteries in other neurovascular bundles. The pterygoid venous plexus lies within and around the lateral pterygoid muscle,[9] and should not be damaged unless the needle is inserted too deeply or laterally. If a positive (venous) aspiration is observed during this procedure, withdrawal will disengage the needle with minimal bleeding resulting – injecting into the friable plexus causes disruption which can lead to haematoma formation and postoperative trismus.

Maxillary nerve block – intraoral approach

The tuberosity approach involves inserting the needle more deeply to enter the pterygomaxillary fissure. If the needle is not kept close to the posterior (infratemporal) aspect of the maxilla, the pterygoid venous plexus can be damaged.[10]

The maxillary artery is close to the injection site. If significant (arterial) aspiration is noted, the needle should be withdrawn a few millimetres, and then aspiration repeated before injecting the solution. If the artery is lacerated, a noticeable haematoma may develop.

With any of the approaches to the maxillary nerve the anaesthetic solution can diffuse to affect orbital structures, usually causing temporary diplopia,[11] and on rare occasions may affect the brainstem[12] (although this is more likely to be after extraoral approaches).

When the nerve is approached via the greater palatine canal, the greater palatine nerve is frequently damaged causing a prolonged period of palatal sensory disturbance. If the canal is narrow, or partially obstructed, attempts to force the needle can result in its fracture. If the canal is very easy to negotiate, insertion of the needle too deeply can result in penetration of the orbit.[6]

Maxillary nerve block – extraoral approach

Inaccuracy is more common when techniques are practised infrequently: in one report, success was reported in 84% of cases.[8]

In addition to the problems mentioned relating to the intraoral techniques, brainstem anaesthesia[12] and orbital involvement[13] have been reported. Temporary paralysis from spread to branches of the facial nerve has also occurred.[8]

MANDIBULAR ANAESTHESIA

It is possible to use supraperiosteal infiltration techniques in the incisor region of the mandible, and injections at the mental foramen are also useful in the anterior region. However, most routine intraoral procedures are undertaken with anaesthesia of the inferior alveolar and lingual nerves, supplemented where necessary with buccal nerve infiltration (for example when surgery involves the posterior aspects of the alveolus).

Nerve blocks in the region of the infratemporal fossa include:

1. The inferior alveolar (and lingual) nerve block: for which two intraoral techniques and one extraoral approach have been described;
2. The mandibular nerve block: one intraoral technique and one extraoral approach have been identified.

INFERIOR ALVEOLAR NERVE BLOCK

Intraoral techniques

Routine open-mouth direct method (Fig. 3.9)

The most secure way of achieving anaesthesia of the inferior alveolar nerve is to use the direct injection technique. (Note that indirect techniques are now not carried out because the fine [27G] needles in common use are deflected and inaccuracy is inevitable.)[14,15]

The patient's mouth is opened widely, and the landmarks palpated with the index finger of the free hand. The entrance to the pterygomandibular space is bounded laterally by the internal oblique ridge of the mandible, and medially by the mucosa over the pterygomandibular raphe (Fig. 4.2). The insertion of the raphe at the posterior end of the mylohyoid line of the mandible creates a triangular fossa. The palpating finger firmly compresses the soft tissues against the retromandibular aspect of the mandible, so that the internal oblique ridge is identified. The syringe is held over the opposite mandibular first premolar. A fine (27G) needle is inserted through the mucosa about

1 cm above the occlusal surface of the posterior teeth. Whilst the soft tissues are held firmly against the internal oblique line of the mandible, the needle is inserted close to the ramus and advanced 1.0–1.5 cm. At this point the medial aspect of the mandible may have been contacted, and if so, the needle is withdrawn slightly, aspiration is performed and 1.5–2.0 ml of solution is slowly injected. Even if the mandible has not been contacted, the injection should be performed at this depth. Inserting the needle further in an attempt to contact the mandible should not be undertaken, as this will encourage the injection to be carried out too deeply where complications can arise.

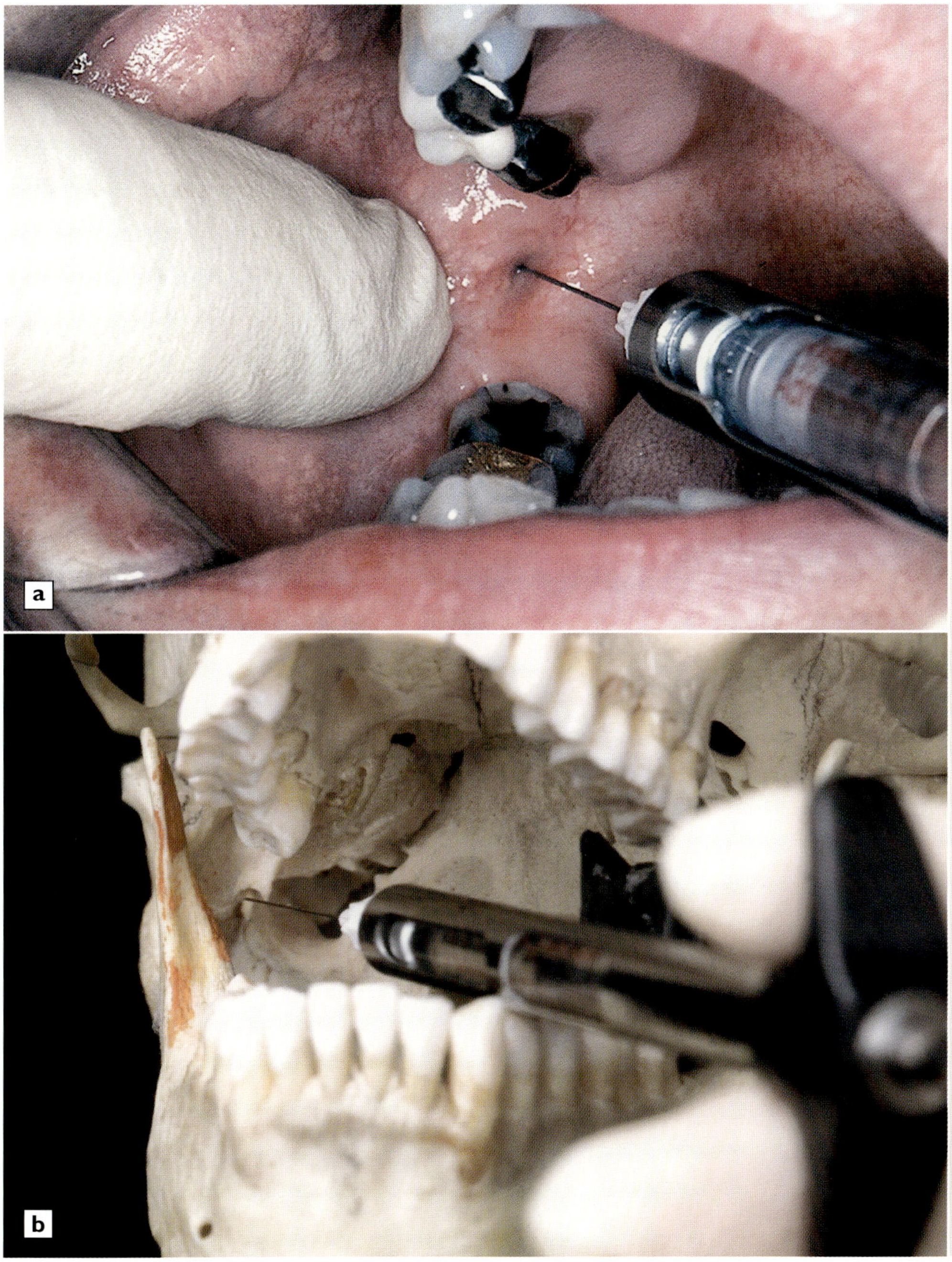

Figure 3.9a, b Inferior alveolar nerve block using the open-mouth direct method.

Closed mouth technique (Fig. 3.10)
When a patient is unable to open the mouth fully, so that the standard direct intraoral technique cannot be undertaken, then this alternative technique may be used.[16,17]

With the index finger, the patient's cheek is retracted laterally, and the retracting finger palpates the external oblique line of the mandible. The syringe is held in the maxillary buccal sulcus, parallel to the gingival margins of the maxillary teeth. The mucosa in the retromolar region is penetrated at a higher site than with the standard direct inferior alveolar nerve block. The needle is inserted about 2–2.5 cm and then, after aspiration, 2.0 ml of solution is slowly injected. The onset of anaesthesia is usually slower than with the direct technique.

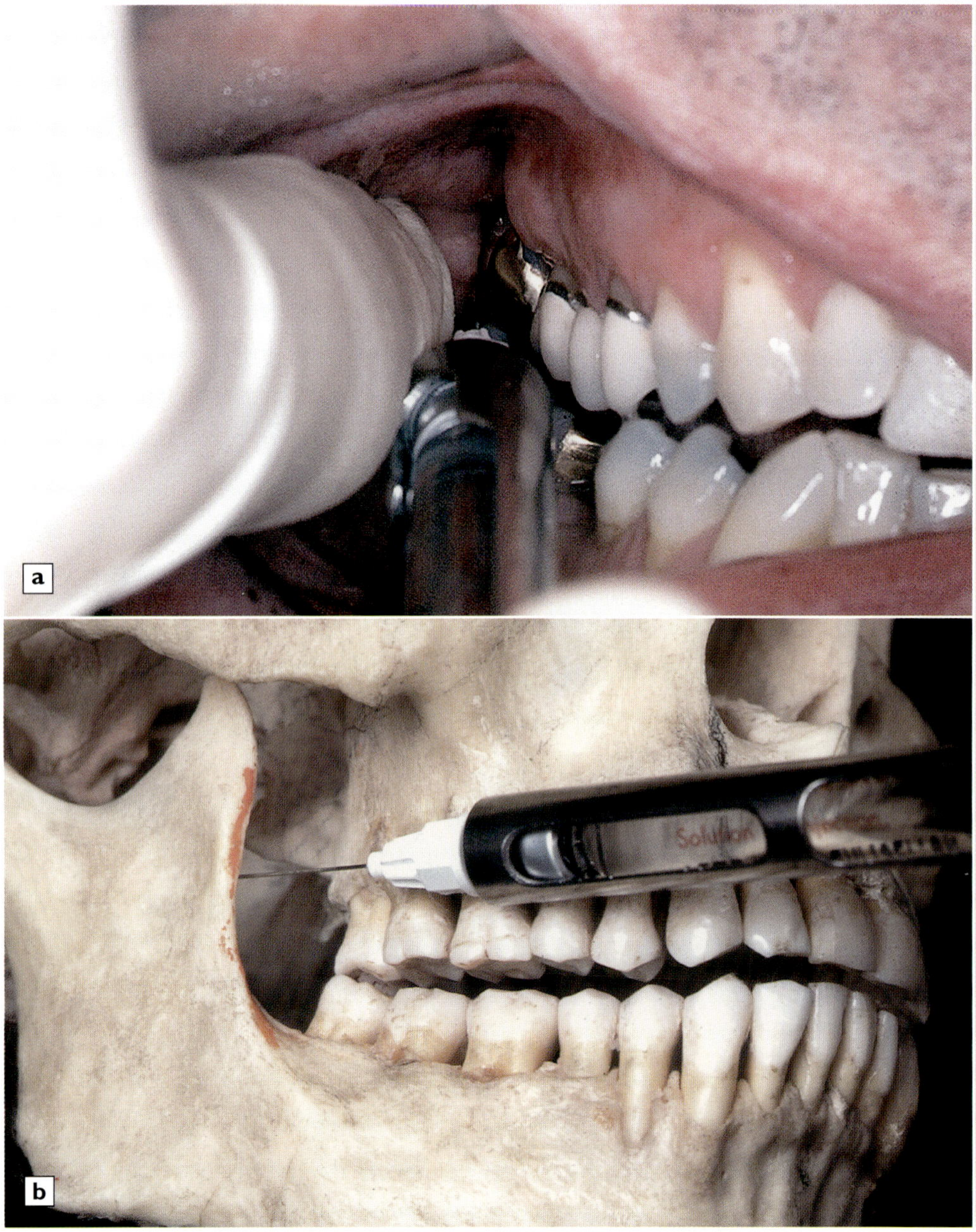

Figure 3.10a, b Inferior alveolar nerve block using the closed mouth technique.

Extraoral approach (Fig. 3.11)
It is usually possible to block the inferior alveolar nerve using an intraoral injection technique. It is therefore rare to use this extraoral approach. There may be occasions when there is acute infection in the intraoral tissues, or when intraoral techniques have failed to produce adequate anaesthesia (usually in the presence of trismus) when this approach could be valuable.

The patient's head is inclined away from the operator: the mouth is closed. A long (6 cm+) heavy gauge (18G) needle is required.

The lower border of the mandible is palpated, and the anterior border of the masseter muscle identified as one landmark. The skin is prepared with a suitable antiseptic, and may be infiltrated with local anaesthetic for the patient's comfort.

The site of penetration is midway between the anterior border of the masseter and the angle of the mandible. Due to the bony ridges associated with muscle attachments, the needle should be inclined towards the medial aspect of the mandible, and inserted parallel to the posterior border of the ramus. At first there may be some resistance due to the insertion of the medial pterygoid muscle, but the needle should be advanced until the tip is judged to be at the level of the crowns of the maxillary molars (if required, this level can be located on the skin before the injection commences). This is usually at a depth of about 4 cm.

During the injection, the needle should have been close to the medial surface of the mandible – which may be concave. Bone might have been contacted during the procedure, but if the operator is confident that the injection was carried out accurately, the needle should not be inserted to a greater depth in an effort to contact the mandible.

After aspirating, 2.0 ml of solution is slowly deposited.

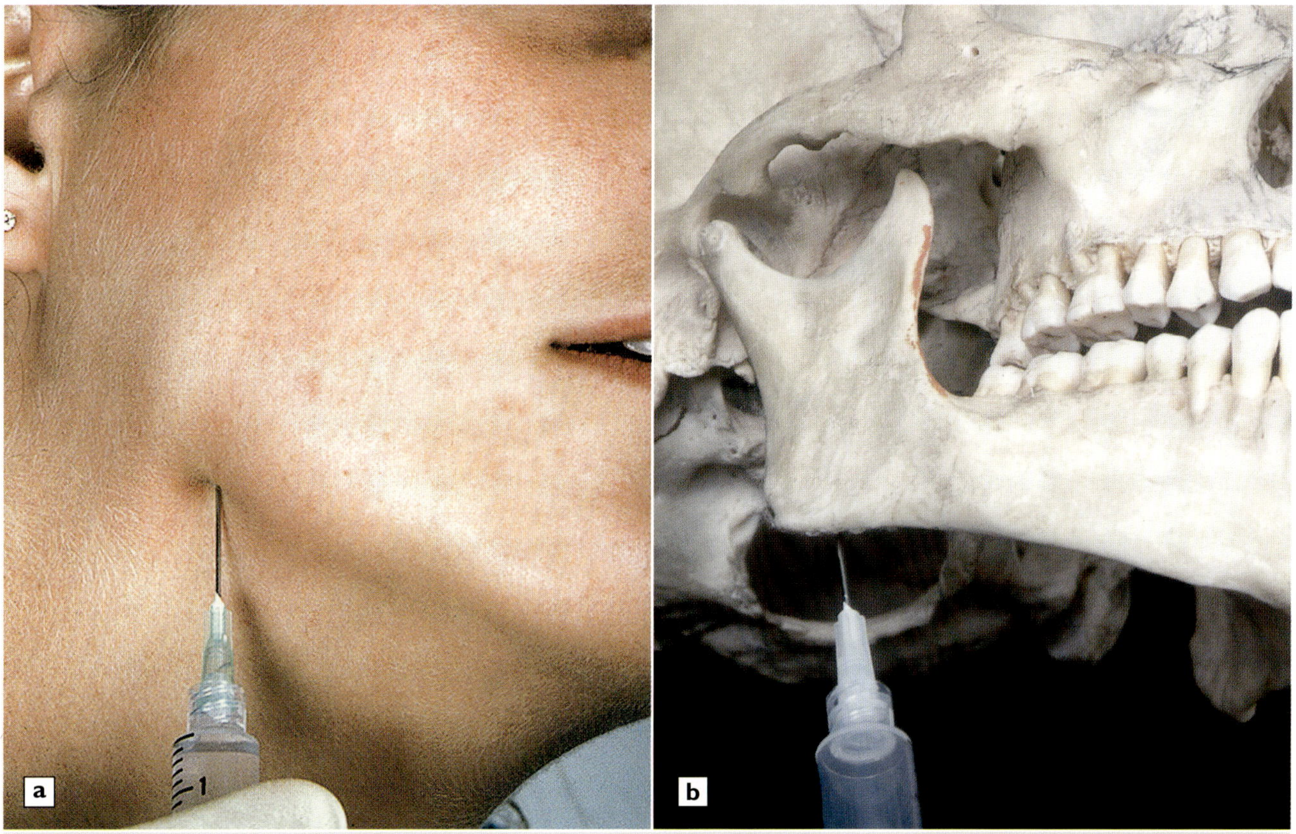

Figure 3.11a, b Inferior alveolar nerve block using the external approach.

Mandibular nerve block

There is one intraoral technique and one extraoral approach.

The intraoral 'high condyle' (Gow–Gates) technique (Fig. 3.12)

This intraoral technique involves the solution being deposited high in the pterygomandibular space, at the level of the neck of the condyle.[18]

The patient's mouth must be widely open, and the index finger of the free hand is used to palpate the anterior border of the ramus. The point of insertion is much higher in the retromolar tissues than that used for the standard (direct) intraoral technique, being at the level of the maxillary molars. With the syringe held over the opposite premolars, the needle is directed towards the mandibular condyle (the extraoral landmark – if required – being the tragus) and advanced until bone is contacted, usually at about 2.5 cm. The target is the neck of the condyle, just beneath the insertion of the lateral pterygoid muscle. After aspiration, 2.0 ml solution is slowly deposited.

It is claimed that this injection blocks all of the branches arising from the mandibular nerve, and in particular anaesthetises the inferior alveolar, lingual and buccal nerves without the need for any additional injections. The solution, however, is deposited further from the nerve trunks than with the standard technique, and inadequate depth of anaesthesia can result.

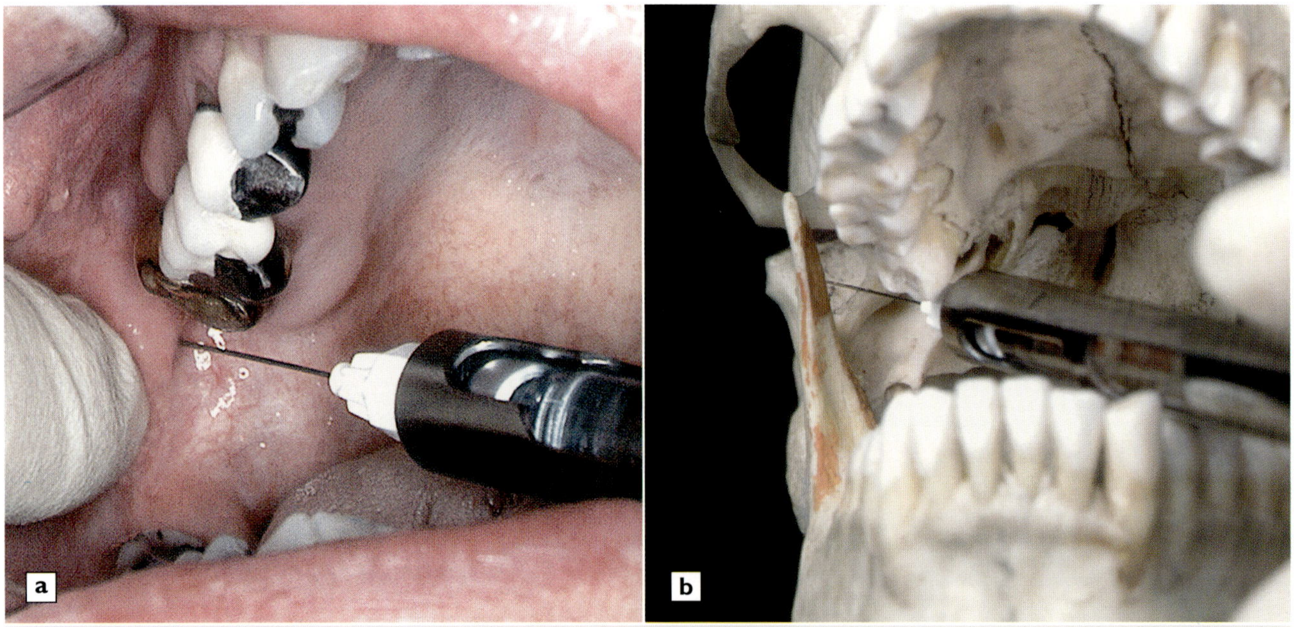

Figure 3.12a, b Mandibular nerve block using internal 'high condyle' (Gow–Gates) technique.

Extraoral approach (Fig. 3.13)
There are very few occasions when an extraoral block of the mandibular nerve would be required, as transoral approaches to most branches of the mandibular nerve can be made even when a patient has trismus. It is sometimes useful in the diagnosis of facial pain or when controlling acute trigeminal neuralgia affecting the mandibular division.

The extraoral landmarks are the same as those used for the maxillary nerve block, being the zygomatic arch and the coronoid process. The skin is prepared, and anaesthetised at the site of penetration.

A long (6 cm+) heavy gauge (18G) needle is again used. The needle is directed at right angles to the surface, and penetrates beneath the arch, behind the coronoid process. It is then advanced until the lateral pterygoid plate is contacted – usually at a depth of about 4 cm. The needle is then withdrawn a short distance, and carefully redirected distally and slowly advanced – this manoeuvre may be undertaken in several small steps until the surgeon feels the needle pass behind the lateral pterygoid plate, at which time the needle is usually about 60° to the sagittal plane. At each manoeuvre to reposition the needle, a small amount of anaesthetic can be deposited to make the procedure more comfortable; and each time aspiration should be performed, and if positive, the needle should be repositioned without injecting to avoid further disruption of the vessels which have been encountered. The needle should then be advanced for a short distance (no more than 5 mm), and after aspiration, 3.0 ml of solution is slowly deposited.

This extraoral approach to the region of the foramen ovale can lead to the same significant complications as with maxillary nerve blocks, so the patient must be managed where appropriate supportive hospital facilities exist.

COMPLICATION OF INFRATEMPORAL NERVE BLOCKS FOR THE MANDIBLE

Inferior alveolar nerve block: intraoral methods

Inaccuracy results in inadequate depth of anaesthesia. Provided that a direct technique was used (avoiding problems arising from attempting to manipulate fine needles within the tissues), the most common cause of inaccuracy from the routine open-mouth procedure is injecting too deeply (which also results in significant complications such as facial paralysis). The distance from the internal oblique line to the mandibular foramen varies, but is never more than 16 mm[15] – a depth of injection of 1.0 cm from the deep tendon of temporalis has been recommended.[6]

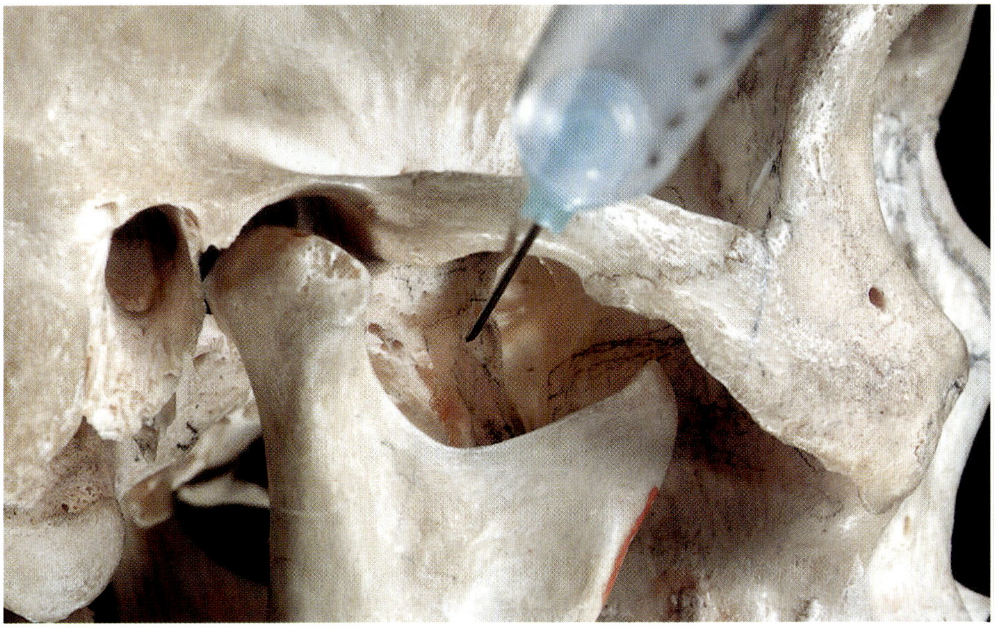

Figure 3.13 Mandibular nerve block using the external approach.

The closed-mouth technique deposits solution higher in the pterygomandibular space. Trismus is less frequent, as the medial pterygoid diverges from the mandible towards its origins, and is less likely to be encountered. As the solution is deposited further away from the neurovascular bundle, there is a reduced rate of positive aspirations, but also a higher failure rate of anaesthesia.[19]

Inferior alveolar nerve block: extraoral approach

The main difficulty with this injection is ensuring that the solution is deposited within the pterygomandibular space, and not within the medial pterygoid muscle. Inadvertent injection into the muscle results in a failure of anaesthesia, and trismus.

Mandibular nerve block: intraoral approach

This injection is made high in the pterygomandibular space, solution being deposited at the neck of the condyle just below the insertion of the lateral pterygoid.

The anaesthetic solution is therefore deposited some distance from the neurovascular bundle, with a slower onset and sometimes inadequate depth of anaesthesia. The main advantage of the injection is said to be simultaneous block of all of the mandibular nerve branches, but this does not always occur.[10]

The maxillary artery or vein may be encountered when injections are directed to the neck of the condyle.[6] Inaccuracy can result in facial paralysis.

Mandibular nerve block: extraoral approach

Similar problems may arise from this injection as described for the extraoral approach to the maxillary nerve. A success rate of 91% has been reported.[8]

References

1. Sicher H. *Oral anatomy*, 3rd edn. Mosby, St Louis, 1960
2. Frommer J, Mele F A, Monroe C W. The possible role of the mylohyoid nerve in mandibular posterior tooth sensation. *J Am Dent Assoc* 1972; 85: 113–117
3. Nortje C T, Farman A G, Grotepass F W. Variations in the normal anatomy of the inferior dental (mandibular) canal: a retrospective study of panoramic radiographs from 3,612 routine dental patients. *Br J Oral Surg* 1977; 15: 55–63
4. Carter R B, Keen E N. The intramandibular course of the inferior alveolar nerve. *J Anat* 1971; 108: 433–440
5. Rood J P. The analgesia and innervation of mandibular teeth. *Br Dent J* 1976: 140: 237–239
6. Jorgensen N B, Hayden J. *Sedation, local and general anaesthesia in dentistry*, 3rd edn. Lea & Febiger, Philadelphia, 1980
7. Mercuri L G. Intraoral second division nerve block. *Oral Surg Oral Med Oral Pathol* 1979; 47: 109–113
8. Stajcic Z, Todorovic L. Blocks of the foramen rotundum and the oval foramen: a reappraisal of extraoral maxillary and mandibular nerve injections. *Br J Oral Maxillofac Surg* 1997; 35: 328–333
9. Last R J. Anatomy. *Regional and applied*, 3rd edn. Churchill, London, 1963
10. Roberts D H, Sowray J H. *Local analgesia in dentistry*, 3rd edn. Wright, Bristol, 1987
11. Seldin H M. *Practical anaesthesia for dental and oral surgery*, 3rd edn Kimpton, London, 1947
12. Nique T A, Bennett C R. Inadvertent brain-stem anaesthesia following extraoral trigeminal V2–V3 blocks. *Oral Surg Oral Med Oral Path* 1981; 51: 468–470
13. Poore T E, Carney M B. Maxillary nerve block: a useful technique. *J Oral Surg* 1973; 31: 749
14. Smith N. An investigation of the influence of gauge on some physical properties of hypodermic needles. *Aus Dent J* 1968; 13: 158–163
15. Rood J P. An investigation into inferior dental nerve block anaesthesia. MDS Thesis, University of Newcastle; 1976
16. Vazirani S J. Closed mouth mandibular nerve block: a new technique. *Dent Dig* 1960; 66: 10–13
17. Akinosi J O. A new approach to the mandibular nerve block. *Br J Oral Surg* 1977; 15: 83–87
18. Gow-Gates G A E. Mandibular conduction anesthesia: a new technique using extraoral landmarks. *Oral Surg Oral Med Oral Path* 1973; 36: 321–328
19. Donkor P, Wong J, Punnia-Moorthy A. An evaluation of the closed mouth mandibular block technique. *Int J Oral Maxillofac Surg* 1990; 19: 216–219

Chapter 4

Infection and the infratemporal fossa and associated tissue spaces

J.D. LANGDON, B.J. MOXHAM AND B.K.B. BERKOVITZ

Most structures in the body are ensheathed by a connective tissue covering of varying thickness. This may be thin and delicate and present little resistance to the spread of infection. Alternatively, the connective tissue layer may be thicker, tendinous and resist the spread of infection, particularly over certain muscles. In these situations, it is associated with the generation of tension, will hold sutures and can be referred to as true fascia. From clinical experience, it is evident that certain predictable pathways exist along which infection may spread. The loose connective tissue uniting fascial planes may be destroyed and the potential space delineated by adjacent structures considerably enlarged as inflammatory exudate accumulates. Such potential spaces are referred to as tissue spaces. Infection may spread from one tissue space to enter another where the spaces are in direct communication or along the side of structures which pass from one tissue space to another (such as blood vessels or nerves). Infection can also invade tissue spaces by directly eroding the intervening fascia. In addition to such direct pathways, infection may also spread through the lymphatics and the blood vessels.

The dissemination of infection in soft tissues is influenced by the natural barriers presented by bone, muscle and fascia. Around the jaws, however, the tissue spaces are primarily defined by muscles, principally the mylohyoid, buccinator, masseter, medial pterygoid, superior constrictor and orbicularis oris muscles.

Because of the occurrence of inflammation in the soft tissues associated with partially impacted mandibular third molars (pericoronitis) and, less commonly, of dental abscesses of these teeth, the infratemporal fossa region is significant in terms of tissue spaces. This is due to the fact that the infratemporal fossa lies in a pivotal position, being intermediate in position between the tissue spaces of the face above and the tissue spaces of the neck below. Unlike infections involving odontogenic tissue that drain directly into the oral cavity (via the buccal or lingual sulci), those involving the tissue spaces around the infratemporal fossa do not drain directly into the oral cavity and have the potential to spread some distance through the head and neck. Of particular relevance are the tissue spaces around the pharynx, as involvement of these spaces may affect the larynx and thus compromise the airway. Symptoms associated with such conditions may include trismus, fever, dysphagia and dyspnoea and such patients must be treated quickly because of the potential development of life-threatening situations. In the most extreme situation, inflammation may eventually spread to the thorax.

The key to understanding the organisation of the tissue spaces in the region of the infratemporal fossa is an appreciation of the attachment of the investing layer of the deep cervical fascia to the mandible and to the muscles of mastication. The investing layer splits to surround the outer and inner surfaces of the body of the mandible, enclosing what is termed the potential space of the body of the mandible. In the region of the ramus, the outer layer (masseteric fascia) covers the outer surface of the masseter muscle while the inner layer surrounds the inner surface of the medial pterygoid muscle. The layers of fascia unite at the anterior and posterior borders of the ramus, merging with the fascia covering the adjacent two remaining muscles of mastication, the temporalis muscle and the lateral pterygoid muscle. The masseteric fascia is attached above to the zygomatic arch, where it is in continuity with the fascia extending upwards to cover the temporalis muscle and which gains attachment to the superior temporal line.

The account of the tissue spaces which follows has been derived from the work of Grodinsky and Holyoke (1938), Laskin (1964), Barker and Davies (1972), Birn (1972), Granite (1976), Tonge and Luke (1981), Hollinshead (1982), Sowray (1985), Koorbusch and Fridrich (1991), Goldberg and Topazian (1994), Lang (1995) and Peterson (1998).[1–12]

TISSUE SPACES ASSOCIATED WITH THE INFRATEMPORAL FOSSA – THE MASTICATOR TISSUE SPACES

The term *masticator tissue space* has been used to describe the space enclosed by the investing layer of the fascia ensheathing the muscles of mastication and the ramus of the mandible. It can be subdivided into the pterygomandibular, infratemporal, temporal and submasseteric tissue spaces (Fig. 4.1).

The *pterygomandibular space* lies between the ramus of the mandible laterally and the medial pterygoid muscle medially (Figs 4.2, 4.3). Above lies the inferior head of the lateral pterygoid muscle. Anteriorly, beneath the overlying oral mucosa, lies fibres of the buccinator muscle that arise from the pterygomandibular raphe. Immediately beneath the buccinator lies the tendon of the temporalis muscle. Posteriorly, the investing layer of deep cervical fascia covering the masseter and medial pterygoid muscles merges with the posterior border of the ramus, behind which lies the parotid gland. The pterygomandibular space is a prominent component of the infratemporal fossa. Between the ramus of the mandible and the medial pterygoid

Figure 4.1 Structures and tissue spaces in the region of the infratemporal fossa.

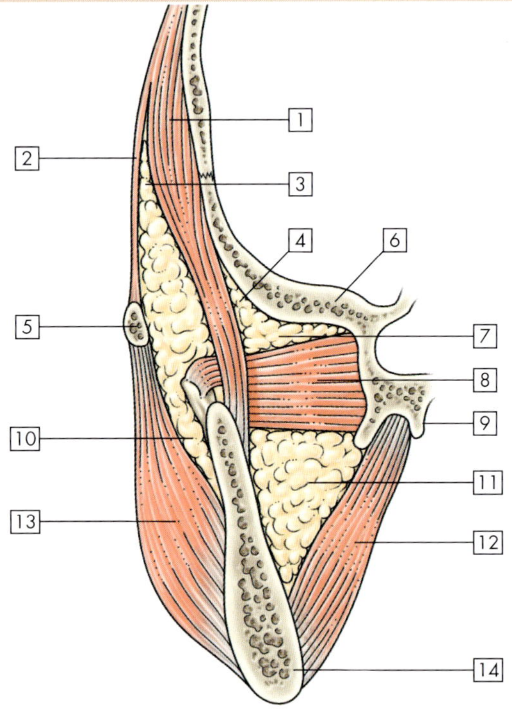

1 Temporalis muscle
2 Temporal fascia
3 Superficial temporal space
4 Deep temporal space
5 Zygomatic arch
6 Sphenoid bone
7 Infratemporal space
8 Lateral pterygoid muscle
9 Pterygoid hamulus
10 Submasseteric spaces
11 Pterygomandibular space
12 Medial pterygoid muscle
13 Masseter muscle
14 Mandible

Figure 4.2 Tissue spaces in the retromandibular region.

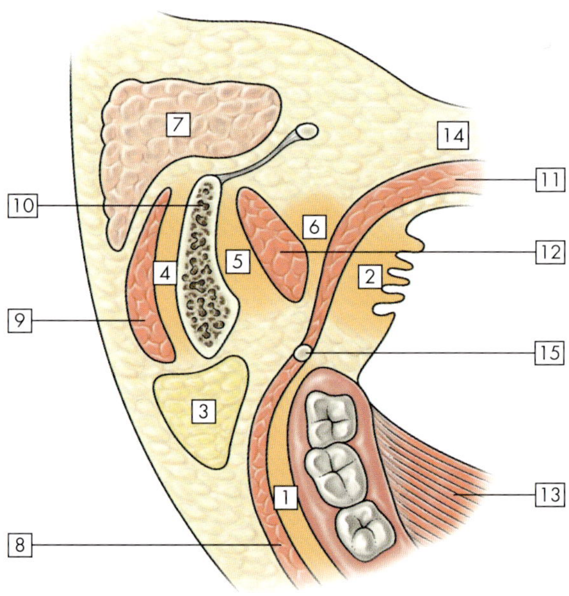

1 Vestibular sulcus
2 Peritonsillar space
3 Buccal space (filled with buccal pad of fat)
4 Submasseteric spaces
5 Pterygomandibular space
6 Parapharyngeal space
7 Parotid space
8 Buccinator muscle
9 Masseter muscle
10 Ramus of mandible
11 Superior constrictor of pharynx
12 Medial pterygoid muscle
13 Mylohyoid muscle
14 Retropharyngeal space
15 Pterygomandibular raphe

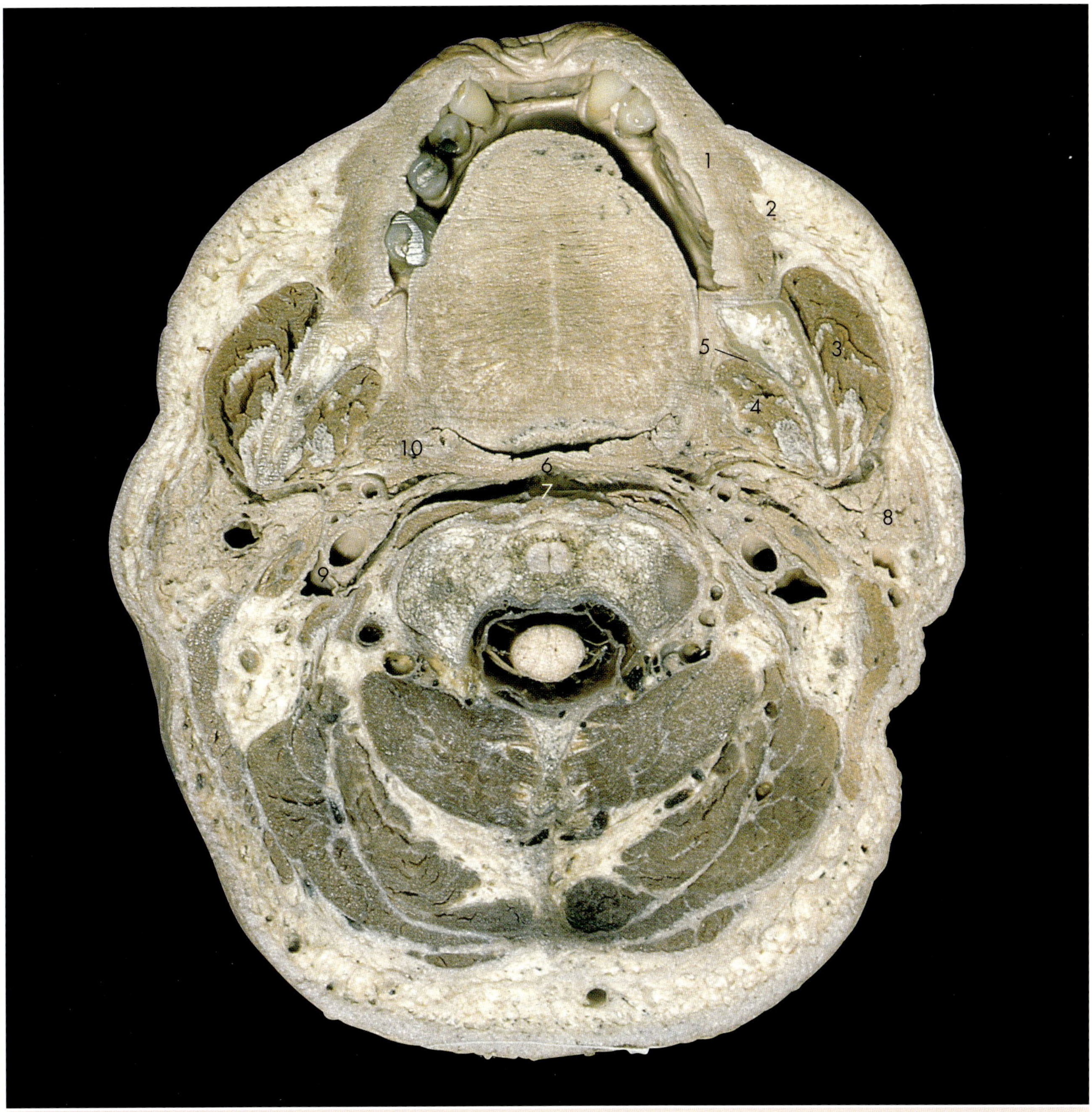

1 Buccinator muscle
2 Buccal space
3 Masseter muscle
4 Medial pterygoid muscle
5 Pterygomandibular space
6 Superior constrictor muscle
7 Retropharyngeal space
8 Parotid gland
9 Carotid sheath
10 Parapharyngeal space

Figure 4.3 A horizontal section of the ramus, showing the pterygomandibular space.

muscle lies the inferior alveolar and lingual nerves, and the pterygomandibular space is therefore the site of injection for an inferior alveolar nerve block. It also contains the maxillary artery and the pterygoid venous plexus (Chapter 1).

The *infratemporal space* is the upper extremity of the pterygomandibular space. It lies behind the maxilla and is bounded medially by the lateral pterygoid plate and above by the base of the skull. It is in continuity with the deep temporal space laterally.

The *temporal space* consists of superficial and deep components that are found in relation to the temporalis muscle. The superficial temporal space lies on the lateral surface of the muscle, beneath the skin and superficial (temporal) fascia. The deep temporal space lies between the medial (deep) surface of the muscle and the adjacent temporal bone.

The *submasseteric space* may take the form of a series of spaces between the lateral surface of the ramus of the mandible and the masseter muscle. These spaces may form because the fibres of the masseter muscle have multiple insertions onto most of the lateral surface of the ramus and may be found between the attachment of the superficial and deep parts of the muscle. Alternatively, they may relate to the passage of the neurovascular bundles. Whatever the true explanation, abscesses may develop between the masseter and the ramus of the mandible.

TISSUE SPACES ADJACENT TO THE INFRATEMPORAL FOSSA

There are several other potential tissue spaces adjoining the infratemporal fossa that are in continuity with the tissue spaces of the infratemporal fossa. Thus, inflammation arising initially within the infratemporal fossa can spread to involve the tissue spaces described below that, with the exception of the submental space, are paired.

The *pharyngeal tissue spaces* can be subdivided into the peripharyngeal spaces around, and external to, the pharynx and the intrapharyngeal space within it. With regard to the peripharyngeal spaces, there is a parapharyngeal space laterally and a retropharyngeal space posteriorly. Some anatomists also include the submental and submandibular spaces as pharyngeal tissue spaces as they lie immediately anteriorly.

Each *parapharyngeal space* (or *lateral pharyngeal space*) passes laterally around the pharynx and is continuous posteriorly with the retropharyngeal space (Fig. 4.2). Unlike the retropharyngeal space, however, it is a space which is restricted to the suprahyoid region. It contains loose connective tissue and is bounded medially by the pharynx (superior constrictor muscle) and laterally by the pterygoid muscles and the parotid gland. Superiorly, it is bounded by the base of the skull. Inferiorly, it does not extend right down the neck but is limited by the suprahyoid structures, such as the fascia associated with the styloid group of muscles, the submandibular gland and mandibulostylohyoid ligament (see page 187). Behind is situated the carotid sheath. The lateral pharyngeal space is partly divided by the styloid process and associated group of muscles into an anterior compartment containing muscle and a posterior compartment containing the carotid sheath and cranial nerves IX–XII.

The *retropharyngeal space* is the area of loose connective tissue lying behind the pharynx and in front of the prevertebral fascia. It extends upwards to the base of the skull and downwards to the retrovisceral space in the infrahyoid part of the neck (Fig. 4.4).

The submental and submandibular spaces are located below the inferior border of the mandible, beneath the mylohyoid muscle, in the suprahyoid region of the neck. The *submental space* lies beneath the chin in the midline, between the mylohyoid muscles and the investing layer of deep cervical fascia and the platysma muscle superficially (Fig. 4.5). It is bounded laterally by the two anterior bellies of the digastric muscles. The submental space communicates posteriorly over the anterior bellies of the digastric muscles with the two submandibular spaces. Occasionally, infection arising from the mandibular incisors, and pointing deep to the origin of the mentalis muscle may produce a localised submental space infection (Figs 4.6, 4.7).

The *submandibular space* (Figs 4.5, 4.8, 4.9) is situated between the anterior and posterior bellies of the digastric muscle and is bounded above and laterally by the body of the mandible. It lies superficial to the mylohyoid muscle anteriorly and superficial to the hyoglossus and styloglossus muscles posteriorly. The space is covered by the platysma muscle and by the investing layer of deep cervical fascia, which explains why

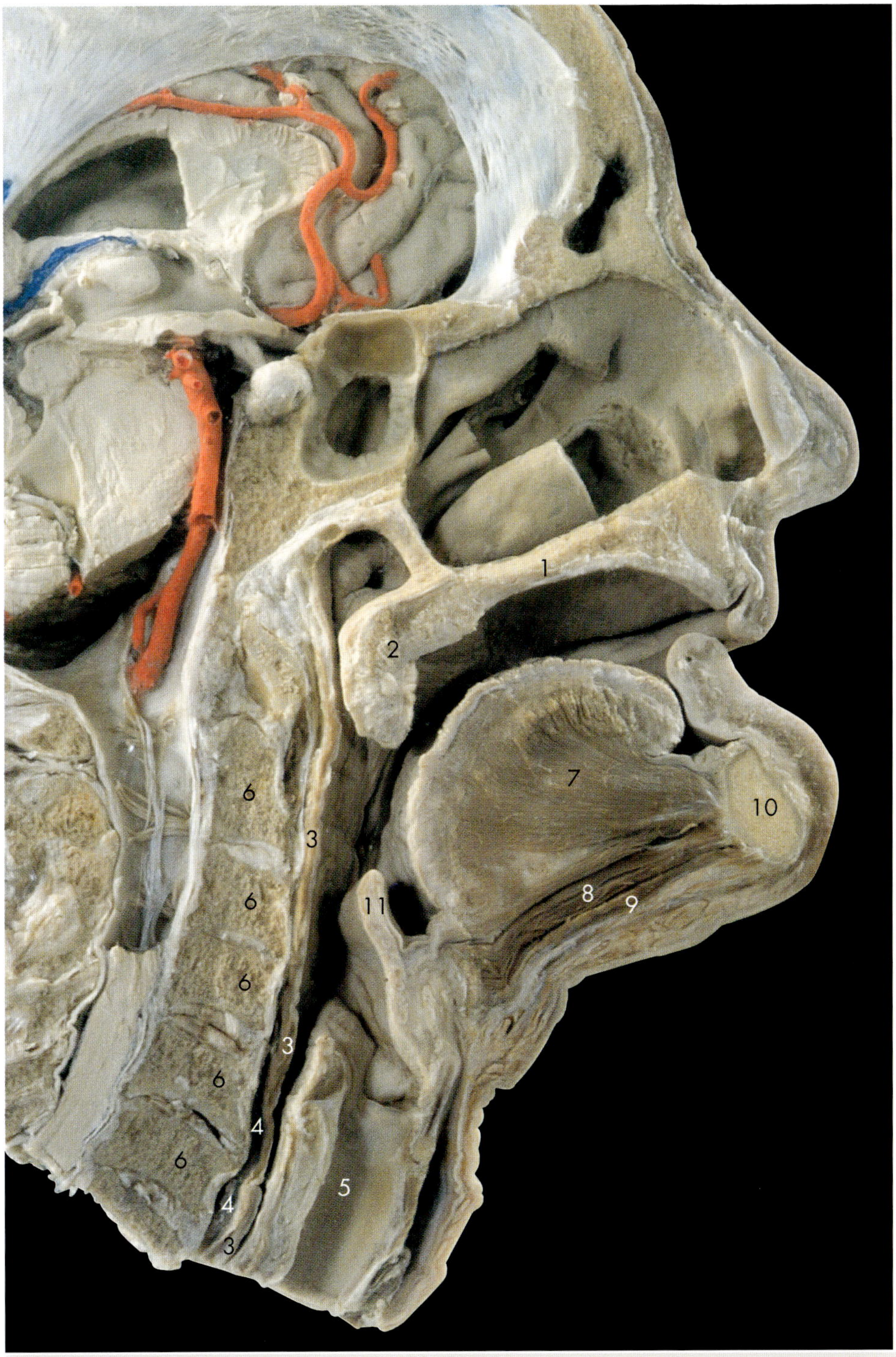

1 Hard palate
2 Soft palate
3 Cut edge of pharynx
4 Retropharyngeal space
5 Larynx
6 Cervical vertebrae
7 Genioglossus muscle
8 Geniohyoid muscle
9 Mylohyoid muscle
10 Mandible
11 Epiglottis

Figure 4.4 A sagittal section of the head showing the retropharyngeal space. (Courtesy Professor S. Standring, GKT School of Biomedical Sciences, London.)

Figure 4.5 Tissue spaces in the floor of the mouth viewed from below.

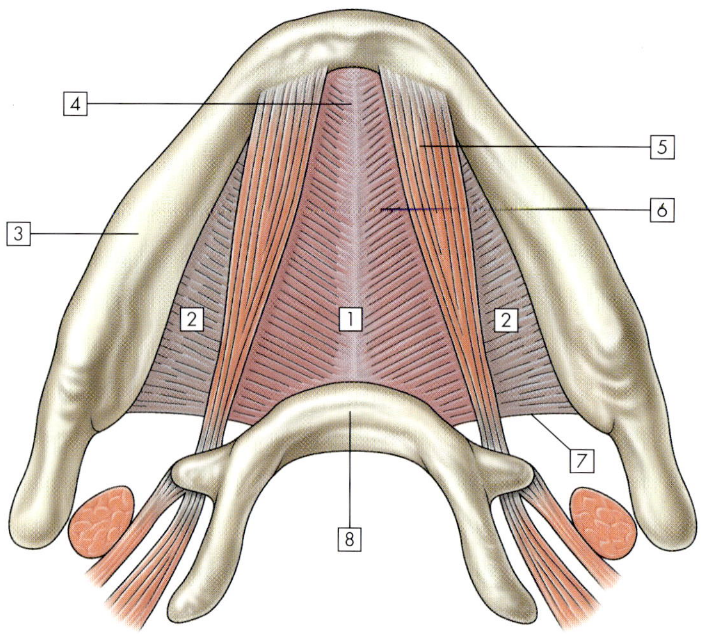

1 Submental tissue space
2 Submandibular tissue spaces
3 Inferior border of mandible
4 Raphè between mylohyoid muscles
5 Anterior belly of digastric muscle
6 Mylohyoid muscle
7 Posterior free border of mylohyoid muscle (communication between submandibular and sublingual spaces)
8 Body of hyoid bone

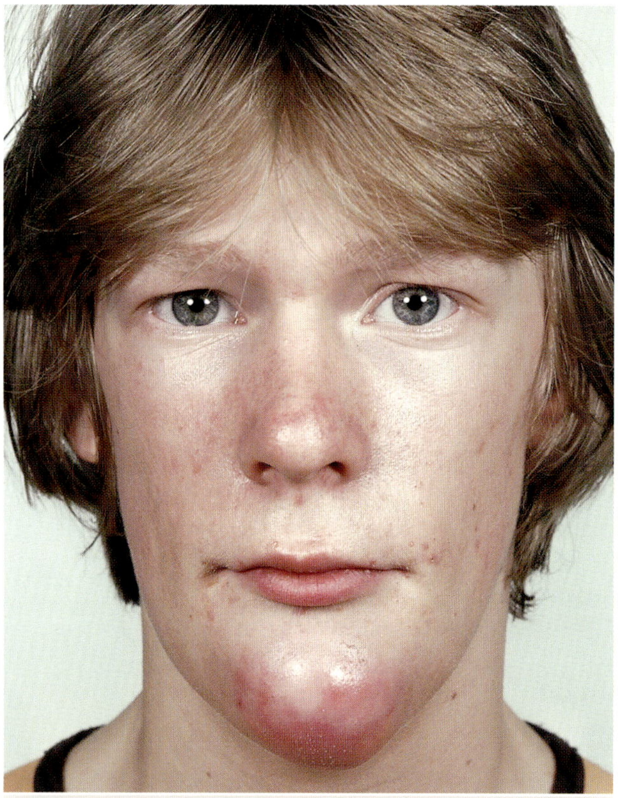

Figure 4.6 A submental space infection from the front.

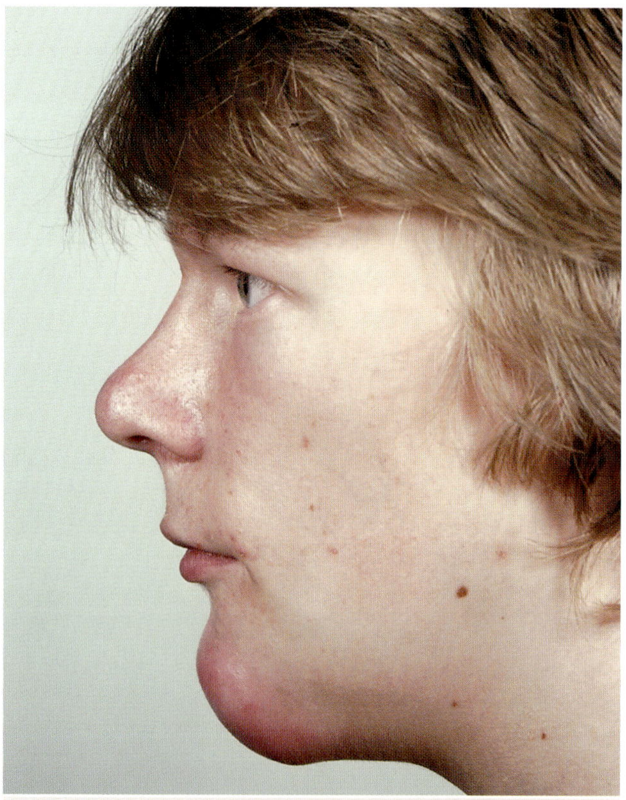

Figure 4.7 A submental space infection from the side.

abscesses in this region do not drain through the skin (Figs 4.8–4.10). The submandibular space communicates with the sublingual space around the posterior free border of the mylohyoid muscle (Figs 4.5, 4.11) and via small deficiencies within the muscular tissue of the mylohyoid muscle. The submandibular space contains the submandibular gland, facial vessels and submandibular lymph nodes (Fig. 4.10).

An *intrapharyngeal space* potentially exists between the inner surface of the constrictor muscles of the pharynx and the pharyngeal mucosa (Fig. 4.2). Infections at this site are either restricted locally or spread through the pharynx into the retropharyngeal or parapharyngeal spaces. An important part of the intrapharyngeal space is the peritonsillar space. This lies around the palatine tonsil, between the pillars of the fauces. Infections here (quinsy) usually spread up or down the intrapharyngeal space, or through the pharynx into the parapharyngeal space.

The *sublingual space* lies in the floor of the mouth, above the mylohyoid muscles and below the oral mucosa (Figs 4.8, 4.9, 4.11). It is delineated in front (and at the sides) by the body of the mandible and behind (and below) by the attachment of the mylohyoid muscle to the hyoid bone. The sublingual space contains the sublingual gland and the submandibular duct (Fig. 4.11).

The *buccal space* is located in the cheek (Figs 4.3, 4.7, 4.9). It has the buccinator muscle (covered by a delicate connective tissue layer called the buccopharyngeal fascia) medially, the skin of the cheek laterally, the pterygomandibular raphe (giving origin to the buccinator muscle) posteriorly, and part of orbicularis oris anteriorly. The buccal space contains the parotid duct accompanied by blood vessels and branches of the facial nerve, as well as the buccal pad of fat.

The *parotid space* surrounds the parotid gland and its contents (Fig. 4.2). It is defined by the parotid capsule. The superficial layer of the parotid capsule is of variable thickness and is not a typical fascia as it contains muscle fibres that parallel those of the platysma muscle. It appears to be continuous with the fascia associated with the platysma muscle. The deep surface of the parotid capsule is derived from the investing layer of deep cervical fascia (see page 182). Above the level of the stylomandibular ligament, the deep surface of the parotid capsule may be thin and may serve as a communicating pathway into the lateral pharyngeal space.

Figure 4.8 Tissue spaces in the floor of the mouth; coronal section.

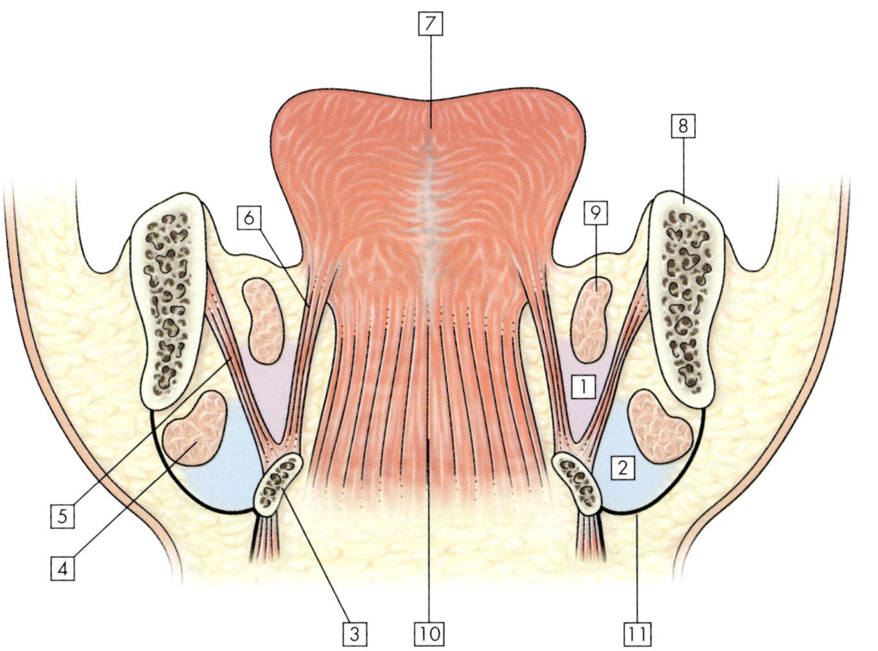

1 Sublingual tissue space
2 Submandibular tissue spaces
3 Greater horn of hyoid bone
4 Submandibular gland
5 Mylohyoid muscle
6 Hyoglossus muscle
7 Tongue
8 Body of mandible
9 Sublingual gland
10 Genioglossus muscle
11 Investigating layer of deep cervical fascia

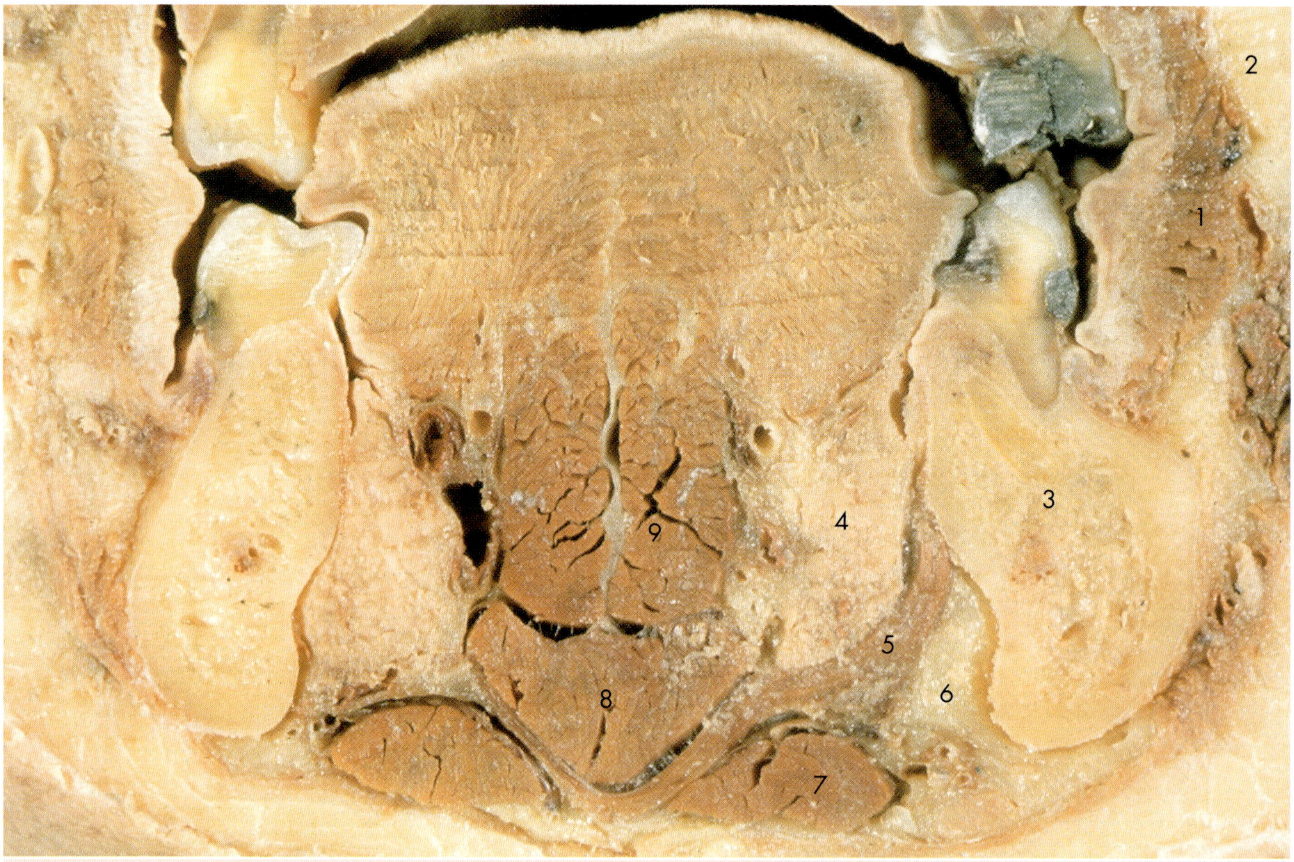

1 Buccinator muscle
2 Buccal tissue space
3 Body of mandible
4 Sublingual gland in sublingual tissue space
5 Mylohyoid muscle
6 Submandibular gland in submandibular tissue space
7 Anterior belly of digastric muscle
8 Geniohyoid muscle
9 Genioglossus muscle

Figure 4.9 A coronal section showing the submandibular and sublingual tissue spaces.

THE FASCIA AND TISSUE SPACES OF THE NECK

Spread of infection from the infratemporal fossa may not be confined to the tissue spaces closely associated with it in the face. Indeed, infection may spread into the region of the neck (and beyond) by way of a further series of tissue spaces.

When considering the tissue spaces of the neck, it is first necessary to consider the arrangement of the fascia. Three main layers can be identified: a superficial (investing) layer of deep cervical fascia, an intermediate (pretracheal) layer and a deep (prevertebral) layer. In addition, a non-membranous layer, the carotid sheath, is also usually described with the cervical fascia. Having outlined the arrangement of the fascial layers, the actual tissue spaces in the neck will then be described.

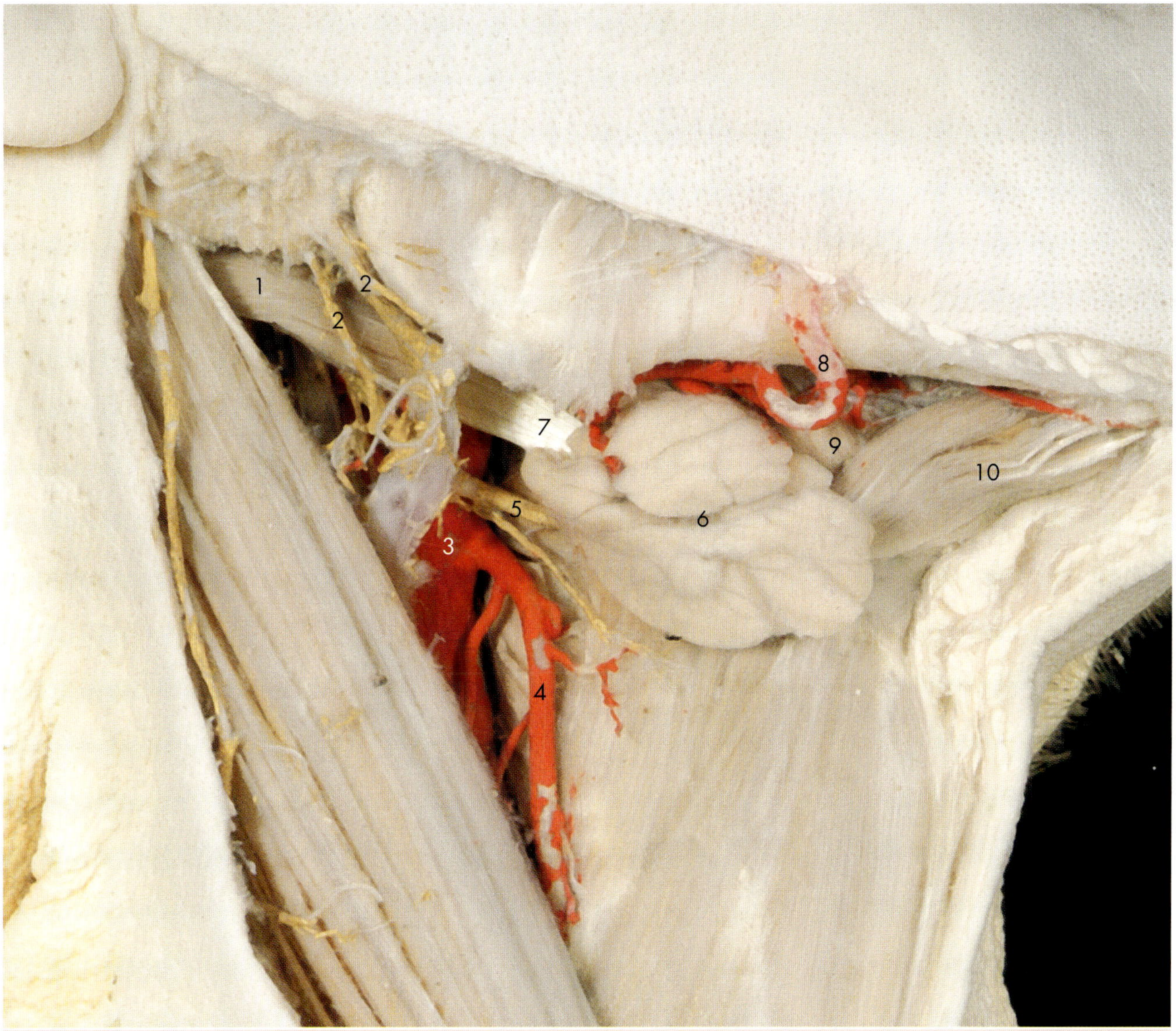

1 Posterior belly of digastric muscle
2 Branches of facial nerve
3 External carotid artery
4 Superior thyroid artery
5 Hypoglossal nerve
6 Submandibular gland (superficial part)
7 Intermediate tendon of digastric muscle
8 Facial artery
9 Mylohyoid muscle
10 Anterior belly of digastric muscle

Figure 4.10 The submandibular gland lying in the submandibular triangle. (Courtesy of Professor C, Dean, Department of Anatomy and Developmental Biology, University College London.)

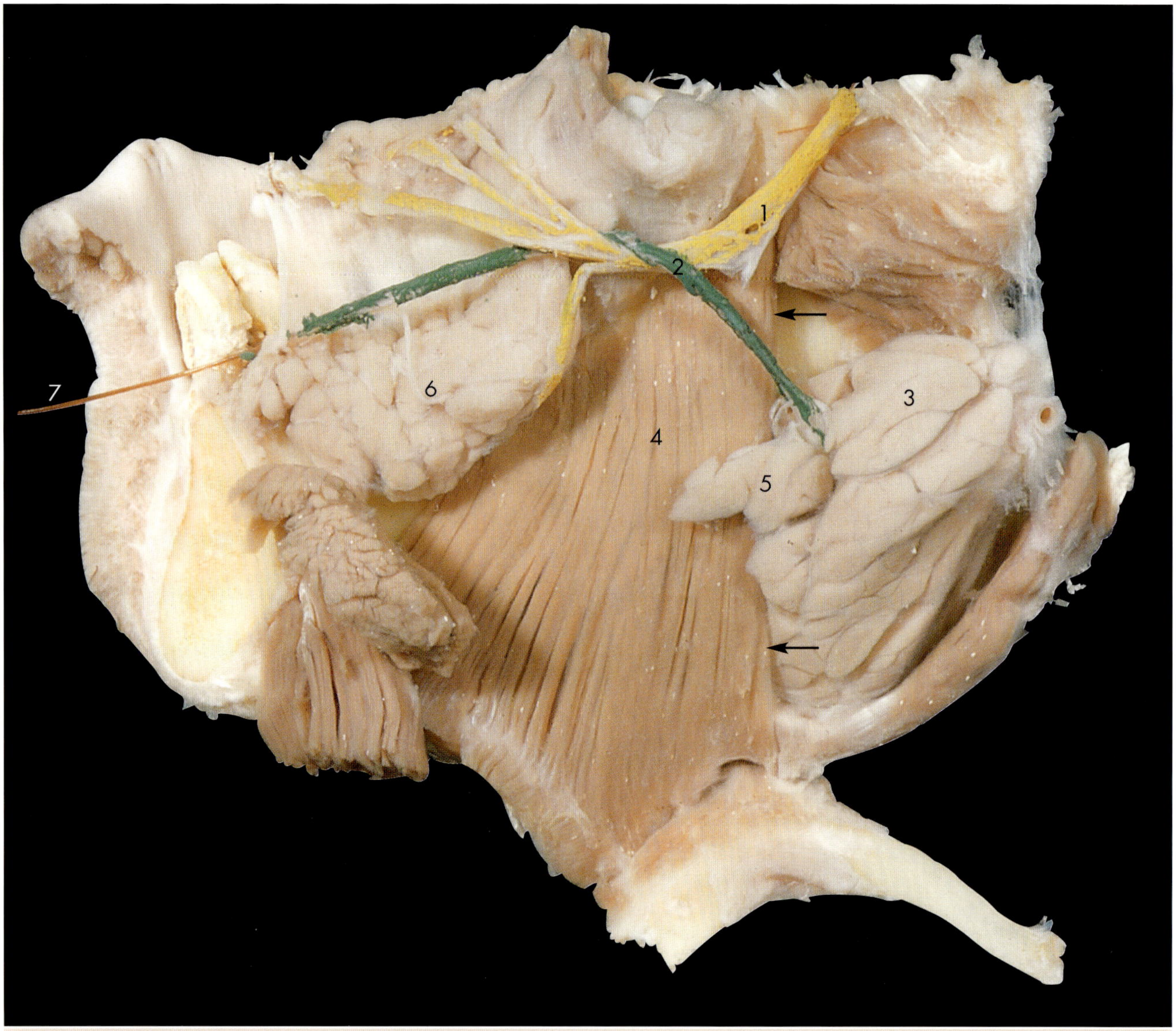

1 Lingual nerve
2 Submandibular duct
3 Submandibular gland (superficial part)
4 Mylohyoid muscle
5 Submandibular gland (deep part) hooking round posterior border of mylohyoid muscle
6 Sublingual gland lying in sublingual tissue space above mylohyoid muscle
7 Bristle in opening of submandibular duct at sublingual papilla

Figure 4.11 Medial view showing sublingual space and its continuity with submandibular space around the free posterior border of mylohyoid muscle (arrows). (Courtesy of Professor C, Dean, Department of Anatomy and Developmental Biology, University College London.)

THE FASCIA OF THE NECK (Figs 4.12–4.14)

The *investing layer of deep cervical fascia* is located just beneath the skin and is the most superficial of the true fascial layers of the neck. It completely encircles the neck like a surgical collar (Fig. 4.12). At the front of the neck, however, the fascia is situated internal to the platysma muscles (which themselves are surrounded by a delicate layer of so-called superficial cervical fascia). Posteriorly, the investing layer is attached in the midline to the spines of the cervical vertebrae and to the ligamentum nuchae. As it passes anteriorly around the neck, it splits to enclose first the trapezius muscle and then the sternocleidomastoid muscle. Between these two muscles, the investing layer forms the roof of the posterior triangle, while its extension in front to the sternocleidomastoid muscle covers the anterior triangle. In the midline anteriorly, the investing layer is attached to the chin, the body of the hyoid bone and to the manubrium sterni (where it splits into superficial and deep layers with the suprasternal space between). Superiorly, the investing layer is attached to the external occipital protuberance, the superior nuchal lines at the back of the skull, the tip of the mastoid process, the tympanic plate, styloid process and across to the mandible. Between the styloid process and the angle of the mandible, the fascia forms the stylomandibular ligament which partially separates the parotid and submandibular compartments. Inferiorly, the investing layer is attached to the sternum, the clavicle (where it also splits into superficial and deep layers between the attachments of the trapezius and the sternocleidomastoid muscles) and to the acromion of the scapula.

In the upper part of the neck, the investing layer of deep cervical fascia roofs over the submandibular and submental tissue spaces. In addition to being covered by this fascia on its outer surface, the submandibular gland has a thin extension on its deep surface, forming a capsule around the gland. The space enclosed by this capsule, which contains the submandibular gland and which is perforated by the submandibular duct, can be referred to as the tissue space of the submandibular gland, thus distinguishing it from the submandibular tissue space.

The *pretracheal fascia* lines the deep surface of the infrahyoid (strap) muscles and surrounds the viscera of the neck, namely the trachea, oesophagus and thyroid gland (Fig. 4.12) (i.e. the visceral compartment). It is found below the hyoid bone and is termed pretracheal because it is particularly prominent in front of the trachea. Superiorly, the pretracheal fascia is attached to the larynx, being limited by the attachments of the overlying infrahyoid strap musculature. Inferiorly, the fascia extends into the superior mediastinum of the thorax and here it is related to the great vessels, merging with the fibrous pericardium. Laterally, the pretracheal fascia merges with the investing layer of deep cervical fascia and with the connective tissues comprising the carotid sheath. The fascia forms a sheath around the thyroid gland.

The *prevertebral fascia* encloses the cervical part of the vertebral column and the prevertebral and postvertebral muscles (i.e. the musculoskeletal compartment) (Fig. 4.12). It is termed the prevertebral fascia because it is particularly prominent in front of the vertebral column. Indeed, here there may be two distinct layers, an outer alar part and an inner prevertebral part (Fig. 4.14). The prevertebral fascia forms the fascial floor of the posterior triangle of the neck. Superiorly, it is attached to the base of the skull in front of the longus capitis and rectus capitis lateralis muscles. Inferiorly, it extends into the thorax to merge with the anterior longitudinal ligament of the third thoracic vertebra at the lower limit of the longus cervicis muscle (Fig. 4.14).

Figure 4.12 Transverse section of the neck at the level of the trachea to show the main layers of the cervical fascia.

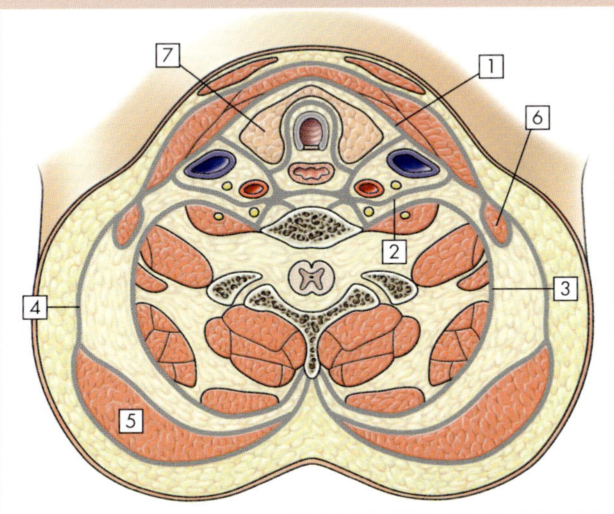

1 Pretracheal fascia
2 Carotid sheath
3 Prevertebral fascia
4 Investing layer of deep cervical fascia
5 Trapezius muscle
6 Sternocleidomastoid muscle
7 Thyroid gland

As the prevertebral fascia passes around the musculoskeletal compartment of the neck, it attaches to the transverse and spinous processes of each cervical vertebra and to the ligamentum nuchae. However, the fascia becomes indistinct posteriorly and often merges with the investing layer of deep cervical fascia.

In the root of the neck, the prevertebral fascia covers the scalene muscles, but tends to thin out beyond this. Thus, the phrenic nerve runs beneath the fascia on the scalenus anterior muscle. Unlike the phrenic nerve, the sympathetic trunk is not covered by the prevertebral fascia, but lies embedded on its surface. As the subclavian artery and the nerves from the brachial plexus emerge from behind the scalenus anterior muscle, the prevertebral fascia forms the axillary sheath. This sheath invests the subclavian and axillary arteries, but not the veins.

The prevertebral fascia provides a base upon which the pharynx, oesophagus and other cervical structures glide during swallowing and neck movements, undisturbed by movements of the prevertebral muscles.

The *carotid sheath* is a distinct, but thin, fascial layer surrounding the carotid arteries (the common and internal carotids, but not the external carotid), the internal jugular vein, and the vagus nerve (Figs 4.12, 4.13). The carotid sheath is thinner over the internal jugular vein. The descendens hypoglossi nerve is embedded in the anterior wall of the carotid sheath, while the sympathetic trunk lies behind the carotid sheath, embedded on the surface of the prevertebral fascia. The sheath is said to be attached superiorly to the base of the skull and inferiorly it merges with the connective tissue around the arch of the aorta. However, most reports suggest that the sheath may be nothing more than the merging of the adjacent investing cervical fascia, prevertebral fascia and pretracheal fascia.

THE TISSUE SPACES OF THE NECK (Figs 4.14, 4.15)

The fascial layers of the neck define a number of tissue spaces. As with those tissue spaces described around the infratemporal fossa, these spaces must be regarded only as 'potential spaces' and not as true anatomical entities. This is because, in the healthy person, the tissues are closely applied to each other or are filled with relatively loose connective tissue. Where there is pathological involvement, particularly with inflammation, this may spread through the tissue spaces, usually taking the line of least resistance. Conversely, the fascia can confine the spread of inflammation.

Below the hyoid bone are located the pretracheal and retrovisceral tissue spaces in the visceral compartment of the neck, the prevertebral space in front of the vertebral column, and a space associated with the carotid sheath. Although a tissue space may be expected between the pretracheal fascia and the investing layer of deep cervical fascia, the area is occupied by the infrahyoid (strap) musculature.

The *pretracheal space* lies behind the pretracheal fascia and the infrahyoid (strap) muscles and in front of the anterior wall of the oesophagus (Figs 4.14, 4.15). It is bounded laterally by the carotid sheath. The space thus immediately surrounds the trachea. It is bounded superiorly by the attachments of the infrahyoid muscles to the thyroid cartilage of the larynx. Inferiorly, it extends down into the anterior portion of the superior mediastinum. Infection usually spreads into the pretracheal space either by perforating the anterior wall of the oesophagus or from the retrovisceral space. The

Figure 4.13 Section through the carotid sheath just below the base of the skull.

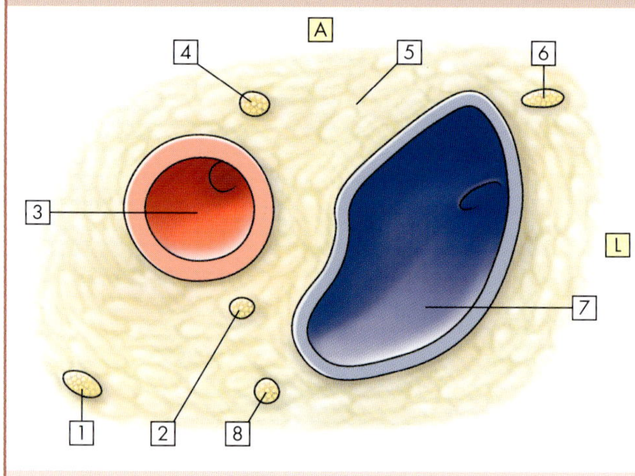

A Anterior
L Lateral
1 Cervical sympathetic trunk
2 Vagus nerve
3 Internal carotid artery
4 Glossopharyngeal nerve
5 Carotid sheath
6 Accessory nerve
7 Internal jugular vein
8 Hypoglossal nerve

space contains the inferior thyroid veins and is delineated from the retrovisceral space behind by connective tissue septa passing from the prevertebral fascia to the lateral aspect of the oesophagus.

The *retrovisceral (retro-oesophageal)* space is continuous superiorly with the retropharyngeal space. It is situated between the posterior wall of the oesophagus and the prevertebral fascia. Inferiorly, the retrovisceral space extends into the superior mediastinum. Should the prevertebral fascia merge with the connective tissue on the posterior surface of the oesophagus (usually at the level of the fourth thoracic vertebra), the retrovisceral space has a distinct inferior boundary.

The *prevertebral space* has been variously described. To some authors, it is the potential space lying between the prevertebral fascia and the vertebral column. Others regard it as the space (the so-called danger space) between the two layers comprising the prevertebral fascia, the alar and prevertebral layers. As the space is closed above, below and laterally (Fig. 4.14), infection usually spreads into it through its fascial walls from the retrovisceral area. Inferiorly, the space extends into the posterior mediastinum.

As already mentioned, the carotid sheath is, in reality, a layer of loose connective tissue demarcated by adjacent portions of the investing layer of deep cervical fascia, the pretracheal fascia and the prevertebral fascia. Nevertheless, there is a potential space, the *carotid space*, into which infections from the visceral spaces may track. However, it has been reported that infections around the carotid sheath are restricted because superiorly (near the hyoid bone) and inferiorly (near the root of the neck) the connective tissues adhere to the vessels.

Figure 4.14 Some of the tissue spaces of the neck seen in longitudinal section. Note that in this diagram the retrovisceral space is limited inferiorly by the merging of a layer of the prevertebral fascia with the oesophagus.

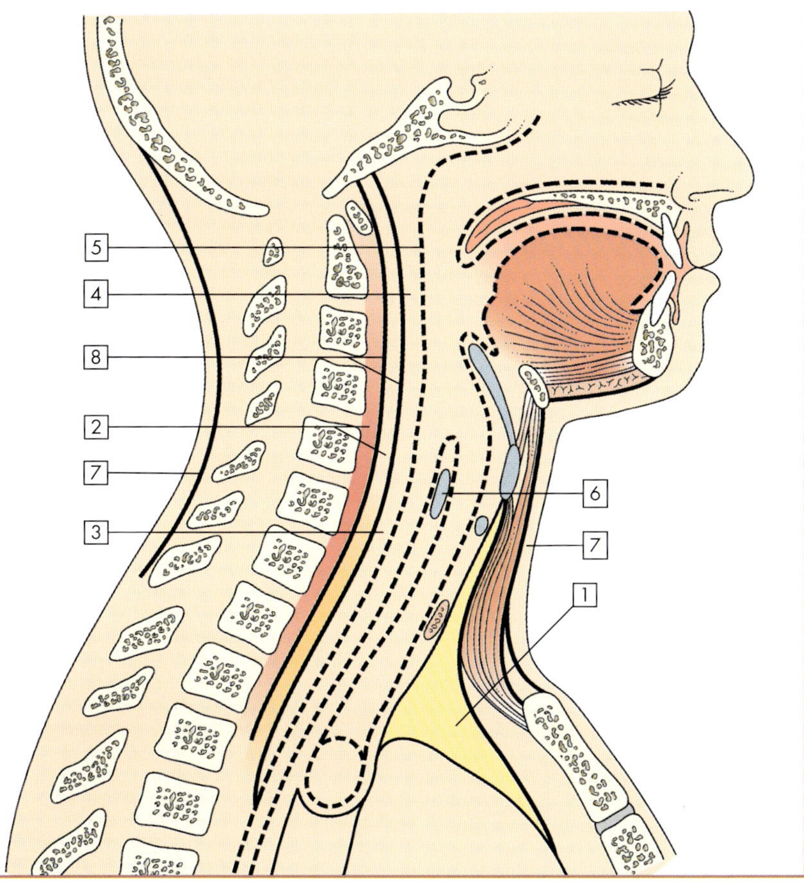

1 Pretracheal space
2 Prevertebral space
3 Retrovisceral space
4 Retropharyngeal space
5 Posterior wall of pharynx
6 Cricoid cartilage (lamina)
7 Investing layer of deep cervical fascia
8 Two layers of prevertebral fascia

Figure 4.15 The tissue spaces of the neck below the hyoid bone, indicated by pale blue regions.

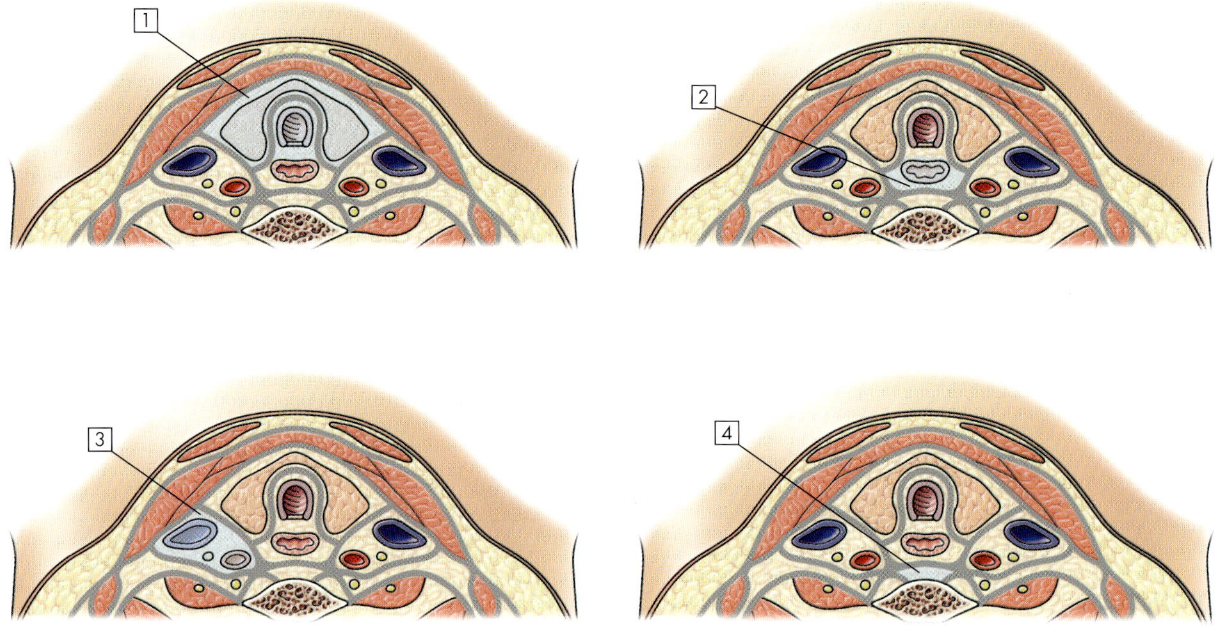

1 Pretracheal tissue space
2 Retrovisceral tissue space
3 Tissue space associated with the carotid sheath
4 Prevertebral tissue space

INFECTION OF THE INFRATEMPORAL FOSSA REGION AND ITS SPREAD

CERVICOFACIAL CELLULITIS

Cellulitis, a spreading infection of connective tissue, is characterised by a gross inflammatory exudate and oedema, together with fever and toxaemia, which may be severe. Before the use of antibiotics, the mortality was high and the condition is still life-threatening if treatment is delayed.

The characteristic features of cellulitis are diffuse brawny swelling, pain, fever and malaise. The swelling is tense and tender with board-like firmness. The overlying skin is characteristically taut and shiny. The pain and swelling result in difficulty in opening the mouth and in swallowing. Constitutional upset becomes severe, with increasing pyrexia, toxaemia and leukocytosis. The regional lymph nodes become swollen and tender. Once oedema extends towards the rima glottidis, as in Ludwig's angina, or the tongue is forced upwards and backwards into the airway, there is increasing respiratory distress that, if not rapidly relieved, quickly results in asphyxia. Tracheostomy may be necessary.

Aetiology and pathology

The organisms mainly responsible are beta-haemolytic streptococci and a variety of anaerobes.

Fascia covering the muscles and other structures are normally in close apposition. If these fascial planes are spread apart a space is created. Such spaces contain little except loose connective tissue and are almost avascular. If, therefore, virulent organisms enter these fascial planes, inflammatory exudate is poured out from nearby vessels but, failure to localise the organisms, opens up the fascial space carrying infection with it. Infection may thus spread through one or more fascial spaces until the natural boundaries are reached.

The main cause of cellulitis of the neck is infection arising from the region of the mandibular molar teeth. Several fascial spaces are accessible from this area and the following factors are contributory:

1. The apices of the second and, more especially, the third mandibular molar teeth are often close to the lingual surface of the mandible.
2. The mylohyoid muscle attachment inclines upwards as it runs forwards; the apices of the roots of the third mandibular molars are usually (and the second molars are often) below this line.
3. The posterior border of the mylohyoid muscle is close to the sockets of the third mandibular molars. At this point, the floor of the mouth consists only of mucous membrane covering part of the submandibular salivary gland.

A virulent periapical infection of a mandibular third molar tooth may therefore penetrate the lingual plate of the mandible and is then at the entrance to several fascial spaces. Anteriorly, there are the submandibular and sublingual spaces, while posteriorly lie the pharyngeal and pterygomandibular spaces. Infections in this area may rarely also spread from an acute pericoronitis, particularly when the deeper tissues are opened to infection by extraction of the tooth during the acute phase.

Cellulitis can be a complication of acute osteomyelitis of the jaws due to spread of an exceptionally virulent infection, but this is virtually never seen nowadays.

In general, cellulitis around the jaw is only likely to develop when the tissues are infected by virulent and invasive organisms at a point where there is access to the fascial spaces. As the predisposing causes do not often coincide, cellulitis is uncommon. Cellulitis in the region of the maxilla is even more uncommon, but fascial space infections may develop in various sites as the result of using infected local anaesthetic needles.

It is evident that there are no barriers running horizontally with respect to the tissue spaces in the neck. Thus, infection entering the third molar region can rapidly spread more or less unhindered down the neck and may enter the thorax.

INFRATEMPORAL FOSSA SPACE INFECTIONS

Infection of the infratemporal fossa is most commonly associated with a pericoronitis affecting a partially impacted mandibular third molar tooth. It may also be associated with a dental abscess of this tooth, or as a result of infection following tooth extraction (dry socket). Rarely, it may result from an infected needle used during an inferior alveolar nerve block. Infection of the infratemporal region may be secondary due to spread from an adjacent infected tissue space.

The main symptom caused by infection of the pterygomandibular space is trismus (painful reflex muscle spasm) generally affecting the medial pterygoid muscle (see page 7). Externally there is usually little evidence of tissue swelling. Other general symptoms of infection (e.g. fever, malaise and lymphadenopathy) may also be evident.

Spread of infection from the infratemporal fossa region to involve the buccal space is characterised by the presence of a swelling of the cheek (Figs 4.16, 4.17). The swelling is bounded above by the zygomatic arch and below by the lower border of the mandible, both landmarks being palpable. The buccal tissue space would also be involved in the case of a dental abscess discharging buccally below the attachment of the buccinator muscle to the alveolar process.

Infection from the infratemporal fossa may also spread directly around the back of the maxillary tuberosity and into the orbit via the inferior orbital fissure. This may result in cavernous sinus thrombosis (see pages 97–98). Once in the orbit, further direct spread of infection through the superior orbital fissure will gain entrance into the cranial cavity.

Spread from the infratemporal fossa via the pterygomaxillary fissure may also involve the pterygopalatine fossa, which contains the maxillary nerve, maxillary artery and the pterygopalatine ganglion and its branches (see pages 42–53). From the pterygopalatine fossa a number of small canals lead into the nose, pharynx and palate. These encourage the spread of infection.

SUBMASSETERIC SPACE INFECTIONS

Sometimes infection around a mandibular third molar tooth tracks backwards, lateral to the mandibular ramus, and pus localises deep to the attachment of the masseter muscle. Such an abscess, deep to this thick muscle, produces little visible swelling, but is accompanied by profound muscle spasm and limitation of jaw opening.

PHARYNGEAL SPACE INFECTIONS

If infection from the masticatory tissue spaces spreads to those associated with the pharynx, additional symptoms will be encountered. If the parapharyngeal space is involved, in addition to pain and trismus, there may

Figure 4.16 A buccal space infection from the front.

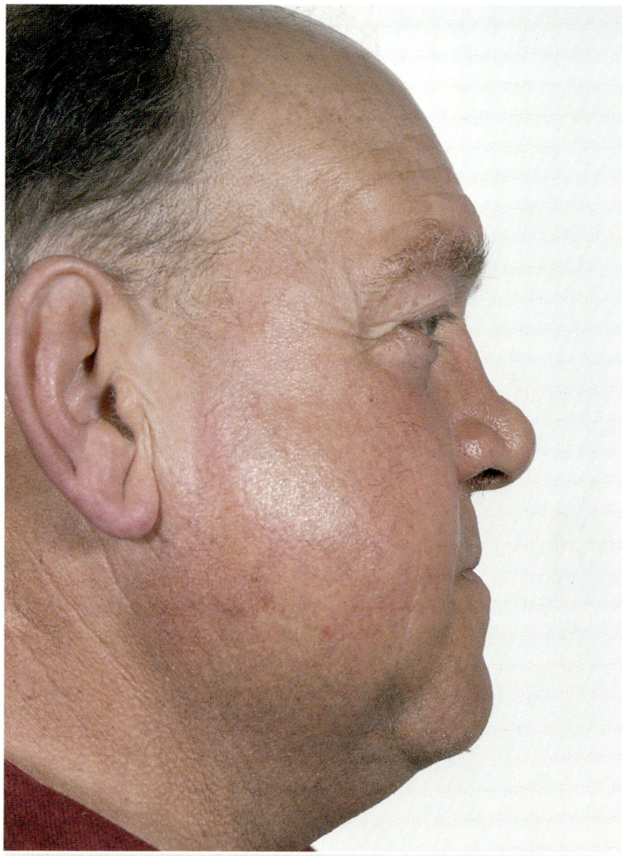

Figure 4.17 A buccal space infection from the side.

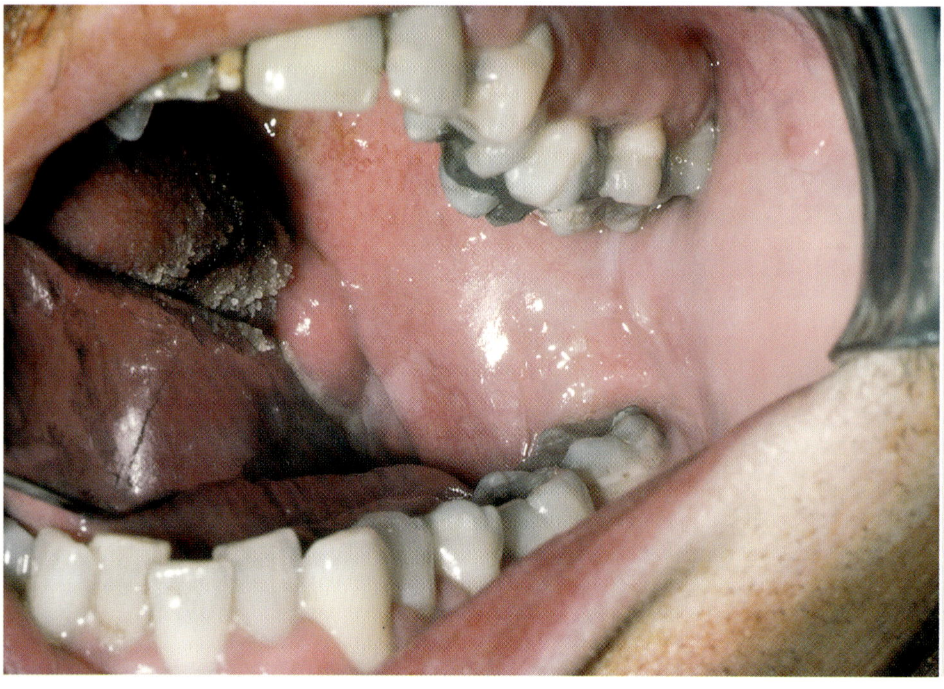

Figure 4.18 An abscess of the parapharyngeal space.

be swelling in the region of the oropharynx which may extend to the soft palate and displace the uvula to the unaffected side (Fig. 4.18). Some swelling may also be evident below the angle of the mandible. The patient may complain of dysphagia. If the infection spreads posteriorly to involve the retropharyngeal space, additional symptoms may be bulging of the posterior pharyngeal wall (Figs 4.19, 4.20), dyspnoea

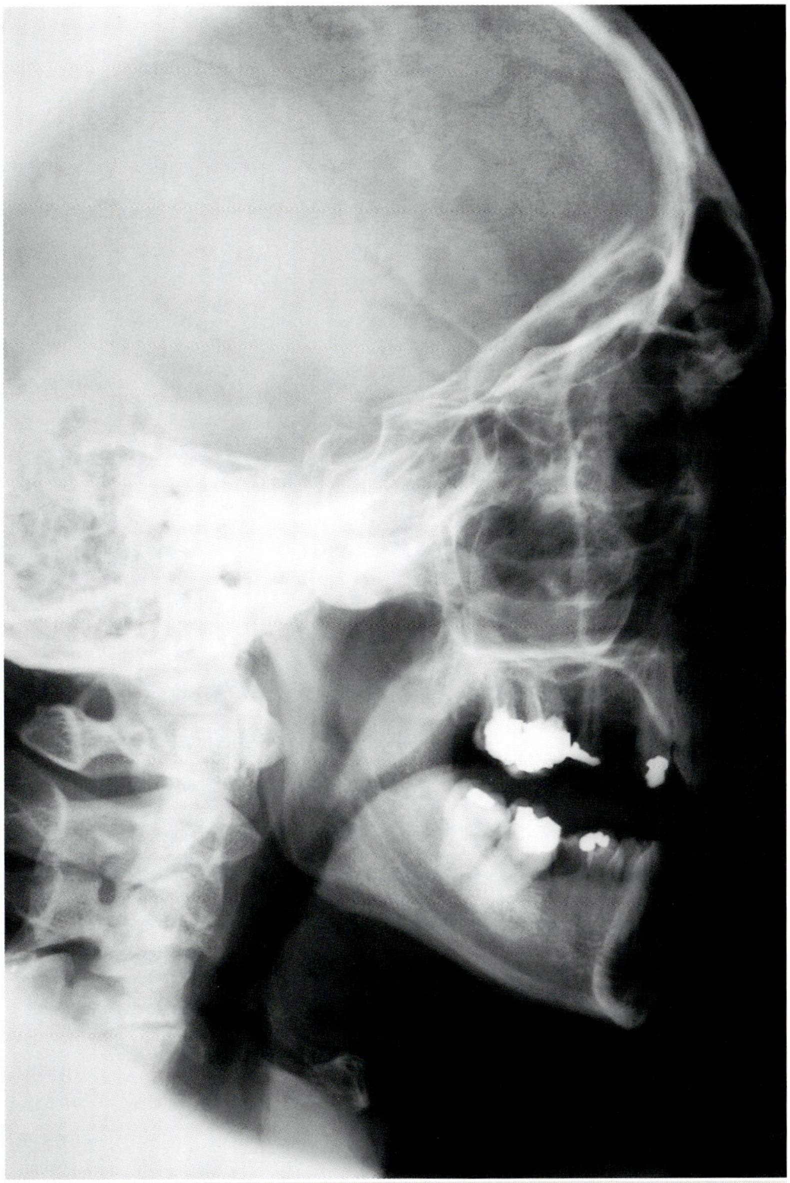

Figure 4.19 A radiograph showing a normal prevertebral soft tissue shadow (compare with Fig. 4.20).

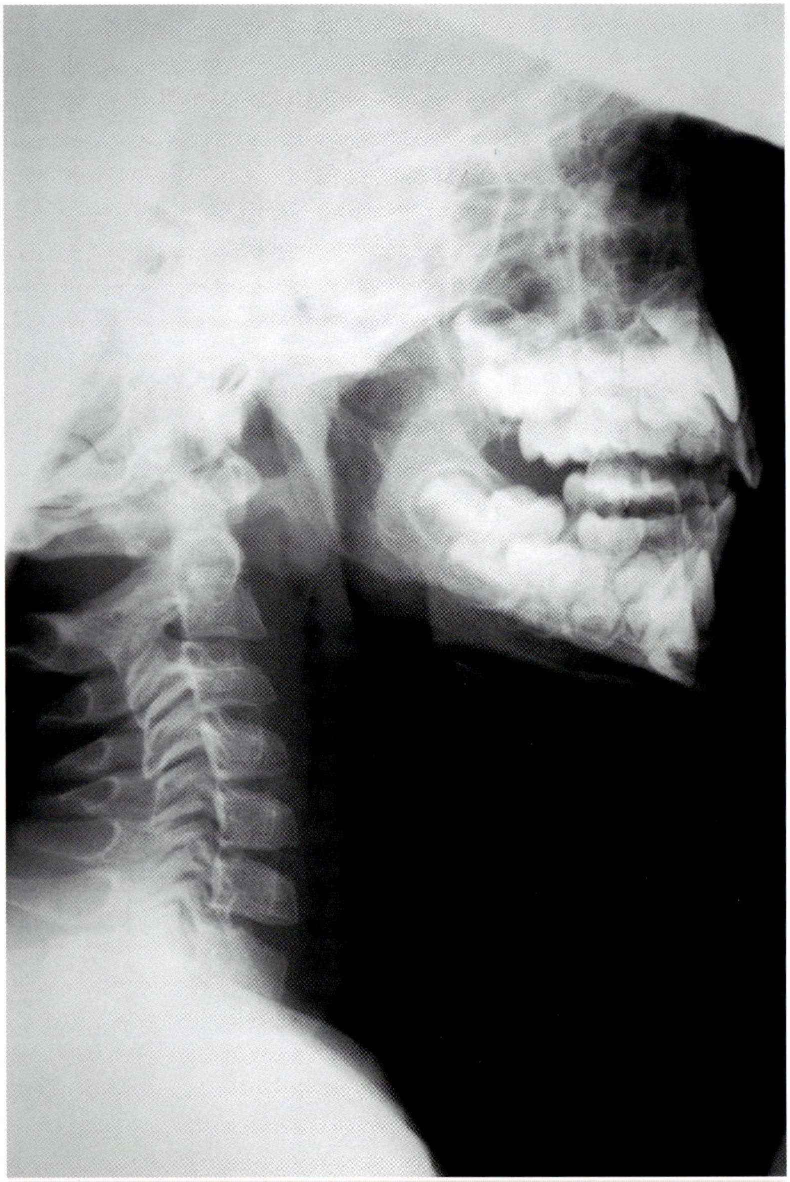

Figure 4.20 A radiograph showing a retropharyngeal abscess. Note the gross thickening of the prevertebral soft tissue shadow (compare with Fig. 4.19).

and nuchal rigidity. Involvement of the carotid sheath and surrounding area may produce symptoms associated with thrombosis of the internal jugular vein and involvement of cranial nerves IX–XII. In the severest of cases, infection may reach the mediastinum, giving rise to the symptoms of mediastinitis. A serious complication associated with infection of the retropharyngeal space is when there is spread through the prevertebral fascia to involve the underlying prevertebral tissue space. Such spread of infection can result in the presence of inflammation in the posterior part of the thorax and into (and even below) the region of the diaphragm. Symptoms may then include chest pain, severe dyspnoea and retrosternal discomfort.

If the peritonsillar space becomes infected, it results in a swelling associated with the lower pole of the tonsil and in acute pain, often resulting in abscess formation (quinsy).

SUBMANDIBULAR CELLULITIS

The submandibular space contains the submandibular salivary gland and is formed by the splitting of the deep cervical fascia above the hyoid bone. The space is bounded laterally and below by the investing layer of the deep cervical fascia, which extends from the hyoid bone to the mylohyoid and forms the superior and medial boundary of the space.

The intercommunications of the three main tissue spaces in this area are related to the deep portion of the submandibular gland. This extends around the free posterior border of the mylohyoid muscle, juts into the entry to the parapharyngeal tissue space and, by curving forwards onto the superior surface of the mylohyoid muscle, comes to lie within the sublingual tissue space.

Clinical features

Submandibular cellulitis is typically caused by infection from the second or third mandibular molar tooth. The swelling is centred upon the upper part of the neck, mainly along the lower border of the mandible (Figs 4.21, 4.22). Infection may spread into the sublingual or parapharyngeal spaces as already indicated.

Incisions in the region of the angle of the mandible must make allowance for important structures situated in this region, whose normal relations may be distorted by tissue swelling. Such structures include the mandibular branch of the facial nerve, submandibular gland, facial artery and vein, and hypoglossal and lingual nerves (Figs 18.18–18.20). For this reason, only the skin is incised and the tissue space opened by blunt dissection using sinus forceps (Hilton's method).

LUDWIG'S ANGINA[13–15]

Ludwig's angina is a severe form of cellulitis which usually arises from the mandibular second or third molar tooth. It involves the sublingual and submandibular tissue spaces bilaterally (and almost simultaneously) and readily spreads into the pharyngeal and pterygomandibular spaces. Spread of the swelling towards the midline of the neck and below the chin would indicate involvement of the submental tissue space.

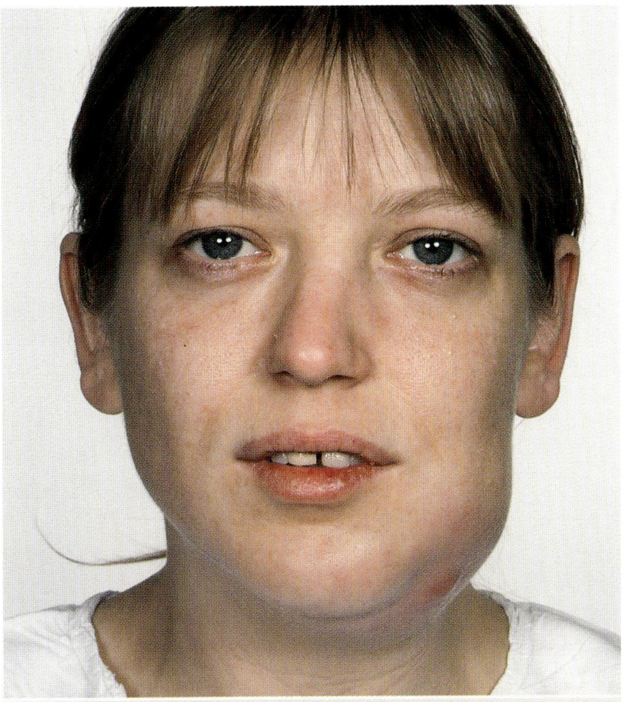

Figure 4.21 A submandibular space infection from the front.

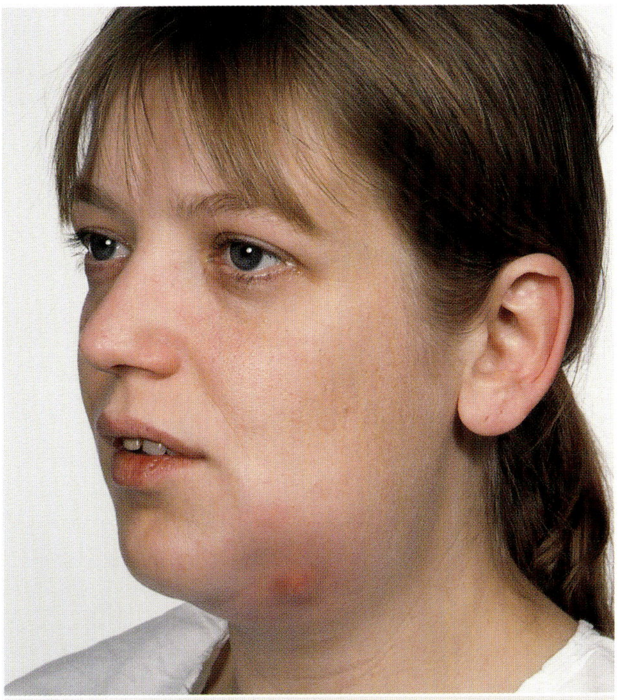

Figure 4.22 A submandibular space infection from the side.

Clinical features

Ludwig's angina is characterised by rapid development of sublingual and submandibular cellulitis with a painful, brawny swelling of the upper part of the neck and the floor of the mouth on both sides (Fig. 4.23). When the parapharyngeal tissue space becomes involved, the swelling tracks down the neck and oedema spreads into the loose connective tissue around the rima glottidis.

There is difficulty in swallowing, opening the mouth may be limited and the tongue may be pushed up against the soft palate (Fig. 4.24). Oedema around the rima glottidis causes increasing respiratory obstruction. The patient soon becomes seriously ill, with fever, headache and malaise.

Respiratory obstruction is suggested by noisy breathing and restlessness, going on to violent efforts at respiration and increasing cyanosis. A patient with cellulitis of the neck may die quickly from this cause, or later from the effects of spread of infection to the mediastinum via the carotid sheath.

Management of cellulitis

A patient with a brawny swelling of the mouth or neck, fever and malaise must be immediately admitted to hospital. The mainstay of treatment of cellulitis is vigorous use of antibiotics. Provided that the patient is not hypersensitive, intravenous penicillin (not less than 600 mg) should be given every 6 hours.

The swelling should be incised at an early stage to relieve the pressure of exudate, particularly when the swelling is so large and tense as to force the tongue into the airway. Little fluid is produced at first but drainage continues in small amounts. When draining such brawny swellings, the neck should be laid open widely, all the tissue spaces opened with sinus forceps and multiple corrugated drains inserted. Often, through and through drainage is required, with bilateral drains through the neck into the oral cavity (Figs 4.25, 4.26).

General anaesthesia is very hazardous in the later stages of Ludwig's angina. The patient will often be

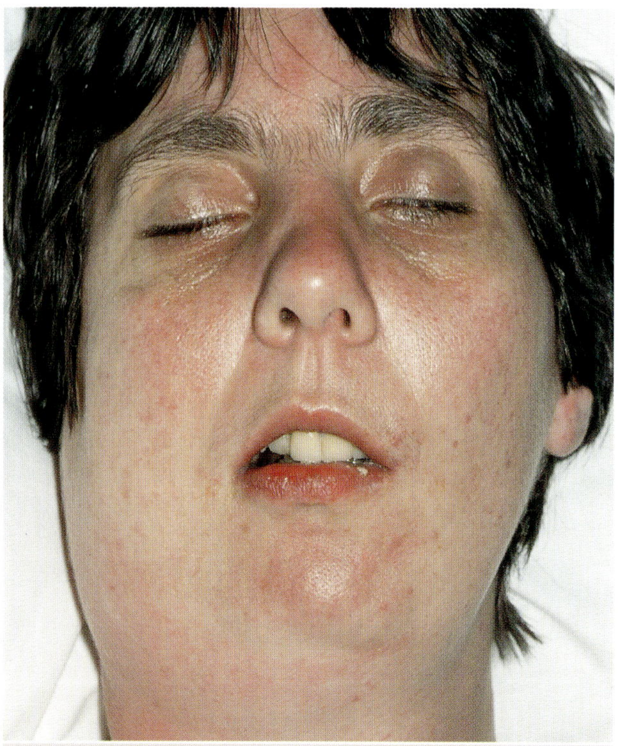

Figure 4.23 Ludwig's angina.

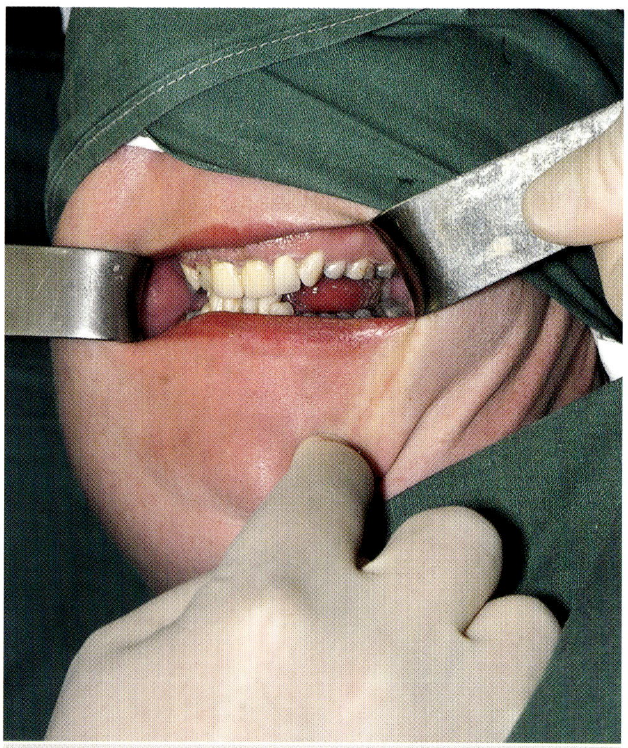

Figure 4.24 Ludwig's angina at surgery.

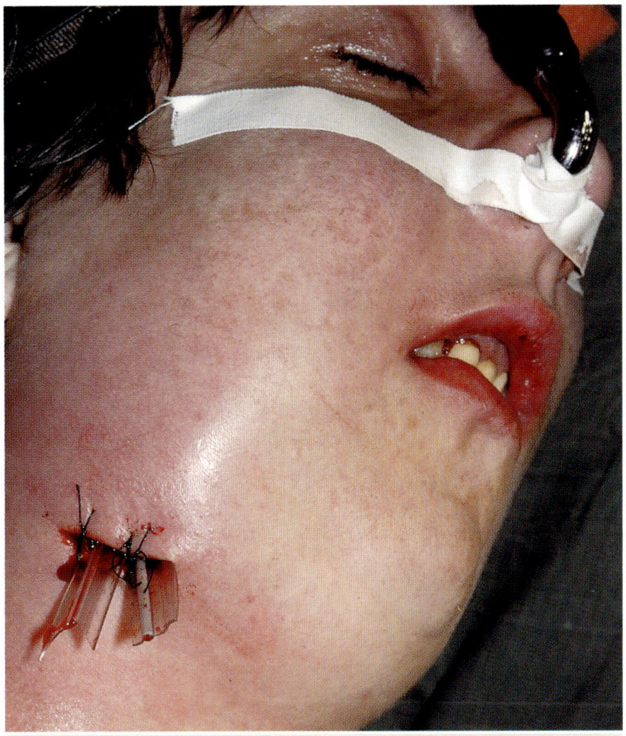

Figure 4.25 External drainage for Ludwig's angina.

relying on a conscious effort to maintain the airway. Immediately a muscle relaxant is given the airway is lost and, if rapid intubation is not possible, an emergency tracheostomy becomes necessary. This in itself opens up further tissue planes to infection. For this reason, surgical drainage should be undertaken early in the course of the infection before respiratory obstruction develops. Should it be necessary at this stage, the patient is given a gaseous induction without muscle relaxants and an attempt is made at either blind nasal intubation or intubation using a fibreoptic laryngoscope. The surgeon must stand by ready to perform an emergency tracheostomy.

The tooth from which the infection started should be extracted as soon as the patient's condition allows.

Cavernous sinus thrombosis[15]

Cavernous sinus thrombosis is a serious complication that can also arise from spread of infection usually from an upper canine tooth (Fig. 4.27). Infected thrombi in the facial vein, or less commonly the pterygoid plexus of veins, communicate with the cavernous sinus

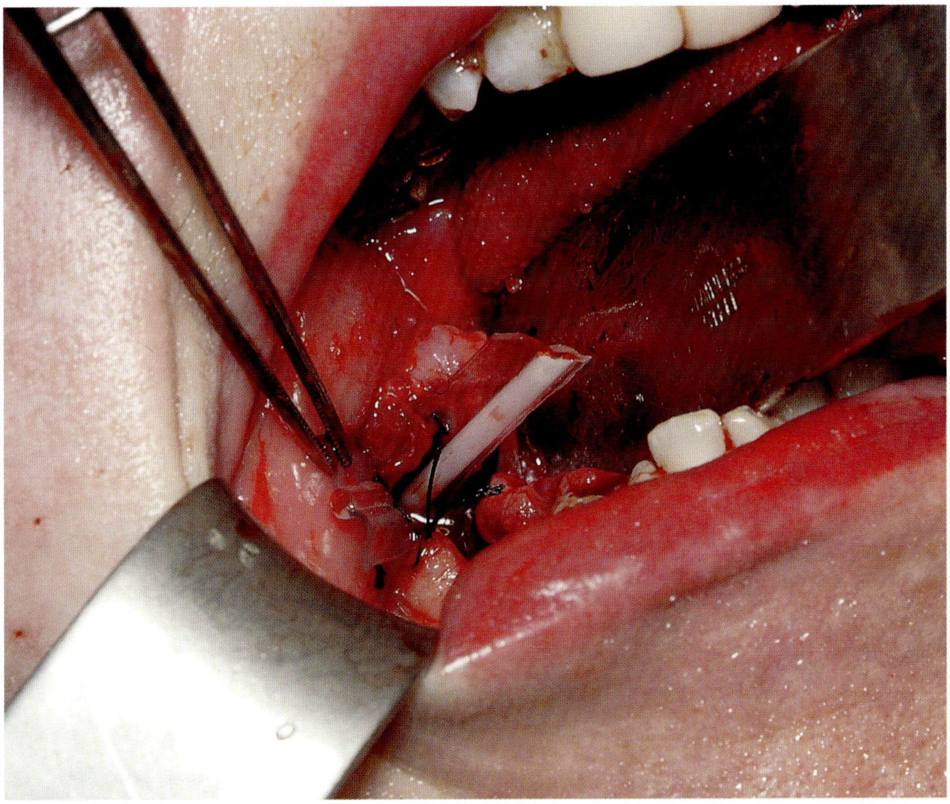

Figure 4.26 Through-and-through drainage for Ludwig's angina.

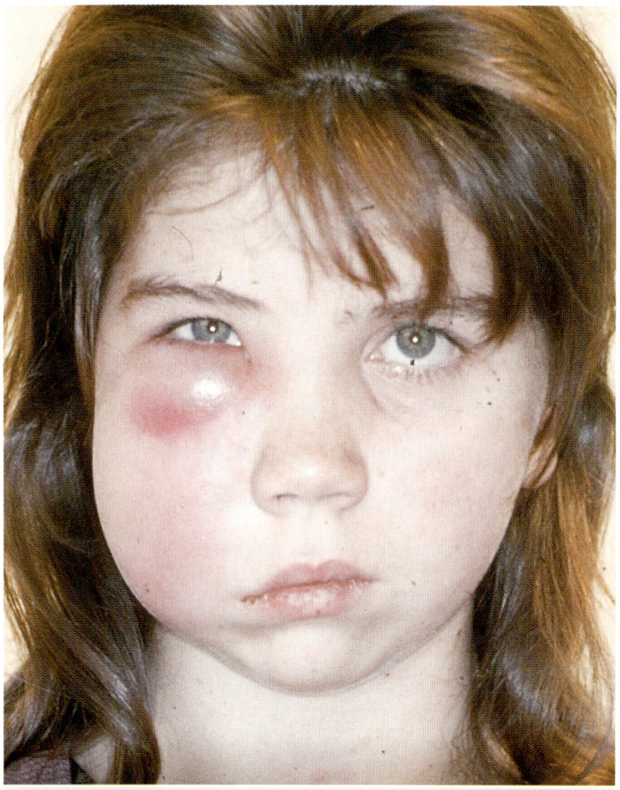

Figure 4.27 An infraorbital facial abscess resulting from a carious upper canine tooth.

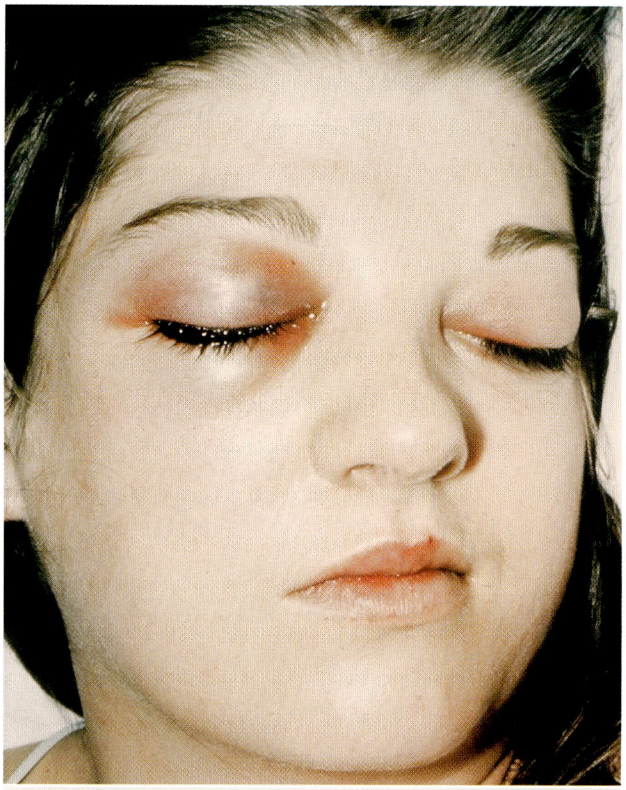

Figure 4.28 Cavernous sinus thrombosis. Note the bilateral orbital signs.

via either the ophthalmic veins or via emissary veins passing through foramen such as the foramen ovale (Fig. 1.14). Infection may also spread via the facial vein from infected spots or boils on the upper lip or in the exterior nares.

Clinical features

There is gross oedema of the eyelids together with pulsatile exophthalmos due to venous obstruction (Fig. 4.28). The venous stasis also leads to cyanosis. The superior orbital fissure syndrome rapidly develops (see Chapter 5). The facial vein is dilated and the conjunctiva oedematous. There is papilloedema and multiple retinal haemorrhages. The patient is seriously ill due to meningitis with rigors and a high swinging pyrexia. Initially, one side is affected but without treatment both sides quickly become infected due to spread via the midline intercavernous sinuses.

Management of cavernous sinus thrombosis

A combination of anticoagulants, antibiotics, drainage of pus and the elimination of the source of infection is essential. There is a 50% mortality and, of those who survive, half will lose the sight of one or both eyes.

SPREAD OF INFECTION BY VASCULAR PATHWAYS AND THE LYMPHATICS

In addition to the spread of infection through interconnecting tissue spaces, infection can spread from the infratemporal fossa through vascular pathways and via lymphatics. The most important vascular route relates to the pterygoid venous plexus, which is particularly prominent around the lateral pterygoid muscle and the maxillary artery. The pterygoid venous plexus communicates with the cavernous sinus via two routes (Fig. 1.14). One pathway is via emissary veins passing through the foramen ovale, foramen spinosum and, where present, the emissary sphenoidal foramen. An additional route is via the deep facial vein, which links the pterygoid venous plexus with the facial vein. The facial vein connects near the medial palpebral ligament with the superior ophthalmic vein, which drains into the cavernous sinus. A further pathway through the orbit is via the inferior orbital vein. Therefore, spread of infection from the infratemporal fossa may reach into the cranial cavity. In this rare situation, there may be both general symptoms leading to meningitis (e.g. fever, headache, vomiting, neck stiffness) as well as specific orbital symptoms due to cavernous sinus thrombosis.

Lymphatic drainage of the infratemporal fossa region is into the submandibular and upper deep cervical group of nodes, so that enlargement of the nodes in this region should alert the clinician to the possibility of infection arising in the infratemporal fossa.

References

1. Grodinsky M, Holyoke E A. The fascia and fascial spaces of the head, neck and adjacent regions. *Am J Anat* 1938; 63: 367–408
2. Laskin D M. Anatomic consideration in diagnosis and treatment of odontogenic infections. *J Am Dent Assoc* 1964; 69: 308–316
3. Barker B C W, Davies P L. The applied anatomy of the pterygomandibular space. *Br J Surg* 1972; 10: 43–55
4. Birn H. Spread of dental infections. *Dent Practitioner* 1972; 22: 347–355
5. Granite E L. Anatomic consideration in infection of the face and neck: review of the literature. *J Oral Surg* 1976; 34: 33–44
6. Tonge C H, Luke D A. Dental anatomy: The spread of infection. *Dent Update* 1981; 291–301
7. Hollinshead W H. Anatomy for surgeons: *The head and neck* (3rd edn). New York: Harper and Row, 1982; 269–289
8. Sowray J H. Acute infections arising from dental causes. In: Moore J R (ed) *The surgery of the mouth and jaws.* Oxford: Blackwell Scientific, 1985; 213–248
9. Koorbusch G F, Fridrich K L. Fascial space infections of the odontogenic origin. In: Zambito R F, Cleri D J (eds) *Immunology and infectious diseases of the mouth, head and neck.* St Louis: Mosby, 1991; 354–356
10. Goldberg M H, Topazian R G. Odontogenic infections and deep fascial space infections of dental origin. In: Topazian R G, Goldberg M H (eds) *Oral and maxillofacial infections* (3rd edn). Philadelphia: Saunders, 1994; 189–250
11. Lang J. *Clinical anatomy of the masticatory apparatus and peripharyngeal spaces.* Stuttgart: Thieme, 1995; 4 91–107
12. Peterson L J. Odontogenic infections. In: Cummings C W *et al.* (eds) *Otolaryngology: head and neck surgery.* Vol. 2, 3rd edn. St. Louis, Mosby Year Book, 1988
13. Patterson H C, Kelly J H, Strome M. Ludwig's angina: an update. *Laryngoscope* 1982; 92: 370–378
14. Linder H H. The anatomy of the fascia of the face and neck with particular reference to the spread and treatment of intraoral infections (Ludwig's) that have progressed into adjacent spaces. *Ann Surg* 1986; 204: 705–714
15. Cawson R A, Langdon J D, Eveson J W. *Surgical pathology of the mouth and jaws.* Oxford: Wright, 1996; 23–27

Chapter 5

The significance of the infratemporal fossa in maxillofacial trauma and orthognathic surgery

D.T. LANIGAN

INTRODUCTION

In order to understand the significance of the infratemporal fossa to maxillofacial trauma and orthognathic surgery the relevant anatomy of the infratemporal and pterygopalatine fossae areas must be briefly reviewed, even though this has been covered thoroughly in Chapters 1 and 2. It is particularly important to be familiar with the boundaries and connections of these fossae, and with the significant neurovascular structures which pass through them.

The infratemporal fossa is a wide irregular space situated behind the maxilla, and deep to the zygomatic arch. It contains the lower part of the temporalis muscle, as well as the lateral and medial pterygoid muscles, the mandibular division of the trigeminal nerve and its branches, the chorda tympani nerve, the otic ganglion, the maxillary artery and its middle meningeal branch, and the pterygoid venous plexus. Anteriorly the infratemporal fossa is bounded by the posterior surface of the maxilla. The roof of the infratemporal fossa is formed by the infratemporal surface of the greater wing of the sphenoid bone. The infratemporal crest on the greater wing of the sphenoid separates the infratemporal fossa below from the temporal fossa above.

The infratemporal fossa is limited medially by the lateral pterygoid plates, which are attached to the greater wing of the sphenoid. The paired pterygoid processes project downwards from the junction of the body and greater wing of the sphenoid bone. The lateral and medial pterygoid plates are in continuity with each other superiorly, but inferiorly are joined together by the pyramidal process of the palatine bone, which is also fused to the maxillary tuberosity. Laterally the infratemporal fossa is bounded by the inner surface of the mandibular ascending ramus and coronoid process.

Two important fissures are located in the depth of the infratemporal fossa, which determines its communication with adjacent areas. The inferior orbital fissure lies horizontally and connects the infratemporal fossa with the orbit. It transmits the infraorbital and zygomatic nerves and the infraorbital artery. The inferior orbital fissure is continuous behind with the pterygomaxillary fissure, which runs vertically and leads medially into the pterygopalatine fossa.

The pterygopalatine fossa is located behind the maxilla, below the apex of the orbit, and lies between the pterygoid plates of the sphenoid bone and the perpendicular lamina of the palatine bone, which forms a posterior part of the lateral wall of the nasal cavity. The anterior surface of the pterygoid processes forms the posterior boundary of the pterygopalatine fossa. The pterygopalatine fossa contains the maxillary artery and its terminal branches, the maxillary nerve and branches, and the pterygopalatine ganglion. In general the plane of the arteries in the pterygopalatine fossa lies anterior to that of the nerves, although the infraorbital nerve crosses superior to the maxillary artery.

At the upper part of its medial wall, the pterygopalatine fossa communicates with the nasal cavity via the sphenopalatine foramen. The two terminal branches of the maxillary artery, the posterior nasal and the sphenopalatine arteries, pass through this foramen. The posterior nasal artery anastomoses with the ethmoid arteries that arise from the ophthalmic branch of the internal carotid artery. The foramen rotundum from the middle cranial fossa, and the pterygoid canal from the anterior wall of the foramen lacerum, open up into the back of the pterygopalatine fossa. The foramen rotundum transmits the maxillary division of the trigeminal nerve, whose branches correspond to those of the third part of the maxillary artery. The pterygoid canal transmits the nerve of the pterygoid canal which runs forwards into the pterygopalatine fossa and ends in the pterygopalatine ganglion.

Maxillofacial trauma and orthognathic surgery, particularly maxillary osteotomies, have the potential to disrupt the soft-tissue contents of the infratemporal and pterygopalatine fossae due to fractures of the bony structures forming their boundaries. These fractures frequently extend to involve the bones immediately adjacent to them (Fig. 5.1). The key to understanding the potential sequelae from bony disruptions involving the infratemporal and pterygopalatine fossae is best gained from a thorough understanding of the sphenoid bone and its connections (Fig. 5.2). Our knowledge of the fracture patterns and soft-tissue changes

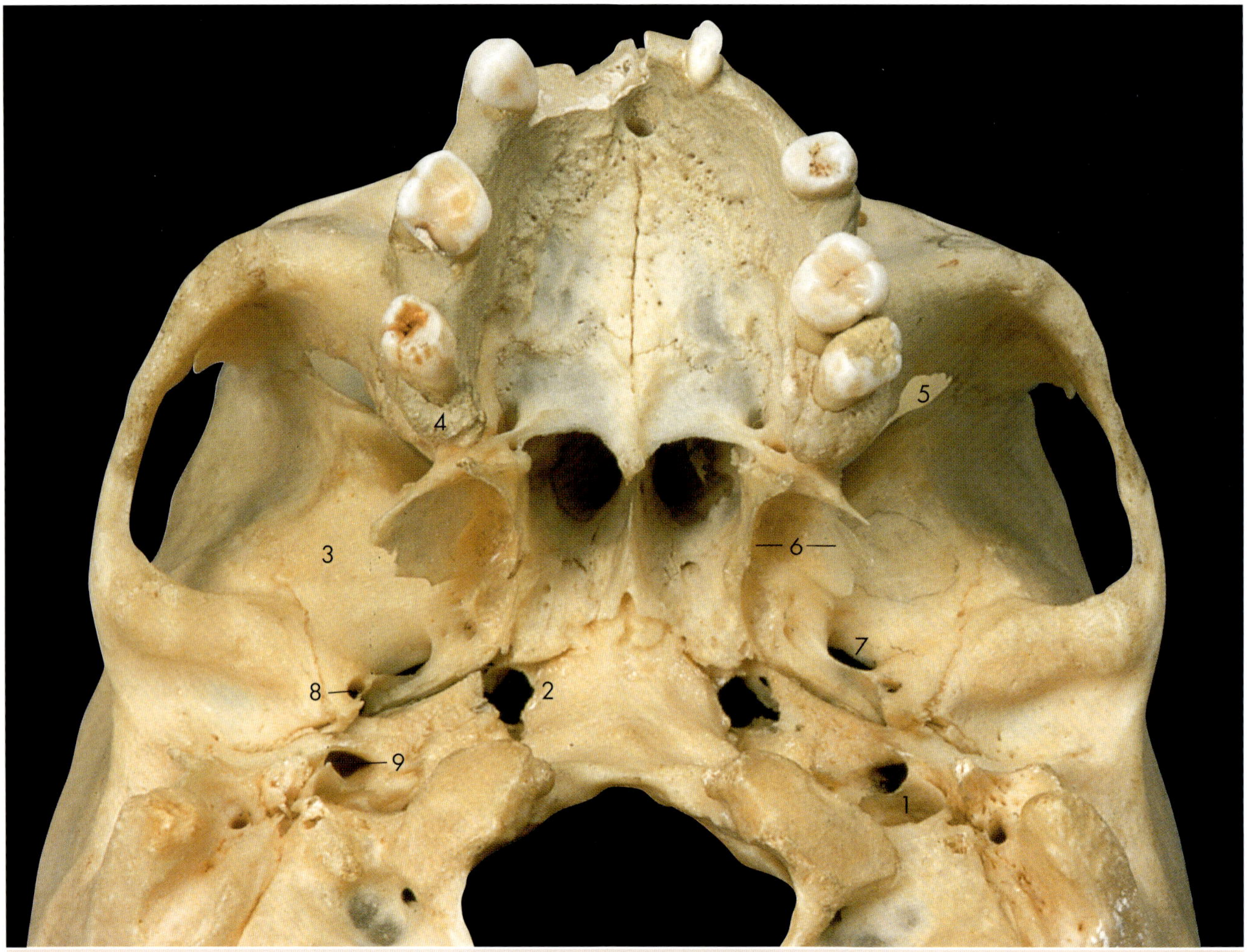

1 Jugular foramen
2 Foramen lacarum
3 Greater wind of the sphenoid
4 Maxillary luberosity
5 Inferior orbital fissure
6 Pterygoid plates
7 Foramen ovale
8 Foramen spinosum
9 Carotid canal

Figure 5.1 The base of the skull. The junction of the pterygoid plates, pyramidal process of the palatine bones, and maxillary tuberosity is in close proximity to the bones forming the base of the skull. Untoward fractures associated with pterygoid plate fractures can damage neurovascular structures coming out through foramina in the skull base, or in the pterygopalatine fossa. (Reproduced with permission from Lanigan.[59])

Figure 5.2 The sphenoid bone: (a) anatomy of the sphenoid bone; (b) coronal section through the sphenoid bone showing fissures, foramina, and neurovascular relationships. (Reproduced with permission from Ghobrial *et al.*[93])

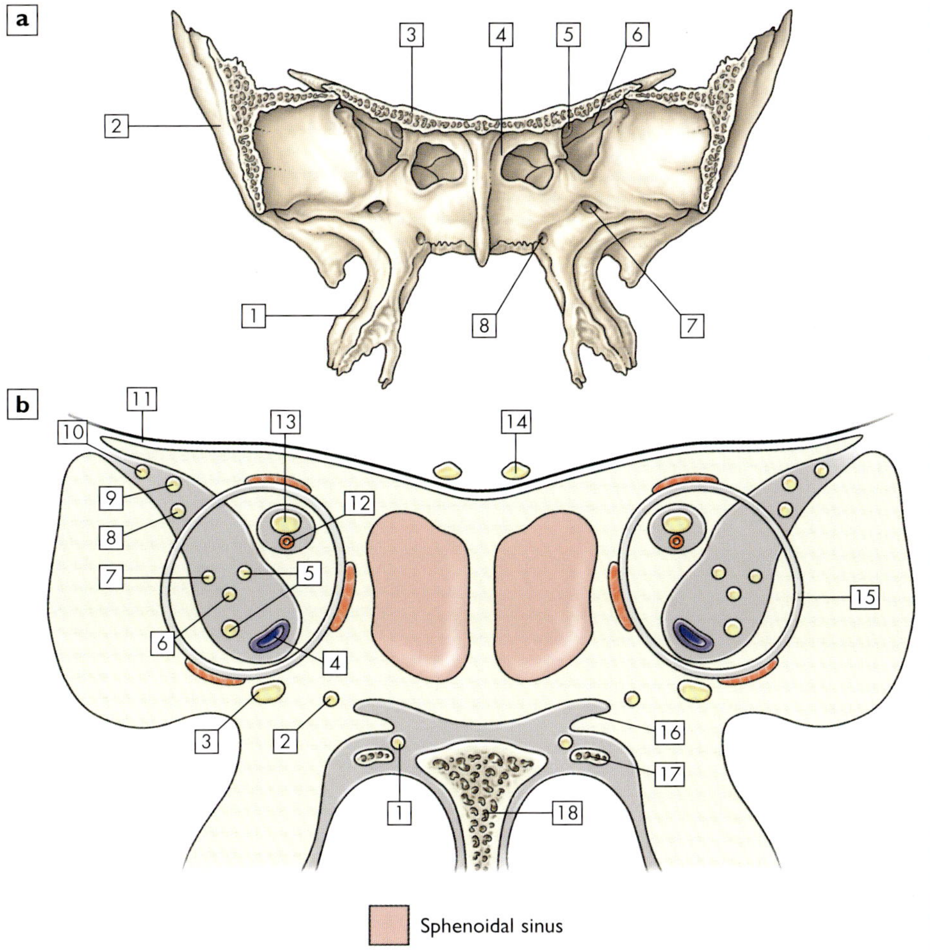

(a)
1 Pterygoid process
2 Greater wing
3 Lesser wing
4 Body
5 Optic canal
6 Superior orbital fissure
7 Foramen rotundum
8 Pterygoid canal

(b)
1 Pharyngeal nerve
2 Nerve of pterygoid canal
3 Maxillary nerve
4 Inferior ophthalmic vein
5 Oculomotor
6 Nasociliary nerve
7 Abducent nerve
8 Trochlear nerve
9 Frontal nerve
10 Lacrimal nerve
11 Dura
12 Ophthalmic artery
13 Optic nerve
14 Olfactory tract
15 Common tendinous ring
16 Vaginal process (medial pterygoid plate)
17 Palatine bone
18 Vomer

involving this area has been greatly facilitated by high resolution computed tomography (CT) scanning.

Fractures of the pterygoid plates, some of which can occur at a high level at or near the base of the skull, have been reported following maxillofacial trauma[1–9] and orthognathic surgery (Figs 5.3–5.6a).[10–15] These pterygoid plate fractures can either be unilateral or bilateral.[2–4,13] If unilateral fractures occur following maxillofacial trauma they largely appear to coincide with the direction of the major force against the midfacial structures.[2] At times, following trauma or orthognathic surgery, it is possible to have fractures of just a lateral or medial plate, but not both simultaneously.[5,13] Pterygoid plate fractures can occur without evidence of Le Fort midfacial fractures, either as isolated fractures[4,8] or as part of more complex facial trauma.[7,8]

Fractures involving the greater wing of the sphenoid can also be of critical importance. The greater wing of the sphenoid must not be thought of in isolation, but must be considered as part of a unit of the bones forming the base of the skull. Not only does the greater wing of the sphenoid bone form the roof of the infratemporal fossa, it also forms the major portion of the lateral wall of the orbit, the anterior part of the temporal surface of the skull, and the anterior surface of the middle cranial fossa. The orbital surface of the greater wing of the sphenoid is limited above by the superior orbital fissure and below by the inferior orbital fissure. The inferior orbital fissure separates the greater wing of the sphenoid from the maxilla. The lesser wing of the sphenoid bounds the optic foramen superiorly, and forms the upper margin of the superior orbital

Figure 5.3 Fractures associated with the pterygomaxillary dysjunction using a curved osteotome: (a) a low level pterygoid plate fracture as well as additional fractures extending up the posterior wall of the maxilla; (b) high level pterygoid plate fractures. (Reproduced with permission from Robinson and Hendy.[12])

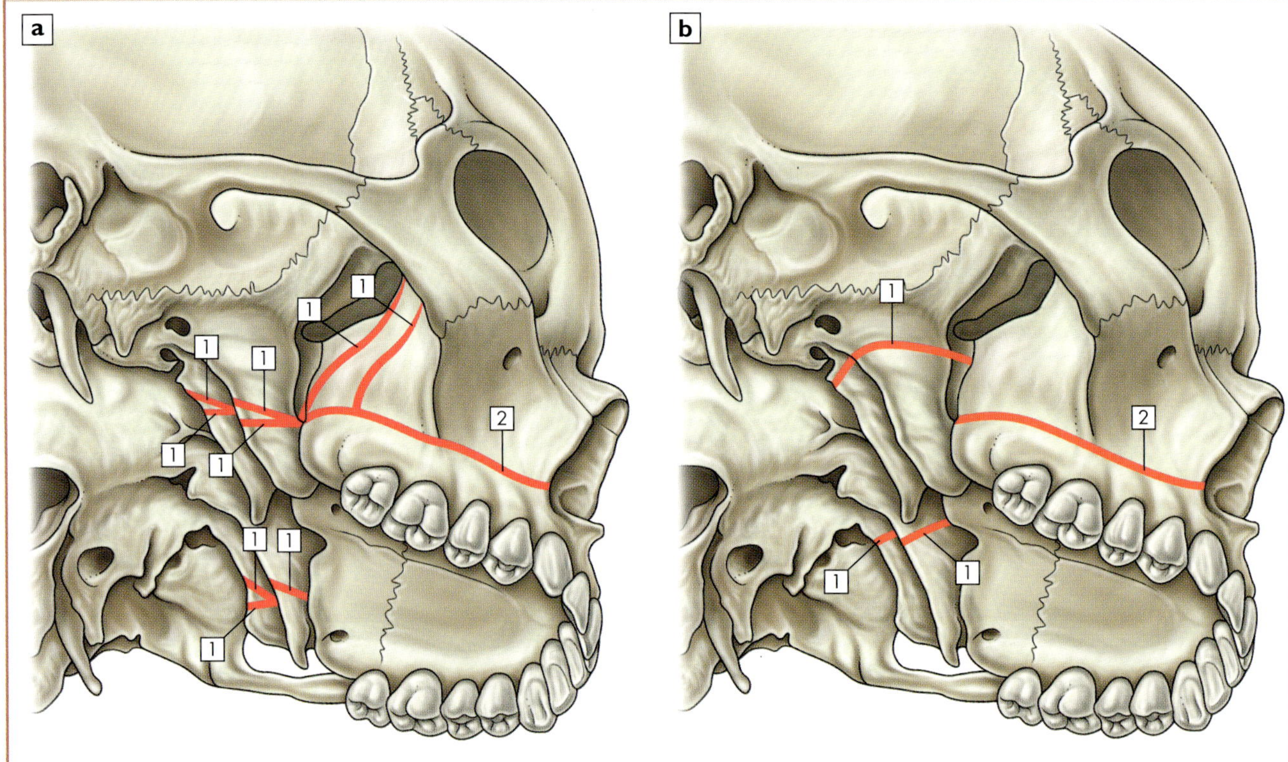

1 Fractures 2 Osteotomy cut

fissure laterally. The greater wing of the sphenoid articulates medially with the sphenoid body and laterally with the lesser wing of the sphenoid and the orbital plate of the frontal bone.

A close anatomic relationship thus exists between the greater and lesser wings of the sphenoid, the sphenoid and ethmoid sinuses, and the inferior and superior orbital fissures.[16] The importance of this relationship in terms of ophthalmic injuries, if fracture lines involve not only the greater wing of the sphenoid but extend to involve contiguous structures, will be examined in more detail later. The majority of fractures involving the sphenoid sinuses also involve the orbital surface of the greater wing of the sphenoid or the pterygoid plates.[6] The sphenoid sinus can extend for a variable distance into the pterygoid process and/or wing of the sphenoid.[17] Ocular injuries are common in patients who have sustained sphenoidal fractures.[9,16]

Sphenoid sinus fractures are generally the result of complex facial trauma,[6,8,9] but can also be associated with orthognathic surgery.[18] Fractures that involve the sphenoid sinus can lead to tears in adjacent soft tissue structures and can result in carotid–cavernous sinus fistulae, carotid dissections, false aneurysms of the carotid artery, and emboli. Sphenoid sinus fractures can also cause damage to the optic nerve leading to decreased visual acuity or blindness, or cranial nerve palsies leading to ophthalmoplegia.[18] The middle cranial fossa is the smallest and structurally most complicated of the three cranial fossae. It is also the most vulnerable part of the skull base to injuries, either related to maxillofacial trauma or orthognathic surgery. Lines of fracture tend to continue in the axis of the striking force,[17] so it is not surprising that, on occasion, untoward fractures occur towards the base of the skull or orbit in association with the pterygomaxillary

Figure 5.4 High level pterygoid plate fractures associated with the use of a curved osteotome. These diagrams illustrate well the potential of these fractures to disrupt the pterygopalatine fossa, regardless of whether the pterygoid plates are loose and not attached to the maxillary tuberosity (1), or remain in continuity with the tuberosity (2). Reproduced with permission from Lanigan and Guest.[13])

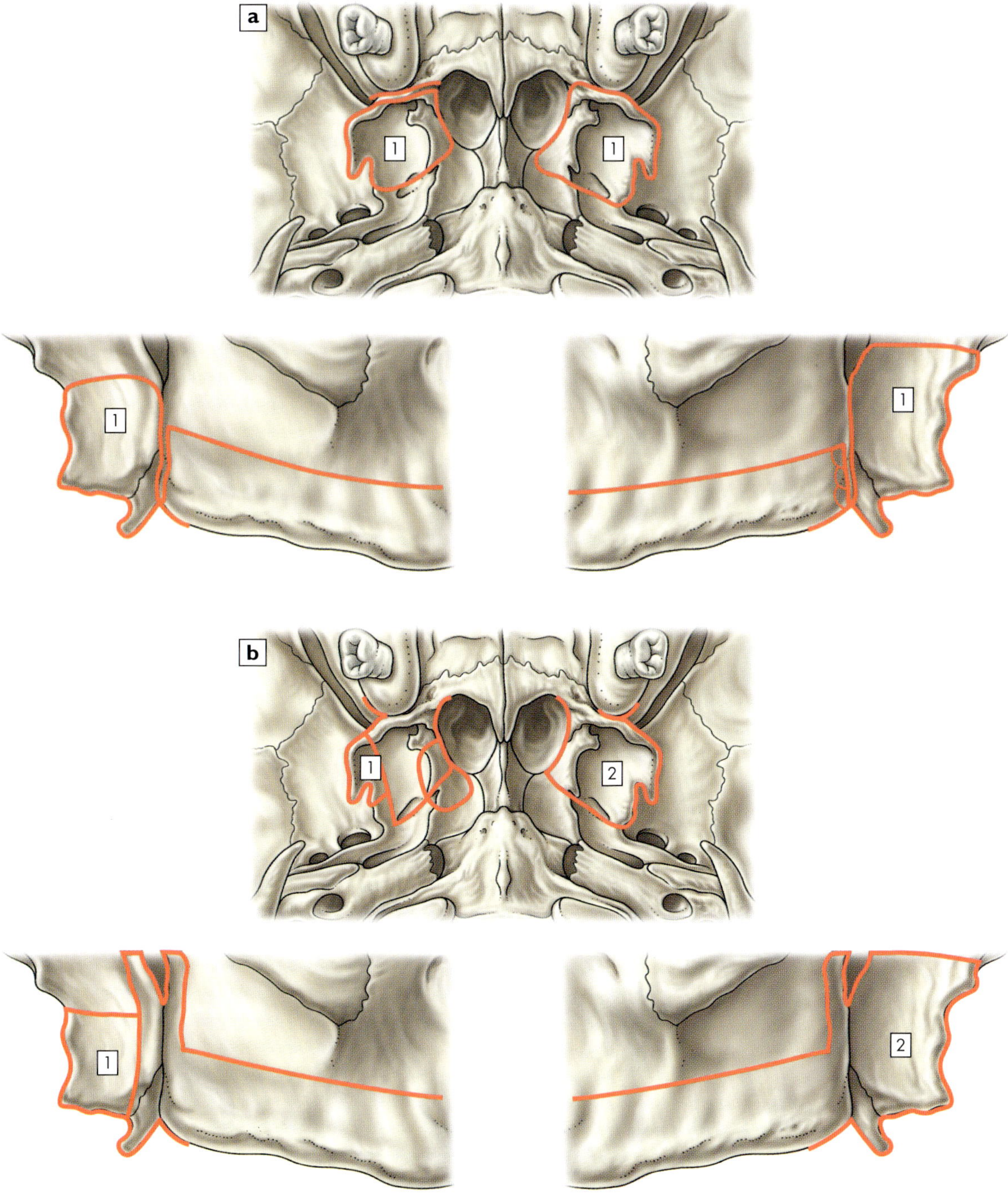

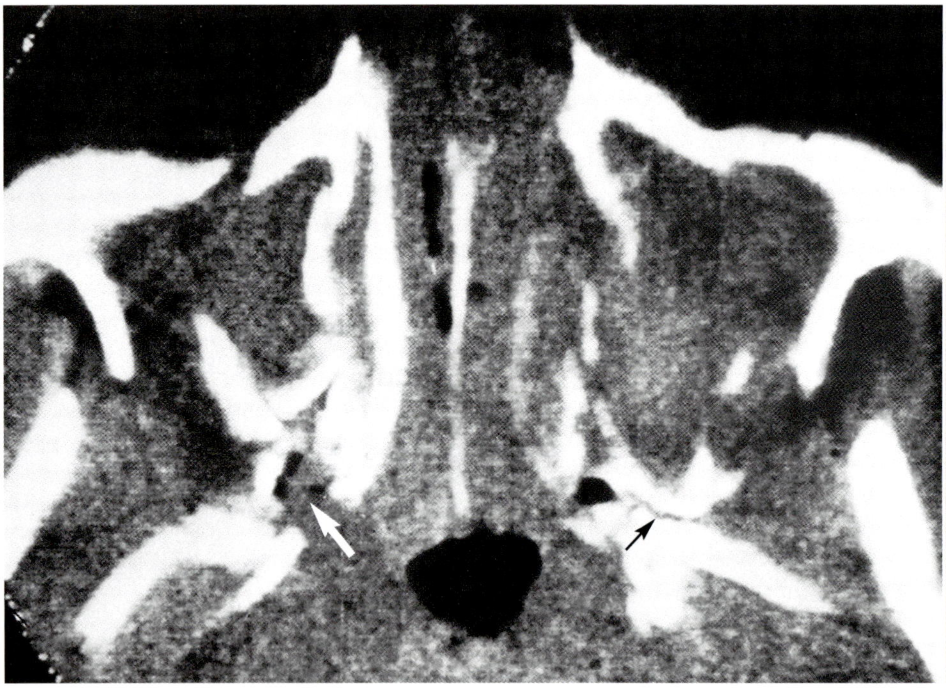

Figure 5.5 Le Fort III fracture. This CT scan shows a post-traumatic maxillary fracture in a 21-year-old woman. The large arrow points out a comminuted fracture of the left pterygoid-palatine-maxillary articulation. The small arrow shows an intact right pterygoid-palatine-maxillary articulation, with the disruption occurring further anteriorly through the maxillary antrum. (Reproduced with permission from Marsh and Gado.[117])

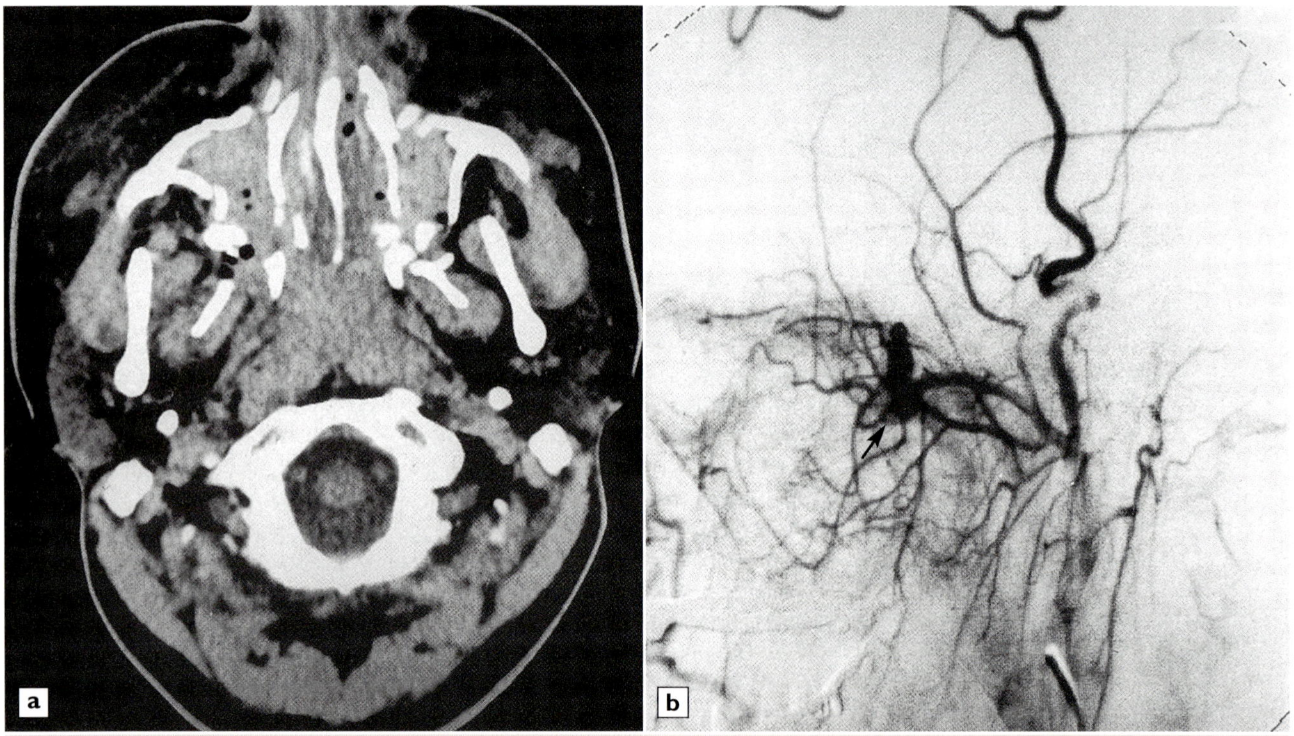

Figure 5.6 False aneurysm of the maxillary artery: (a) CT scan showing maxillofacial injuries, including a grossly comminuted Le Fort III fracture, in a 20-year-old man following a motor vehicle accident. The patient developed recurrent epistaxis in the postoperative period; (b) selective angiography showing a false aneurysm (arrow) of the right maxillary artery with leakage of contrast into the nasal cavity. (Reproduced with permission from Rogers *et al.*[118])

dysjunction during maxillary osteotomies. Unfavourable anatomic variants at the base of the skull, such as bony defects or incomplete ossification, could put a patient at increased risk.[19,20] These anatomic abnormalities may be more common in patients with craniofacial malformations (Fig. 5.7).[21–23]

Fractures of the sphenoid bone, either associated with complex facial fractures or representing primarily fractures of the skull base itself, appear to be more common than were once thought. Studies utilising CT scan imaging have shown that the incidence of sphenoidal fractures exceeds that of the other bones of the base of the skull, including the temporal bone.[8] The majority of these sphenoidal fractures involved the greater wing, body and pterygoid plates, whereas fractures of the lesser wings were relatively uncommon.[8] The greater wing of the sphenoid may be predisposed to fracture due to its anterolateral location, its large surface area, and its developmentally thin bone, especially when subjected to forces that exceed the protective effects provided by the facial buttresses.[8,24] The pterygoid plates may be subject to fracture due to their largely unprotected location, inherent fragility, and functional relationship to the facial skeleton. The walls of the sphenoid sinus may be prone to fracture due to their relative thinness, while the lesser wing of the sphenoid may be protected by its thickness, density, and central location.[8] Fractures of the sphenoid bone are most commonly associated with maxillary fractures.[9,24] This is primarily due to the high incidence of pterygoid plate fractures associated with maxillary trauma, rather than specifically being able to relate these injuries to fractures of the greater and lesser wings and body.[9]

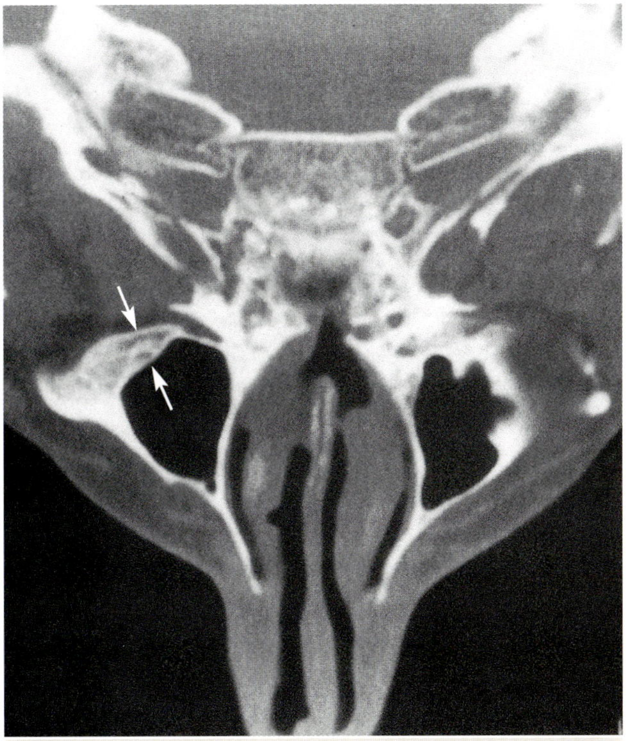

Figure 5.7 Altered anatomy in patients with craniofacial anomalies. The arrows on this coronal CT scan point to the excessive thickness of the posterior maxillary walls in a 14-year-old boy with mandibulofacial dysostosis. Significant difficulty was encountered in achieving the maxillary downfracture following standard Le Fort I osteotomy cuts. Reproduced with permission from Reaume and MacNicol.[22])

Since the early research of Le Fort it has been known that the bones of the face must be considered as a functional unit and not merely as a series of individual bones.[1,25] The bones of the midface appear to transmit the forces of impact directly to the cranium.[24] Although the bony fracture patterns created by facial trauma or orthognathic surgery may be grossly similar, the complications that can occur due to damage to the underlying soft tissues can be markedly different, depending on the exact pathway and nature of the resultant fracture lines.[25,26] In terms of the infratemporal and pterygopalatine fossae, the most important of the bony facial struts is the posterior coronal or 'pterygomaxillary' strut.[25] Fractures involving this strut have the potential to extend elsewhere, including to the cranial base and orbit, which must be taken into account when considering the potential sequelae of fractures in this area.

Soft-tissue injuries from fractures involving the infratemporal and pterygopalatine fossae can result in damage to nerves, blood vessels, and/or muscles, which can be reversible or permanent. Injuries to the sensory nerves of the second or third division of the trigeminal nerve or the chorda tympani nerve could result in decreased sensation to the oral cavity, face, or jaws or in an impaired ability to taste.[2] Fractures extending to the orbit could result in decreased visual acuity and ophthalmoplegia. Injuries to motor nerves and/or muscles could potentially compromise chewing, deglutition, speech, middle ear function, and eye movements.[2,4,9] Injuries to the pterygopalatine or otic ganglia supplying glandular structures could interfere with lacrimation, nasal secretions, and salivation.[2] Injuries to blood vessels could result in haemorrhage, thrombosis, emboli, and the formation of false aneurysms or arteriovenous fistulae.

HAEMORRHAGE

HAEMORRHAGE ASSOCIATED WITH MAXILLOFACIAL TRAUMA AND MIDFACIAL OSTEOTOMIES

A great deal has been written about haemorrhage following maxillofacial injuries, perhaps due to its potential to be life-threatening.[27,28] Disruptions in the pterygopalatine fossa arising secondary to pterygoid plate fractures could result in damage to the maxillary artery and the branches of its third part, which include the posterior superior alveolar artery, the infraorbital artery, the descending palatine artery, the artery of the pterygoid canal, the posterior nasal artery, the sphenopalatine artery, and the pharyngeal artery (Fig. 5.8). Veins, in general, tend to be scanty in the pterygopalatine fossa, although occasionally a sphenopalatine vein is found passing diagonally from the sphenopalatine foramen to the lower end of the pterygomaxillary fissure and then into the pterygoid plexus. Following facial trauma massive haemorrhage can also occur from the anterior and/or posterior ethmoidal arteries, which are branches of the ophthalmic artery,[27] or even from the internal carotid artery itself secondary to a basal skull fracture.[29,30] Other arteries that can be injured if fractures extend to the base of the skull include the artery of the pterygoid canal and the ascending pharyngeal artery.[31,32] The case reported by Kurata *et al.*[32] of bleeding from the left ascending pharyngeal artery was associated with a fracture of the posterior wall of the left maxillary sinus, the left pterygoid plate, and the greater wing of the left sphenoid bone.

Most of the significant treatable bleeding following facial trauma arises from the maxillary artery and its branches, particularly the sphenopalatine or descending palatine branches, as they course through the pterygopalatine fossa,[26] or from injuries to the pterygoid venous plexus in the infratemporal fossa. The same holds true for bleeding, either intraoperative or postoperative, following midfacial osteotomies,[33–35] although

Figure 5.8 The branches of the maxillary artery. (Reproduced with permission from Lanigan and West.[33])

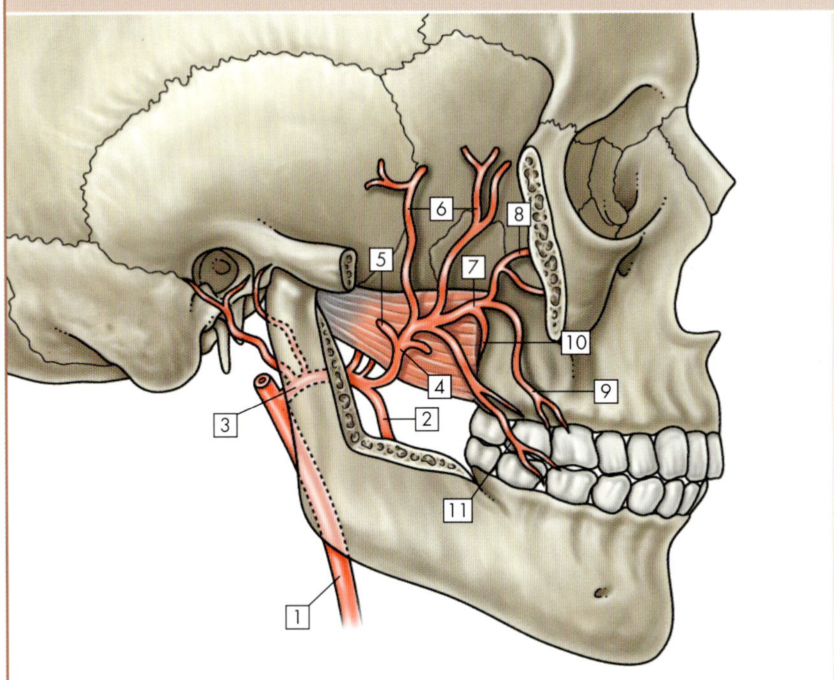

1 External carotid
2 Inferior alveolar
3 Maxillary
4 Pterygoid
5 Masseteric
6 Deep temporal
7 Sphenopalatine
8 Infraorbital
9 Posterior superior alveolar
10 Descending palatine
11 Buccal

sometimes bleeding has been reported from more unusual sites such as the internal carotid artery[36] or the infraorbital artery.[37] Marked venous and/or arterial bleeding can also occur secondary to tears in the pterygoid muscles associated with pterygoid plate fractures.[34] Patients with craniofacial malformations can present with vascular anomalies that can increase their susceptibility to vascular damage and haemorrhagic complications.[22]

The maxillary artery and its branches are most vulnerable to damage in their course through the pterygopalatine fossa when the maxillary tuberosity is separated from the pterygoid plates and pyramidal process of the palatine bone with an osteotome, or during the maxillary downfracture procedure.[33–35,38] The descending palatine artery in particular is vulnerable to damage during these manoeuvres due to its location in the posteromedial wall of the maxillary sinus.[38] Damage to the maxillary artery during the pterygomaxillary dysjunction is unlikely to be from the osteotome itself if the osteotome is correctly positioned.

Turvey and Fonseca[38] have shown that the margin of safety from the superior edge of the osteotome to the maxillary artery is approximately 10 mm in the adult patient. Injuries to the maxillary artery and its branches during the pterygomaxillary dysjunction therefore primarily occur indirectly via high level pterygoid plate fractures caused by the use of the curved osteotome which has conventionally been used to achieve the pterygomaxillary separation.[10–14] The potential disruptions of the pterygopalatine fossa caused by the use of a curved osteotome are well delineated in the diagrams presented in Lanigan and Guest's paper on alternative approaches to the pterygomaxillary separation (Fig. 5.4).[13] Braun and Sotereanos[39] have shown in Macaca monkeys that disruption of the maxillary artery in the pterygopalatine fossa does occur after craniofacial dysjunction surgery.

Although the potential exists for massive bleeding following maxillary orthognathic surgery, significant bleeding occurs less than 1% of the time following Le Fort I osteotomies.[33–35,40] Arterial bleeding requiring up to 20 units of blood for transfusion has been reported, however.[34] Freihofer and Brouns[41] reported blood loss in excess of 3000 ml in 15% of cases of high midfacial osteotomies not operated on under hypotensive anaesthesia.

Bleeding after Le Fort I osteotomies or maxillofacial trauma primarily presents as epistaxis, regardless of which vessel is responsible for the haemorrhage. This epistaxis can be anterior, posterior, or both.[33] A number of measures have been advocated to try to control this haemorrhage, including anterior or posterior nasal packing, packing of the maxillary sinus, surgical exploration of the fracture or osteotomy site, specific artery ligation, and angiography and embolisation.[33–35] Venous bleeding may respond to anterior and posterior nasal packing, as may arterial, although arterial haemorrhage tends to be more persistent and can be recurrent.[33,34]

Exploration of the operative site by re-downfracturing the maxilla after a Le Fort I osteotomy may be useful treatment if bleeding occurs early in the postoperative course, or if angiography has pinpointed the source of the haemorrhage as being easily accessible and embolisation is not feasible because of technical reasons or because of a lack of expertise on the part of the radiologist.[34,37] At the time of reoperation, depending on the findings from surgical exploration or from angiography, suspected vessels can be electrocoagulated or have vascular clips applied to them, depending on their size. It can be difficult, however, in this relatively inaccessible area, even with the maxilla downfractured, to find all the branches of the maxillary artery or its collaterals that could perpetuate postoperative haemorrhage.[34,42] In addition it can be extremely difficult at times, due to the sheer volume of bleeding, to get a good enough view of the surgical field to ascertain the specific vessel(s) responsible for the haemorrhage, especially if the anatomy is grossly distorted from postoperative swelling.

The pattern of branching of the maxillary artery within the pterygopalatine fossa can vary from simple to very complex (Fig. 5.9).[17,43,44] Tortuosity of the vessels also gets worse with increasing age.[17] The variability in the arterial supply within the pterygopalatine fossa makes clipping all the vessels that could perpetuate epistaxis difficult. Even if the descending palatine artery and the maxillary artery at its terminal bifurcation into the sphenopalatine and posterior nasal branches are identified and all three vessels are ligated, variations still exist in the arterial supply which theoretically could still perpetuate bleeding.[17] If one attempts to ligate the terminal branches of the maxillary artery the relation of one vessel to another can become confused, and an important branch can easily be overlooked, especially if an early bifurcation occurs in a vessel like the sphenopalatine or descending palatine artery.[17] Arterial ligation can therefore fail occasionally, even when the suspected vessels have been well identified and adequately ligated.[45]

The classical transantral approach for the control of epistaxis[17,44] is not generally useful in cases of bleeding following maxillofacial trauma or orthognathic surgery because of multiple pieces of fractured bone, a blood-filled maxillary antrum or excessive bleeding obscuring vision, and altered anatomy.[33,45,46] Rosnagle *et al.*,[46] however, described a case where a patient with massive bleeding from an extensive maxillary fracture had the bleeding controlled by the maxillary artery being ligated through the fracture site. Maceri and Makielski[47] have described an intraoral approach for the ligation of the maxillary artery between its first and second parts to control haemorrhage following facial trauma. Although anomalous branches or extensive collateral supply could have resulted in persistent epistaxis requiring more distal ligation, no additional surgical procedures were required in any of their patients in this small series.

Ligation of the external carotid artery has also been suggested for the control of epistaxis.[48] It has proved successful in the management of bleeding following maxillary orthognathic surgery.[33,34] Ligation of the maxillary artery and/or its terminal branches, however, is more in keeping with the surgical principle of controlling the source of haemorrhage as close to the bleeding point as possible.[48] Even after an external carotid artery ligation procedure, collateral arterial supply to the maxillary artery or its branches distal to the point of ligation, or across the midline, could allow bleeding to continue. Not only are there multiple collateral pathways between the various branches of the external carotid artery, but also between the external and internal carotid arterial systems.

Rosenberg *et al.*,[49] in an experimental study on baboons, found that ligation of the external carotid

Figure 5.9 The variability in branching of the maxillary artery within the infratemporal fossa. This branching can range from simple to complex, which can make surgical control of bleeding difficult to achieve. M marks the beginning of the third part of the maxillary artery. (Reproduced with permission from Pearson *et al.*[119])

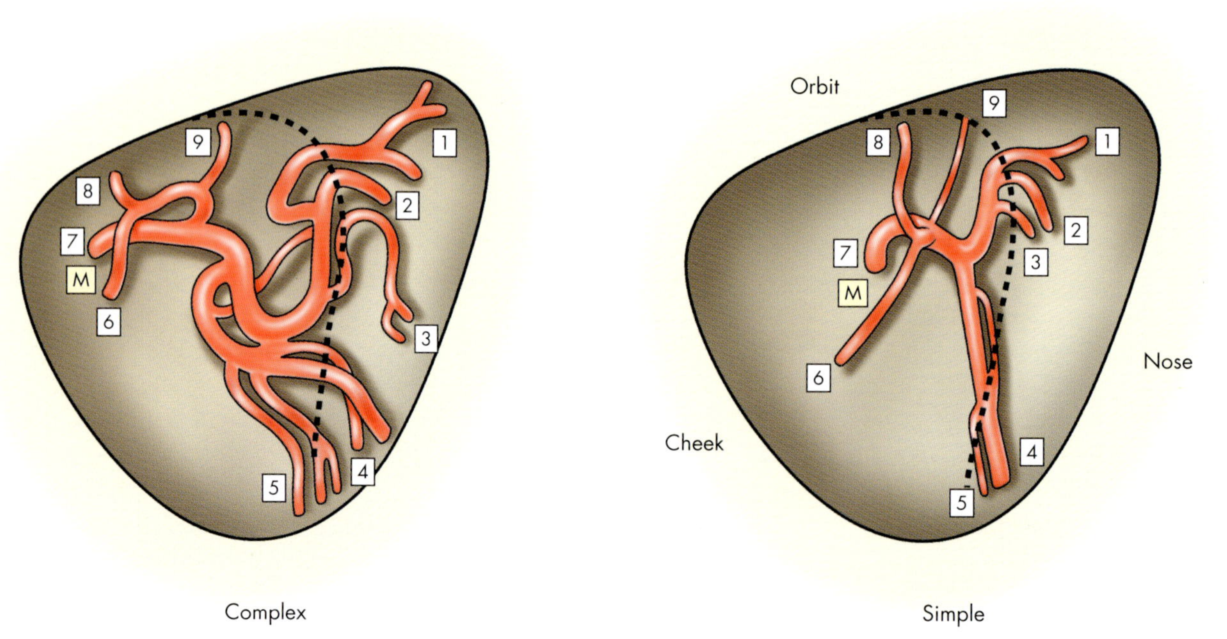

1 Sphenopalatine
2 Posterior nasal
3 Pharyngeal
4 Descending palatine
5 Lesser palatine
6 Posterior superior alveolar
7 Maxillary
8 Infraorbital
9 Branch to foramen rotundum

artery close to the carotid bifurcation, both below and above the origin of the lingual and facial arteries, decreased maxillary artery blood flow by only 40% and 73% respectively due to collateral blood flow. Ligation of the external carotid artery above the origin of the lingual and facial vessels, combined with ligation of the posterior auricular occipital trunk, reduced maxillary artery blood flow by 99%. It was suggested, therefore, that haemorrhage from the maxillary artery might be most effectively controlled in man by ligation of the external carotid artery in the retromolar fossa, distal to the origin of the posterior auricular artery, combined with ligation of the superficial temporal artery at the root of the zygoma.[49] Lownie *et al.*[50] reported a case where Rosenberg's suggested technique was used successfully to control severe haemorrhage from the maxillary artery following a handgun injury to the face.

Angiography and embolisation of the maxillary artery and its terminal branches has also proven to be a useful technique for the control of epistaxis following maxillofacial trauma or orthognathic surgery (Figs 5.6b, 5.10).[32–35,42,45,51,52] Digital subtraction angiography with superselective embolisation is the technique currently utilised.[32,42] Angiography can not only be useful in delineating the source of bleeding, but also variations in normal and abnormal anatomy, such as anomalous vessels or false aneurysms, and in pointing out the development of collateral blood flow that could contribute to the haemorrhage. Although angiography can only localise the site of haemorrhage when active bleeding is occurring, embolisation can be carried out even if haemorrhage is not occurring. Under these circumstances, however, more extensive embolisation will be required than if the bleeding point is visualised angiographically, when selective occlusion of the involved vessel may be all that is required.

Angiography can be particularly useful in determining and controlling the source of haemorrhage if previous attempts at surgical control of the bleeding have not been successful.[45] Even if superselective embolisation of the bleeding vessel is not possible due to technical reasons, the additional information gained about the vascular supply from the angiograms will generally allow a subsequent surgical procedure to be planned and executed in a more precise and safe manner, and increase the chances of finding and dealing with the source of haemorrhage.[37,53] Angiography and embolisation should be considered the treatment of choice for haemorrhage following orthognathic surgery if the bleeding is recurrent in nature, or if it occurs later in the postoperative course where reoperation to downfracture the maxilla would interfere with the healing process.[34,35]

HAEMORRHAGE ASSOCIATED WITH MANDIBULAR OSTEOTOMIES

The potential also exists with mandibular orthognathic surgery to injure the maxillary artery in its course through the infratemporal fossa. The maxillary artery runs posteriorly around the mandibular condyle, and then medially in a superior direction through the infratemporal fossa towards the pterygopalatine fossa.[38] Although injuries to the maxillary artery have been reported following sagittal split mandibular ramus osteotomies,[54,55] this vessel is much more apt to be

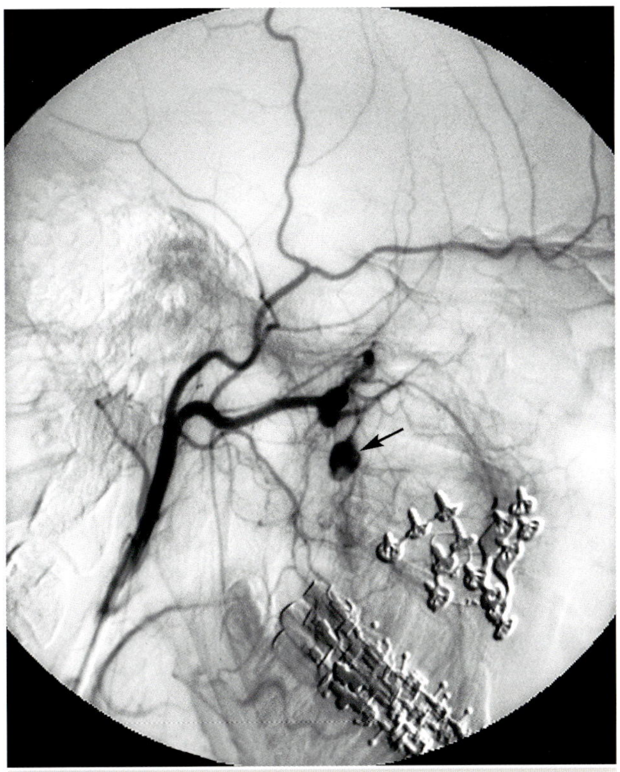

Figure 5.10 False aneurysm of the sphenopalatine artery (arrow). A 16-year-old girl developed recurrent epistaxis that was not controlled by nasal packing following a Le Fort I osteotomy. The bleeding was arrested by angiography and embolisation.

damaged during the vertical oblique mandibular ramus osteotomy (Fig. 5.11). Due to the design of the vertical oblique osteotomy the maxillary artery can be severed when the bone cut is made near the sigmoid notch. Significant bleeding from the maxillary artery following vertical oblique ramus osteotomies has been reported.[55–57] It is also possible that the maxillary artery could be injured by fractures of the mandibular condylar neck.[58]

FALSE ANEURYSMS, ARTERIOVENOUS FISTULAE, AND THROMBOSIS

Arterial damage can result from penetrating or blunt damage. An intimal laceration can act as a site for occlusion of a vessel, or for the development of a false aneurysm, while a more severe disruption of the vessel wall could result in haemorrhage or the formation of an arteriovenous (A-V) fistula.[59] A false aneurysm is generally the result of a tangential, incomplete tear of the arterial wall, whereas an incomplete laceration of an artery in conjunction with a concomitant injury to its accompanying vein can lead to an abnormal communication between them resulting in the formation of an arteriovenous fistula. False aneurysms or A-V fistulae are rare in the facial region because the small size of most facial blood vessels makes their partial transection unlikely. The deep well-protected location of the maxillary artery and its terminal branches also undoubtedly contributes to the rare occurrence of false aneurysms or arteriovenous fistulae involving these vessels.

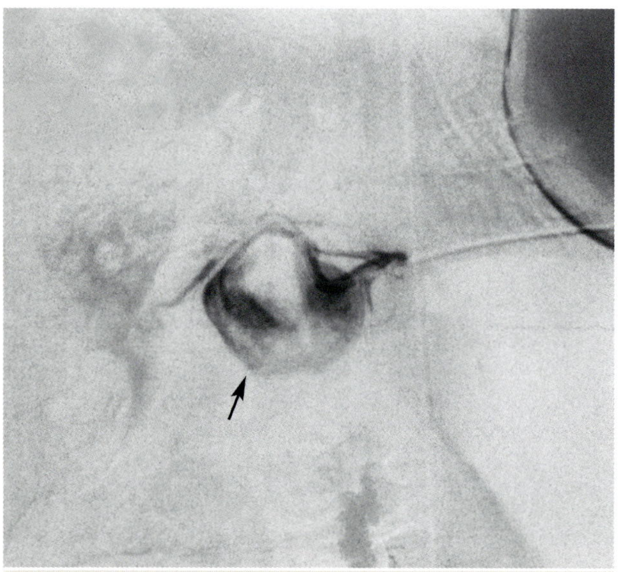

Figure 5.11 False aneurysm of the maxillary artery. A 15-year-old boy developed swelling over the left mandibular ramus secondary to a false aneurysm of the maxillary artery following intraoral vertical oblique ramus osteotomies for the correction of mandibular prognathism. (Reproduced with permission from Clark *et al.*[6])

FALSE ANEURYSMS

High pterygoid plate fractures can disrupt the maxillary artery and its branches within the pterygopalatine fossa,[4,12,13,39] while osteotomies or fractures involving the mandibular condylar neck region could injure the maxillary artery in the infratemporal fossa.[58,60,61] False aneurysms of the maxillary artery have been reported after maxillofacial trauma,[2,58,62,63] Le Fort I osteotomies[60,64] and mandibular vertical oblique ramus osteotomies (Fig. 5.11).[60,61] Following maxillary orthognathic surgery false aneurysms have also been reported involving the sphenopalatine artery (Fig. 5.10),[60,65] the descending palatine artery,[37] and the infraorbital artery.[37] False aneurysms involving the maxillary artery and its terminal branches were treated either with angiography and embolisation,[58,60,61,63–65] or angiography followed by surgical resection or clipping of the aneurysm.[37,60,61]

False aneurysms of the internal carotid artery have been described after a head injury associated with basal skull fractures involving the sphenoidal wings.[66] False aneurysms of the cavernous portion of the internal carotid artery have also been reported after maxillofacial trauma.[30,67] A false aneurysm of the internal carotid artery was reported secondary to a basal skull fracture extending to the foramen lacerum region after a difficult downfracture associated with a Le Fort I osteotomy.[68]

ARTERIOVENOUS FISTULAE

Arteriovenous fistulae of the maxillary artery caused by blunt trauma associated with maxillofacial injuries appears to be rare.[69,70] Albernaz and Tomsick[71] reported two cases of arteriovenous fistulae involving the maxillary artery after Le Fort I osteotomies.

Arteriovenous fistulae after maxillofacial trauma or orthognathic surgery are more apt to involve larger vessels, particularly the internal carotid artery.[59,71] If facial fractures extend from the pterygopalatine fossa region the internal carotid artery can be damaged as it enters the base of the skull at the carotid canal, within the apex of the petrous temporal bone, within the cavernous sinus or sphenoidal sinus, or remotely from the orbital apex.[29,72] Carotid–cavernous sinus fistulae have been reported following midfacial trauma with associated skull base fractures,[73–75] or following maxillary orthognathic surgery, including Le Fort I osteotomies[60,68,76] (Fig. 5.12) and a modified Le Fort II osteotomy.[77] Carotid–cavernous fistulae are currently primarily treated by angiography in conjunction with detachable balloon catheters.[68,72] An A-V fistula between the internal carotid artery and the jugular vein has also been reported after a Le Fort I osteotomy,[36] probably secondary to fractures extending to the skull base in the region of the carotid canal and jugular foramen (Fig. 5.13).[59]

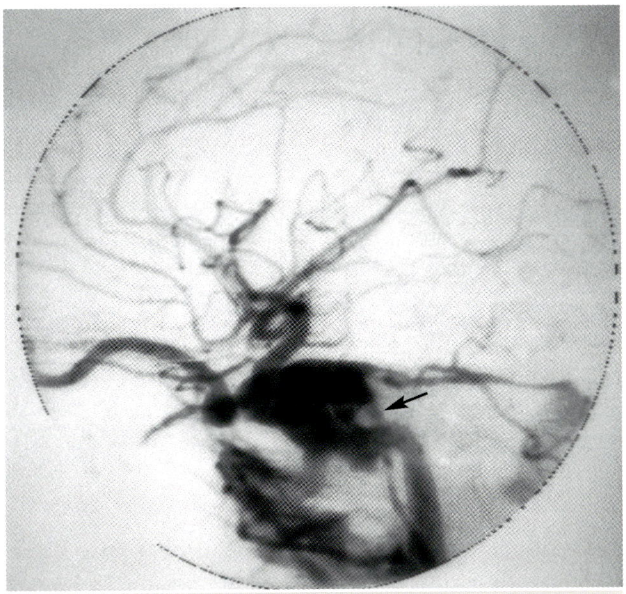

Figure 5.12 Carotid–cavernous sinus fistula (arrow). Lateral view of an internal carotid arteriogram showing a left carotid–cavernous sinus fistula following a difficult maxillary downfracture associated with a Le Fort I osteotomy in a 23-year-old man. (Reproduced with permission from Lanigan and Tubman.[68])

THROMBOSIS

Thrombosis of blood vessels after maxillofacial trauma or orthognathic surgery likely occurs frequently, but because the face has such a good collateral blood supply it seldom results in significant complications. Unger and Unger[2], for instance, have discussed the possibility of thrombosis of the maxillary artery

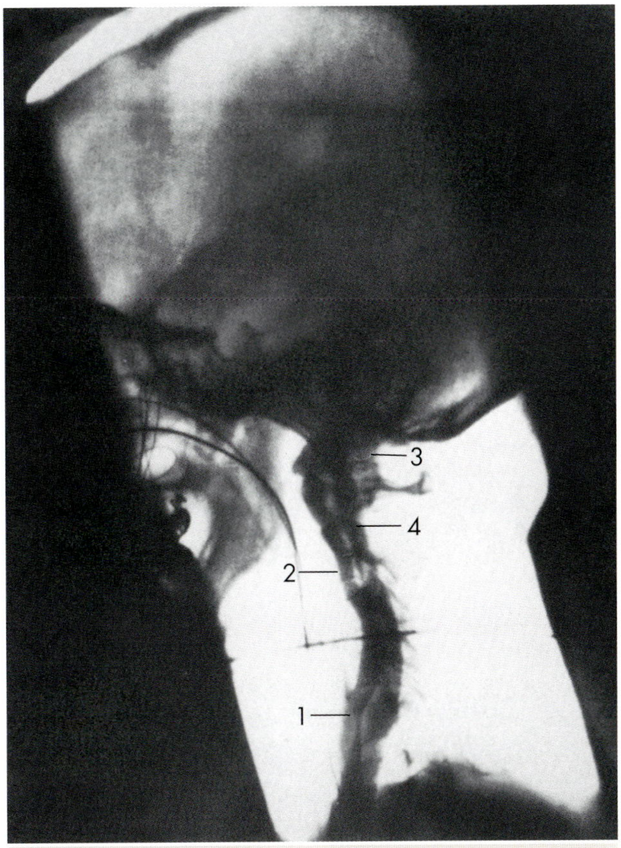

Figure 5.13 Arteriovenous fistula. A 32-year-old woman developed life-threatening haemorrhage from the right posterior maxillary region following the maxillary downfracture associated with a Le Fort I osteotomy. This lateral view of a right common carotid subtraction arteriogram demonstrates a traumatic arteriovenous fistula between the right internal carotid artery (ICA) and the right internal jugular vein (IJA) at the base of the skull. Contrast was injected into the common carotid artery (1), travelled via the ICA (2) to the A-V fistula (3) at the base of the skull, and was immediately shunted down the IJA (4). No intracranial flow could be seen from the right internal carotid artery. (Reproduced with permission from Newhouse *et al.*[36])

secondary to it being injured in the pterygopalatine fossa area. It is only when thrombosis occurs to important vessels with a limited collateral supply, such as the internal carotid artery or the ophthalmic artery, that complications relating to this occlusion become obvious. Severe neurological consequences are not unexpected following maxillofacial trauma associated with significant head injuries. Many cases of carotid artery thrombosis have undoubtedly been missed, however, because the concomitant head injury has drawn attention away from possible arterial damage.[78]

Stroke following maxillary orthognathic surgery is an unexpected event, but it has been reported in association with fractures which have extended to the base of the skull.[23,36,79] In the case presented by Willmar[23] it resulted in the death of the patient.

OPHTHALMIC COMPLICATIONS

The potential ophthalmic complications that can be associated with maxillary osteotomies and blunt maxillofacial trauma include decreased visual acuity, extraocular muscle dysfunction, neuroparalytic keratitis, and decreased or increased tearing.

KERATITIS SICCA

Of major importance when disruptions of the pterygopalatine fossa occur, associated with high level pterygoid plate fractures, is possible damage to the pterygopalatine ganglion or to the postganglionic parasympathetic fibres running with the maxillary nerve in the pterygopalatine fossa. From the pterygopalatine ganglion postganglionic fibres pass to the lacrimal gland and parasympathetic fibres pass to the nasal and palatine glands. Disruptions involving the pterygopalatine ganglion or its postganglionic fibres can produce unilateral diminution of tearing and dryness of the nasal mucosa, and is frequently associated with paraesthesia or hyperaesthesia in the area innervated by the maxillary division of the trigeminal nerve.[80]

The nerve supply involved with lacrimation is complex (Fig. 2.22). The preganglionic parasympathetic fibres run initially with the facial nerve, but leave at the geniculate ganglion to form the greater petrosal nerve. The greater petrosal nerve exits the middle cranial fossa at the foramen lacerum (Fig. 8.25), where it joins the deep petrosal nerve to form the nerve of the pterygoid canal, which runs to the pterygopalatine ganglion. The postganglionic fibres involved with lacrimation then join the maxillary nerve and form part of its zygomatic branch. The zygomatic nerve enters the orbit through the inferior orbital fissure but, before it leaves the orbit, the zygomatic nerve anastomoses, probably via its zygomatico-temporal branch, with the lacrimal nerve. The lacrimal nerve ultimately carries secretory fibres to the lacrimal gland. Direct branches from the maxillary nerve to the lacrimal nerve have also been described.

The complication of decreased tearing following Le Fort I maxillary osteotomies associated with pterygoid plate fractures has been reported by Tomasetti *et al.*[81] and Lanigan *et al.*[82] Of the three cases of keratitis sicca reported by Lanigan *et al.*[82] one patient also had anaesthesia of the left cornea from damage to the nasociliary branch of the first division of the trigeminal nerve (see page 61), while another patient had ipsilateral malar hyperaesthesia and dryness of the nasal turbinates.

Interestingly, as far as I have been able to discover, keratitis sicca has not been reported following maxillofacial trauma. Walsh and Hoyt[80], in their textbook on clinical neuro-ophthalmology, mention diminished tear secretion as a theoretical possibility after facial trauma, but give no supporting references. They feel this complication could most commonly result from trauma involving the posterior lateral orbital wall, leading to damage to the zygomaticotemporal nerve producing postganglionic denervation of the lacrimal gland.[80] Despite the paucity of reports in the literature, a dry eye may occur more commonly following maxillofacial trauma than is suspected, but it is possible that it could be overlooked if the trauma has resulted in a more serious complication such as visual loss.

SECRETOMOTOR RHINOPATHY AND LACRIMAL HYPERSECRETION

Marais and Brookes[83] reported a case of severe secretomotor rhinopathy and lacrimal gland hypersecretion after a Le Fort I maxillary osteotomy. This complication was believed to have resulted from marked autonomic neural imbalance to the nasal and lacrimal glands secondary to an injury to the pterygopalatine ganglion.

Decreased visual acuity or blindness

A loss of visual acuity is one of the most devastating complications that can occur as a consequence of maxillofacial trauma or orthognathic surgery. There is a close anatomical relationship between the optic canal and the greater and lesser wings of the sphenoid bone and the sphenoid and ethmoid sinuses.[16] Manfredi *et al.*[16] studied facial trauma associated with blindness using CT scans and found the most significant findings in the demonstration of indirect injury to the optic nerve were a fracture of the lesser wing of the sphenoid, or a subdural haematoma in the optic nerve sheath. Signs of sphenoethmoidal haemorrhage noted on conventional radiographs should also arouse suspicion of the possibility of an optic canal fracture with associated optic nerve injury.[16]

Manfredi *et al.*[16] reviewed 547 cases of facial fractures in 379 patients in which 21 patients (6%) had complete loss of vision in at least one eye secondary to the facial trauma. Holt *et al.*[84] did a retrospective study of 1436 cases of maxillofacial injuries, of which 727 patients had received a formal ophthalmic evaluation. Of these 727 patients 3% had sustained injuries that resulted in permanent blindness, either from optic nerve injury, retinal detachment, or corneal-scleral rupture. All the blinding injuries were due to midfacial or frontal fractures. Al-Qurainy *et al.*[85] did a prospective study of 363 patients over 2 years to look at the association between ocular injuries and midfacial fractures. Permanent blindness occurred in 2.5% of patients, the result of a traumatic optic neuropathy in each case.

Midfacial fractures most apt to be associated with visual loss are comminuted zygomatico-maxillary complex fractures, orbital floor fractures, naso-ethmoidal–frontal fractures, and Le Fort II or Le Fort III fractures.[16,84–88] Babajews and Williams[89] reported, however, that there appeared to be no direct correlation with the severity of the maxillofacial injury and disturbances in visual acuity. No cases of blindness have been reported after Le Fort I fractures, where injuries occur in an uncontrolled fashion, and yet blindness has been reported after Le Fort I osteotomies[82] or following craniofacial surgery.[90]

Untoward fractures associated with midfacial fractures or maxillary orthognathic surgery, that extend towards the orbit or skull base, have the potential to cause blindness. It is well established, for instance, that zygomatico-maxillary complex fractures have the potential to extend to the optic foramen, the superior orbital fissure, and the orbital plates of the frontal bone.[16,17,91–94] The medial displacement of a fractured zygoma can cause a secondary fracture of the greater wing of the sphenoid, fragments of which can in turn be displaced medially with the potential for a bone fragment to compress the optic nerve.[86,91] The optic nerve in its intraorbital position may thus be injured by a bone spicule, even in the absence of a fracture of the optic canal or a retrobulbar haemorrhage.[91] Funk *et al.*[95] have described a case where a man sustained fractures of the right superior and lateral orbital rims in a motor vehicle accident. A CT scan showed medial and posterior impaction of the entire lateral orbital wall, with the medial edge of the orbital plate of the greater wing of the sphenoid pushed across the superior orbital fissure into the orbital apex.

Lanigan *et al.*[82] have reported extensive fractures around the orbit following a Le Fort I osteotomy. A CT scan postoperatively showed multiple facial and basal skull fractures, including a fracture through the medial wall of the right orbit posteriorly. A fracture of the right sphenoid bone extended through the floor of the middle cranial fossa. A fracture through the right lesser wing of the sphenoid bone extended just lateral to the optic foramen. A small bony spicule, probably from a fracture of the roof of the right maxillary sinus, was in close proximity to, and perhaps even in contact with, the right optic nerve in the region of the optic foramen, close to the superior orbital fissure (Fig. 5.14).

A carotid–cavernous sinus aneurysm,[30,67] or a carotid–cavernous sinus fistula,[73,74] secondary to facial trauma can also result in decreased visual acuity or blindness. Fujii *et al.*[96] studied the optic canal/sphenoid relationship in cadavers. They found that 4% of optic nerves were devoid of medial osseous cover, with only the optic sheath and the sinus mucosa separating the nerve from the sinus, while 78% were covered by 0.5 mm or less of bone.

Traumatic optic neuropathy

A traumatic optic neuropathy is the most common cause of permanent visual loss following blunt midfacial trauma. Although injuries to the optic nerve can

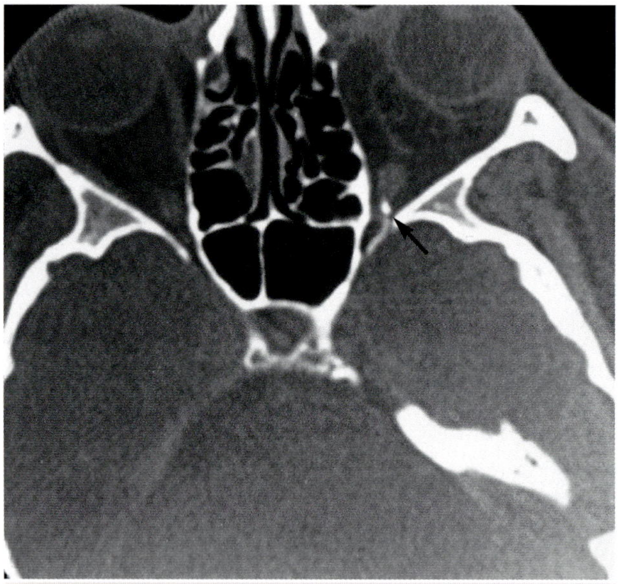

Figure 5.14 Optic nerve injury. This CT scan demonstrates a fracture at the back of the right orbit, associated with a Le Fort I osteotomy, that resulted in permanent blindness of the right eye in a 33-year-old woman. The arrow points towards a small fragment of bone impinging on the right optic nerve. (Reproduced with permission from Lanigan *et al.*[82])

occur anywhere along its length, the canalicular portion of the optic nerve is more vulnerable since it is encased in a tight bony canal to which it is partially adherent. The orbital portion of the optic nerve can withstand fairly strong shearing forces since it is somewhat loose and therefore able to stretch with eye movements or eyeball compression.[92] An injury to the optic nerve, however, is seldom the result of direct injury by osseous compression, laceration, or haemorrhage into the nerve itself.[16] The most common mechanism of injury would appear to be indirect, as a result of haemorrhage into the optic nerve sheath, or via contusion of the nerve with resulting oedema and compression leading to a secondary compromise of the vascular supply to the optic nerve resulting in nerve infarction.[16] A fracture through the optic canal can cause a direct compression or contusion of the optic nerve, or it could have a shearing effect on the meningeal layers of the optic nerve or on the nerve itself.

A fracture of the optic canal is obviously not necessary for damage to the optic nerve to occur, and its presence is merely indicative of the excessive energy that was delivered to the region of the optic canal and subsequently transmitted to the optic nerve.[97] Hairline fractures of the optic canal can occur that may be too small to be visualised radiographically, even with high resolution CT scanning. Other mechanism(s) may be involved as well.

Anderson *et al.*[98] have suggested that indirect optic nerve injury could occur secondary to the stretching, tearing, torsion, or contusion to the nerve and its blood supply, caused not only from the momentum of the eyeball and orbital contents being absorbed by the fixed canalicular portion of the optic nerve, but from skeletal distortion caused by forces remote from the initial impact. During maxillary osteotomies this impact could occur during the pterygomaxillary dysjunction using a curved osteotome. During maxillary orthognathic surgery, strains related to the pterygomaxillary dysjunction are widely distributed throughout the craniomaxillofacial complex.[99] This could lead to compression, traction, or *contre-coup* injuries to neurovascular structures supplying the orbit.[82] These strains could be intensified if the correct technique to achieve the pterygomaxillary separation with an osteotome is not used, or if unusual anatomical features are present.[82,99]

Retrobulbar haemorrhage

Retrobulbar haemorrhage is another potential cause of blindness that has been reported following maxillofacial trauma[93,94,100,101] and orthognathic surgery (Fig. 5.15).[82,102] Most cases of retrobulbar haemorrhage after maxillofacial trauma have been associated with zygomatico-maxillary complex fractures or orbital floor fractures. In addition to direct vascular damage from fractured bones, shearing or traction forces could lead to disruption of a vessel within the orbit. The central retinal artery and ophthalmic artery are seldom injured.[92] The central retinal artery, which provides blood supply to the retina, is a branch of the ophthalmic artery. Although occlusion of the branches of the central retinal artery would cut off all circulation to the retina, this is a rare cause for visual loss since the anatomical location of the central retinal artery within the optic nerve tends to protect it.

Visual loss from retrobulbar haemorrhage is more likely to be due to compromise in the circulation of the posterior ciliary arteries which supply the optic nerve head. The long and short ciliary arteries lie within the extraocular muscle cone, and enter the eye around the

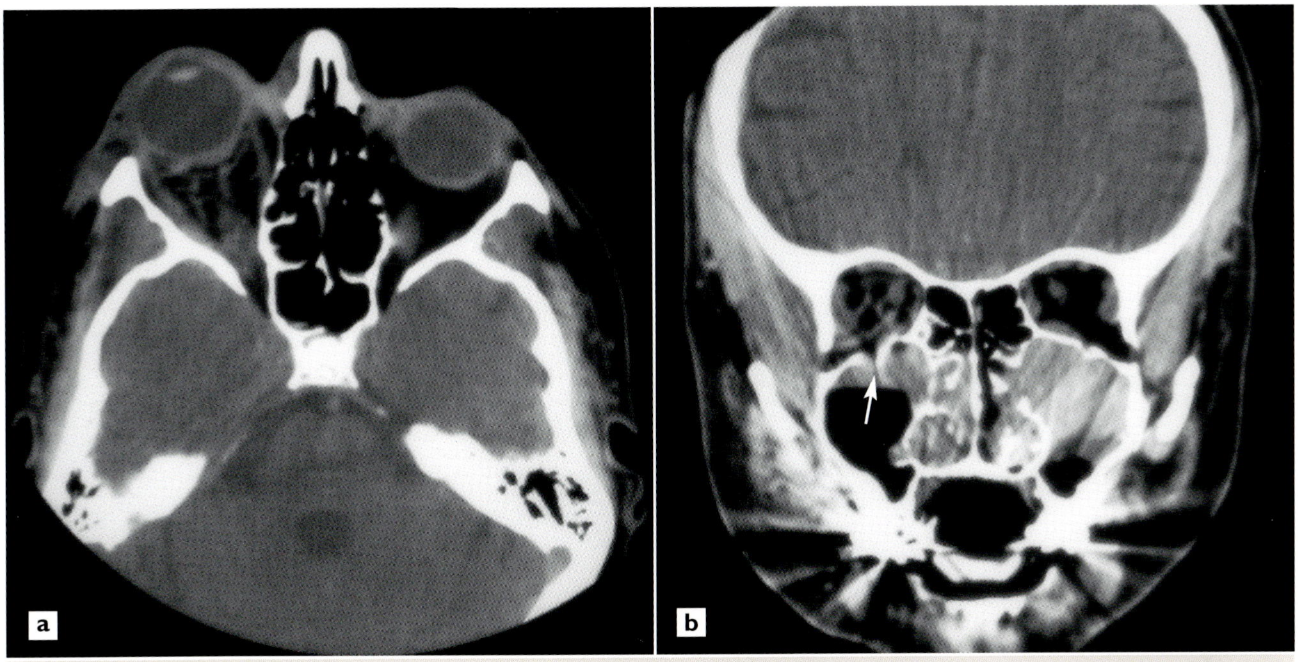

Figure 5.15 Retrobulbar haemorrhage: (a) CT scan showing proptosis of the left eye and soft tissue densities in the retrobulbar space consistent with blood. A 34-year-old woman developed a left retrobulbar haemorrhage leading to an orbital compartment syndrome resulting in permanent blindness of the left eye following surgically assisted rapid maxillary expansion. (b) CT scan showing a fracture of the left orbital floor (arrow). There was no evidence of an optic canal fracture or optic nerve sheath haematoma. (Reproduced with permission from Li *et al.*[102])

optic nerve to supply the uveal tract and anterior optic nerve. An injury within the confines of the extraocular muscles, or penetration by bone spicules from the orbital floor or elsewhere into this space, may result in the rupture of vessels posterior to the globe, particularly the short posterior ciliary arteries. This intraconal space is almost completely closed. A retrobulbar haemorrhage into the muscle cone leads to increased pressure which could compress, and eventually occlude, the posterior ciliary arteries leading to anterior optic nerve head ischaemia and visual loss.[94,101]

Other potential sources for a retrobulbar haemorrhage following maxillofacial trauma or maxillary osteotomies include the infraorbital artery, the sphenopalatine artery, or the anterior or posterior ethmoidal arteries which are branches of the ophthalmic artery. The infraorbital artery could easily be a source of bleeding following midfacial fractures, and damage to its perforating branches could lead to extraconal haemorrhage.[100] The sphenopalatine artery, which can be injured during the pterygomaxillary dysjunction and maxillary downfracture, is one of the major branches involved in postoperative haemorrhage following maxillary osteotomies.[33–35] Haemorrhage from the sphenopalatine artery could perhaps result in bleeding into the orbital cavity via a connection with the pterygopalatine fossa via the inferior orbital fissure.[82]

INDIRECT INJURIES FROM OEDEMA OR HAEMATOMA

Injuries to nerves can occur not only from direct compression from fractures, but also indirectly from oedema and/or haematoma formation. Ascending oedema and haematoma formation from the pterygopalatine fossa area has been postulated to have resulted in blindness and/or ophthalmoplegia.[103–105] If one looks at the anatomy of the pterygopalatine fossa area the foramina of the pterygoid canal and the foramen rotundum are located within millimetres of each other, while directly above these is the superior orbital fissure, with the optic foramen being further superior and medial.[105] There is free communication between the pterygopalatine fossa and the orbit through the infraorbital groove and foramen. Postoperative oedema and haematoma formation, therefore, could possibly transmit pressure to the superior orbital fissure and optic canal via the inferior orbital fissure, leading to the superior orbital fissure or orbital apex syndromes.[105]

EXTRAOCULAR MUSCLE DYSFUNCTION

Ophthalmoplegia has been reported following maxillofacial trauma and maxillary osteotomies in conjunction with fractures of the base of the skull.[17,18,82,106] In the case reported by Goubran[17] of bilateral abducent nerve palsies associated with severe craniofacial trauma, the patient had sustained a comminuted fracture through the sella turcica. Antoniades *et al.*[106] reported three cases of patients who had sustained craniofacial injuries which included a transverse fracture of the middle cranial fossa through the sphenoid sinus, which extended to the petrous apex and led to abducent, facial, and eighth nerve dysfunction.

Reiner and Willoughby[18] described a right abducent nerve palsy after a Le Fort I osteotomy. A postoperative CT scan demonstrated a comminuted fracture through the sella turcica with lateral displacement of a fragment on to the medial surface of the right cavernous sinus. A comminuted fracture of a portion of the greater wing of the sphenoid bone in the middle cranial fossa just lateral to the foramen rotundum was also noted. Lanigan *et al.*[82] reported a case of partial oculomotor palsy following a Le Fort I osteotomy that was associated with multiple facial and basilar skull fractures. The diplopia resolved spontaneously within a short period of time. Abducent nerve function also usually resolves spontaneously over time. The abducent nerve is generally injured by being either stretched or compressed at its attachment to the skull base or to the pons.[17,106]

Ophthalmoplegia has also been reported as part of the clinical picture of a carotid–cavernous sinus fistula following maxillofacial trauma,[73,74] or maxillary osteotomies,[68,76] or secondary to a traumatic carotid–cavernous sinus aneurysm.[67]

Paresis of the extraocular muscles resulting from involvement of the abducent and oculomotor nerves, that can not be specifically related to injuries from basal skull fractures, has also been reported following maxillary orthognathic surgery.[107,108] The superior oblique muscle and the four rectus muscles arise from a common tendinous ring, which surrounds the optic foramen and a portion of the superior orbital fissure (Fig. 5.2). As these muscles are fixed posteriorly to this tendinous ring they will be put on a stretch, in conjunction with their neurovascular structures, if the orbital contents are subjected to traction, compression, or *contre-coup* forces during the pterygomaxillary dysjunction with an osteotome.[82]

Cranial nerve injuries leading to ophthalmoplegia have been reported following maxillofacial trauma.[9,84,85,87,109,110] Rucker[110] reviewed 1000 cases of ocular nerve palsies, of which 17% were traumatic in origin. Of the cases related to trauma 34% involved the abducent (VI) nerve, 30% the oculomotor (III) nerve, 14.5% the trochlear (IV) nerve, and 21.5% were combinations. In addition to cranial nerve injuries, ophthalmoplegia can be due to injuries to the extraocular muscles themselves or to muscle trapping.[9,109,110]

SUPERIOR ORBITAL FISSURE AND ORBITAL APEX SYNDROMES

The superior orbital fissure syndrome, characterised by a fixed dilated pupil, ptosis and proptosis of the eye, ophthalmoplegia, and anaesthesia of the upper eyelid and forehead, has been reported following facial trauma, especially zygomatico-maxillary complex and Le Fort II and III midfacial fractures.[9,111,112] Varying degrees of ocular palsy can be noted.[111] If the optic nerve is involved as well it is known as the orbital apex syndrome. The superior orbital fissure is bounded medially by the body of the sphenoid bone, inferolaterally by the greater wing of the sphenoid, superiorly by the lesser wing of the sphenoid, and laterally by the frontal bone. The optic foramen lies superomedially. The superior orbital fissure leads from the middle cranial fossa to the orbit and transmits the oculomotor, trochlear, and abducent nerves, the ophthalmic division of the trigeminal nerve, and some sympathetic fibres from the cavernous plexus (Fig. 5.2).[111] If fractures with displacement occurs to the bones surrounding the superior orbital fissure, the nerves within the fissure can be compressed.[9] The prognosis for a slow spontaneous partial or complete recovery in sensory and motor nerve function is generally good, unless the nerves have been seriously injured.[112]

SENSORY DEFICITS OF THE SECOND AND THIRD DIVISIONS OF THE TRIGEMINAL NERVE

It is common for patients to have numbness over the distribution of the infraorbital nerve following midfacial fractures, due to fracture lines which extend to the infraorbital canal in the floor of the orbit. This complication generally occurs following Le Fort II or III or zygomatico-maxillary complex fractures, rather than with Le Fort I fractures which generally occur below the level of the infraorbital foramen. It is also possible that damage to the maxillary division of the trigeminal nerve and its branches could occur in the pterygopalatine fossa, which could result in sensory changes to the midface and oral cavity.[2,4,9] It is unusual to have a case of facial trauma where the injuries are so localised as to only involve the pterygopalatine fossa area. It becomes easier in this instance, however, to sort out where the possible damage to neurovascular structures could have occurred.

Eriksson and Hakansson[4] have described a case in which a woman was struck in the left cheek by a bicycle handle just below the anterior part of the zygomatic arch. The patient complained of numbness over the infraorbital nerve distribution to the left cheek, as well as numbness of the left maxillary teeth, the left posterior buccal gingivae, and the left palatal mucosa and gingivae. The only facial fracture that could be demonstrated radiographically was a non-displaced fracture through the base of the left pterygoid process, which also involved the posterior wall of the left maxillary sinus. The floor of the left orbit and the anterior wall of the maxillary sinus were not affected. The authors speculated that bleeding or oedema in the pterygopalatine fossa may have led to pressure on the maxillary nerve as it entered the superior part of the pterygopalatine fossa through the foramen rotundum. This theory is supported by the fact that, in addition to cheek numbness, the patient complained of numbness of the posterior part of the left palatal mucosa, and the buccal gingivae of the left maxillary molars. These areas are innervated by the greater palatine and the posterior superior alveolar nerves, which are branches of the maxillary nerve given off in the pterygopalatine fossa. These nerves would not normally be affected by injuries that result in infraorbital nerve anaesthesia from fractures involving the infraorbital nerve canal.

Damage to the third division of the trigeminal nerve is also possible in its course through the infratemporal fossa, possibly secondary to subcondylar fractures.[113] The anatomical pull of the lateral pterygoid muscle tends to displace a subcondylar fracture anteromedially. Law[113] feels this displaced fragment could compress the inferior alveolar and lingual nerves against the sphenomandibular ligament resulting in paraesthesia. This compression could also be caused by oedema and haemorrhage. In the three cases of anaesthesia involving the third division of the trigeminal nerve following mandibular subcondylar fractures reported by Law,[113] one patient had numbness involving the left mental, long buccal, and auriculotemporal branches, but not the lingual nerve. The second patient had initial anaesthesia that involved the left inferior alveolar, mental, lingual, long buccal, and auriculotemporal branches. The third patient had anaesthesia only over the mental nerve distribution.

Goubran[17] feels that the cases of paraesthesia reported by Law[113] were not due to extracranial damage to the trigeminal nerve, but to a transverse fracture of the base of the skull. A violent upward displacement of the condylar process into the mandibular fossa has the potential to lead to a fracture of the middle cranial fossa. If this fracture line radiates medially to involve the foramen ovale it can give rise to a neuropraxia of the third division of the trigeminal nerve.[9,17]

MUSCLE DYSFUNCTION

Fractures that involve the bones surrounding the infratemporal and pterygopalatine fossae can result in soft tissue injuries to muscles or their nerve supply leading to muscle dysfunction. Despite the fact that these injuries have the potential to affect important stomatognathic functions, they are rarely discussed as complications following maxillofacial trauma, and there is little information available in this area in the literature.

Ghobrial *et al.*[9] have noted that interference with muscle attachments following injuries to the sphenoid bone can result in important functional sequelae. The lateral and medial pterygoid muscles take their origin from the lateral pterygoid plates and insert into the mandible. The superior constrictor muscle of the pharynx takes its origin from the medial pterygoid plate. The pterygoid hamulus, which is an extension of the medial pterygoid plate, acts as a pulley for the tensor veli palatini muscle. Fractures of the sphenoid bone can lead to changes in these muscle attachments and result in problems with jaw opening, mastication, swallowing, speech, and Eustachian tube function.

Ghobrial *et al.*[9] reported that trismus was a frequent early finding following sphenoid bone fractures. Jaw function generally returned to normal over a several week period. One patient, however, developed fusion of the pterygoid plate to the coronoid process of the mandible and the ankylosis was not resolved until the pterygoid plate was surgically resected.

Following pterygoid plate fractures protrusion and lateral movements of the mandible may be limited, at least temporarily.[9] Archer and Sundaram[5] reported a patient who developed an inability to close his mouth following fractures of both medial pterygoid plates, with sparing of the lateral pterygoid plates. The open-bite deformity was attributed to spasm, and possible laceration, of some fibres of the medial and lateral pterygoid muscles. A fracture of the pterygoid hamulus can also compromise the actions of the tensor veli palatini muscle.[2,9]

Ericksson and Hakansson[4] have described a patient who complained of difficulty in swallowing following an isolated unilateral fracture through the posterior wall of the left maxillary sinus and the base of the left pterygoid process. This difficulty in swallowing was felt to be secondary to a post-traumatic interference in function of muscles in the infratemporal fossa, particularly the superior constrictor and tensor veli palatini muscles. Damage to the process tubarius, a bony prominence which projects from the posterior edge of the medial pterygoid plate, can lead to Eustachian tube dysfunction and the possible accumulation of blood and fluid in the middle ear.[2,9]

SUMMARY AND CONCLUSIONS

It is important for maxillofacial surgeons to be knowledgeable about the boundaries and contents of the infratemporal and pterygopalatine fossae in order to understand the potential sequelae of disruptions in this area associated with maxillofacial trauma and orthognathic surgery. This is particularly so due to the potential for fractures in this area to extend into contiguous regions, such as the skull base and orbit, with the potential for catastrophic complications.

In terms of midfacial osteotomies the potential for the occurrence of untoward fractures is highest in association with the pterygomaxillary dysjunction and maxillary downfracture.[120] Lanigan and Guest[13] investigated five different methods of achieving the pterygomaxillary separation in fresh cadavers. The safest and most predictable separations were found using a micro-oscillating saw. This technique had previously been suggested by Kinnebrew and Dzyak[114] and Juniper and Stajcic.[115] The efficacy and safety of this technique in patients was confirmed in a CT scan study by Lanigan and Loewy (Fig. 5.16).[116]

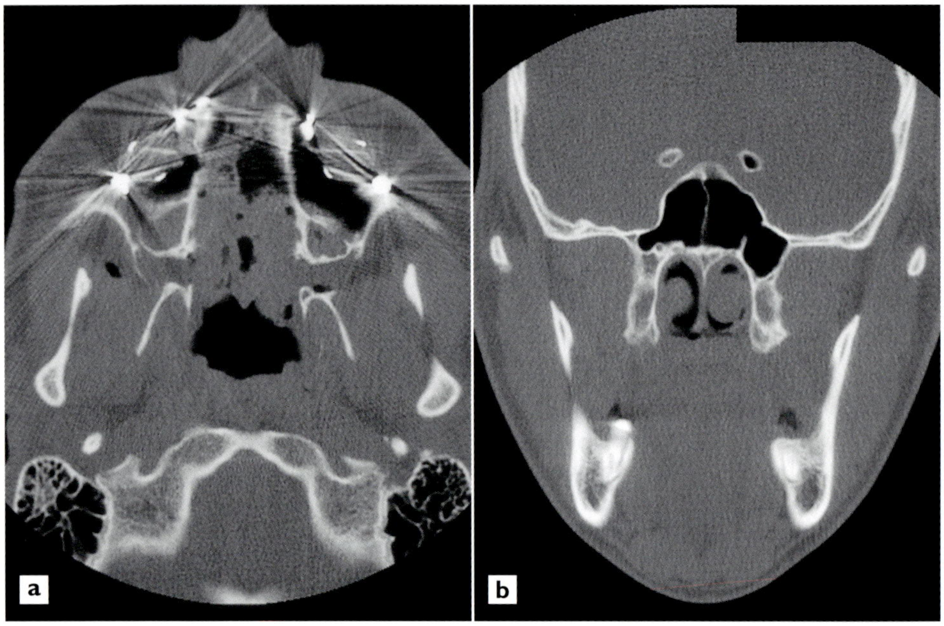

Figure 5.16 Pterygomaxillary separations using a micro-oscillating saw. Axial (a) and coronal (b) CT scans showing ideal separations bilaterally between the maxillary tuberosities, pyramidal processes of the palatine bones, and the pterygoid plates with the use of a micro-oscillating saw. (Reproduced with permission from Lanigan and Loewy.[116])

REFERENCES

1. Le Fort R. Etude experimentale sur les fractures de la machoire superieure. *Rev Chir* 1901; 23: 208–227; 360–379; 479–507
2. Unger J, Unger G F. Fractures of the pterygoid processes accompanying severe facial bone injury. *Radiology* 1971; 98: 311–316
3. Potter G D. The pterygopalatine fossa and canal. *Am J Roentgenol Radium Ther Nucl Med* 1969;107: 520–525
4. Eriksson L, Hakansson H. Unilateral fracture of the pterygoid process. *Oral Surg Oral Med Oral Pathol* 1979; 47: 127–130
5. Archer C R, Sundaram M. Uncommon sphenoidal fractures and their sequelae. *Radiology* 1977; 122: 157–161
6. Laine F J, Conway W F, Laskin D M. Radiology of maxillofacial trauma. *Curr Probl Diagn Radiol* 1993; 22: 147–188
7. West O C, Mirvis S E, Shanmuganathan K. Transsphenoid basilar skull fracture: CT patterns. *Radiology* 1993; 188: 329–338
8. Unger J M, Gentry L R, Grossman J E. Sphenoid fractures: prevalence, sites, and significance. *Radiology* 1990; 175: 175–180
9. Ghobrial W, Amstutz S, Mathog R M. Fractures of the sphenoid bone. *Head Neck Surg* 1986; 8: 447–455
10. Wikkeling O, Koppendraaier J. In vitro studies on lines of osteotomy in the pterygoid region. *J Max-Fac Surg* 1973; 1: 209–212
11. Wikkeling O, Tacoma J. Osteotomy of the pterygomaxillary junction. *Int J Oral Surg* 1975; 4: 99–103
12. Robinson P, Hendy C. Pterygoid plate fractures caused by the Le Fort I osteotomy. *Br J Oral Maxillofac Surg* 1986; 24: 198–202
13. Lanigan D T, Guest P. Alternative approaches to the pterygomaxillary separation. *Int J Oral Maxillofac Surg* 1993; 22: 131–138
14. Renick B M, Symington J M. Postoperative computed tomography study of pterygomaxillary separation during the Le Fort I osteotomy. *J Oral Maxillofac Surg* 1991; 49: 1061–1065
15. Precious D, Goodday R, Bourget L *et al.* Pterygoid plate fracture in Le Fort I osteotomy with and without pterygoid chisel: a computed tomography scan evaluation of 58 patients. *J Oral Maxillofac Surg* 1993; 51: 151–153
16. Manfredi S J, Raji M R, Sprinkle P M *et al.* Computerized tomographic scan findings in facial fractures associated with blindness. *Plastic Reconstr Surg* 1981; 68: 479–490
17. Goubran G F. Traumatic bilateral abducent nerve palsies. *Br J Oral Surg* 1978; 15: 268–275
18. Reiner S, Willoughby J H. Transient abducens nerve palsy following a Le Fort I maxillary osteotomy: report of a case. *J Oral Maxillofac Surg* 1988; 46: 699–701
19. Renn W H, Rhoton A L. Microsurgical anatomy of the sellar region. *J Neurosurg* 1975; 43: 288–298
20. Feuerman T F, Hieshima G B, Bentson J R *et al.* Carotid–cavernous fistula following nasopharyngeal biopsy. *Arch Otolaryngol* 1984; 110: 412–414
21. Ullik R. Erweiterte osteotomie des oberkiefers. *Österr Z Stomat* 1970; 4: 122–128
22. Reaume C E, MacNicol B M. Complications encountered during Le Fort I osteotomy in a patient with mandibulofacial dysostosis. *J Oral Maxillofac Surg* 1988; 46: 1003–1004
23. Willmar K. On Le Fort I osteotomy. *Scand J Plast Reconstr Surg* 1974; 12: Suppl, 1–68
24. Haug R H, Adams J M, Conforti P J *et al.* Cranial fractures associated with facial fractures: a review of mechanism, type, and severity of injury. *J Oral Maxillofac Surg* 1994; 52: 729–733
25. Gentry L R, Manor W F, Turski P A *et al.* High-resolution CT analysis of facial struts in trauma: 2. Osseous and soft-tissue complications. *AJR* 1983; 140: 533–541
26. Frable M A, El Roman N, Lenis A *et al.* Hemorrhagic complications of facial fractures. *Laryngoscope* 1974; 84: 2051–2057
27. Murakami W T, Davidson T M, Marshall L F. Fatal epistaxis in craniofacial trauma. *J Trauma* 1983; 23: 57–61
28. Thaller S R, Beal S L. Maxillofacial trauma: a potentially fatal injury. *Ann Plast Surg* 1991; 27: 281–283
29. Kleid M S, Millar H S. Internal carotid artery epistaxis. *Otolaryngol Head Neck Surg* 1986; 94: 480–486
30. Teal J S, Bergeron R T, Rumbaugh CI *et al.* Aneurysms of the petrous or cavernous portions of the internal carotid artery associated with nonpenetrating head trauma. *J Neurosurg* 1973; 38: 568–574
31. Lasjaunias P B O, Marsot-Dupuch R, Doyon D. The radio-anatomical basis of arterial embolization for epistaxis. *J Neuroradiol* 1979; 6: 45–53
32. Kurata A, Kitahara T, Miyasaka Y *et al.* Superselective embolization for severe traumatic epistaxis caused by fracture of the skull base. *Am J Neuroradiol* 1993; 14: 343–345
33. Lanigan D T, West R A. Management of postoperative hemorrhage following the Le Fort I maxillary osteotomy. *J Oral Maxillofac Surg* 1984; 42: 367–375
34. Lanigan D T, Hey J H, West R A. Major vascular complications of orthognathic surgery: hemorrhage associated with Le Fort I osteotomies. *J Oral Maxillofac Surg* 1990; 48: 561–573
35. Lanigan D T. Hemorrhage associated with orthognathic surgery. *Oral Maxillofac Surg Clin North Am* 1990; 2: 887–899
36. Newhouse R, Schow S R, Kraut R A *et al.* Life-threatening hemorrhage from a Le Fort I osteotomy. *J Oral Maxillofac Surg* 1982; 40: 117–119
37. Tiner B D, Van Sickels J E, Schmitz J P. Life-threatening delayed hemorrhage after Le Fort I osteotomy requiring surgical intervention: report of two cases. *J Oral Maxillofac Surg* 1997; 55: 91–93

38. Turvey T A, Fonseca R J. The anatomy of the internal maxillary artery in the pterygopalatine fossa: its relationship to maxillary surgery. *J Oral Surg* 1980; 38: 92–95
39. Braun T W, Sotereanos G C. Vascular changes in the pterygopalatine fossa after craniofacial dysjunction surgery. *J Oral Surg* 1979; 37: 88–92
40. Freihofer H P. Latitude and limitation of midface movements. *Br J Oral Maxillofac Surg* 1984; 22: 393–413
41. Freihofer H P, Brouns J. Midfacial movements: a reappraisal. *Oral Maxillofac Surg Clin North Am* 1990; 2: 761–773
42. Schilstra S H A, Marsman J W P. Embolization for traumatic epistaxis. *J Cranio-Max-Fac Surg* 1987; 15: 28–30
43. Lofgren R H. Surgery of the pterygomaxillary fossa. *Arch Otolaryngol* 1971; 94: 516–524
44. Montgomery W W, Katz R, Gable J F. Anatomy and surgery of the pterygomaxillary fossa. *Ann Otol Rhinol Laryngol* 1970; 79: 606–618
45. Rosnagle R, Allen W, Kiev E *et al.* Use of selective arteriography in the treatment of epistaxis. *Arch Otolaryngol* 1980: 106: 137–142
46. Rosnagle R S, Yanagisawa E, Smith H W. Specific vessel ligation for epistaxis: survey of 60 cases. *Laryngoscope* 1973; 83: 517–525
47. Maceri D R, Makielski K H. Intraoral ligation of the maxillary artery for posterior epistaxis. *Laryngoscope* 1984; 94: 737–741
48. Hunter K, Gibson R. Arterial ligation for severe epistaxis. *J Laryngol Otol* 1969; 83: 1099–1103
49. Rosenberg I, Austin J C, Wright P G *et al.* The effect of experimental ligation of the external carotid artery and its major branches on haemorrhage from the maxillary artery. *Int J Oral Surg* 1982; 11: 251–259
50. Lownie J F, Shakenovsky B N, Berezowski B M *et al.* Haemorrhage from the maxillary artery: a case report. *S Afr Med J* 1987; 71: 781–782
51. Stein B R, Kerber C W. Therapeutic arterial embolization for posttraumatic hemorrhage: report of case. *J Oral Surg* 1981; 39: 439–441
52. Sakamoto T, Yagi K, Hiraide A *et al.* Transcatheter embolization in the treatment of massive bleeding due to maxillofacial injury. *J Trauma* 1988; 28: 840–843
53. Lanigan D T. Discussion. Life-threatening hemorrhage after Le Fort I osteotomy requiring surgical intervention: report of two cases. *J Oral Maxillofac Surg* 1997; 55: 93–94
54. Behrman S J. Complications of sagittal osteotomy of the mandibular ramus. *J Oral Surg* 1972; 30: 554–561
55. Lanigan D T, Hey J H, West R A. Major vascular complications of orthognathic surgery: hemorrhage associated with mandibular osteotomies. *J Oral Maxillofac Surg* 1991; 49: 713–724
56. Tuinzing D B, Greebe R B. Complications related to the intraoral vertical ramus osteotomy. *Int J Oral Surg* 1985; 14: 319–324
57. Quinn P D, Wedell D. Complications from intraoral vertical subsigmoid osteotomy: review of literature and report of two cases. *Int J Adult Orthodont Orthognath Surg* 1988; 4: 189–196
58. Field P L, Tress B M, O'Brien M D *et al.* Embolization of traumatic aneuysm of the maxillary artery. *Aust NZ J Surg* 1979; 49: 244–246
59. Lanigan D T. Injuries to the internal carotid artery following orthognathic surgery. *Int J Adult Orthodontics Orthognathic Surg* 1988; 4: 215–220
60. Lanigan D T, Hey J H, West R A. Major vascular complications of orthognathic surgery: false aneurysms and arteriovenous fistulas following orthognathic surgery. *J Oral Maxillofac Surg* 1991; 49: 571–577
61. Clark R, Lew D, Giyanani V L *et al.* False aneurysm complicating orthognathic surgery. *J Oral Maxillofac Surg* 1987; 45: 57–59
62. Schwartz H C, Kendrick R W, Pogorel B S. False aneurysm of the maxillary artery: an unusual complication of closed facial trauma. *Arch Otolaryngol* 1983; 109: 616–618
63. Miller F J, Entzminger L B, Coleman L L *et al.* Successful catheter embolization of a false aneurysm of the internal maxillary artery. *Arch Otolaryngol* 1975; 101: 517–519
64. Solomons N, Blumgart R. Severe late-onset epistaxis following Le Fort I osteotomy: angiographic localization and embolization. *J Laryngol Otol* 1988; 102: 260–263
65. Hemmig S B, Johnson R S, Ferraro N. Management of a ruptured pseudoaneurysm of the sphenopalatine artery following a Le Fort I osteotomy. *J Oral Maxillofac Surg* 1987; 45: 533–536
66. Burton R. Massive epistaxis from a ruptured traumatic internal carotid aneurysm. *Med J Australia* 1973; 1: 692–694
67. Guest P G, Schnetler J. Traumatic cavernous aneurysm resulting in blindness. *Br J Oral Maxillofac Surg* 1992; 30: 395–397
68. Lanigan D T, Tubman D E. Carotid–cavernous sinus fistula following Le Fort I osteotomy. *J Oral Maxillofac Surg* 1987; 45: 969–975
69. Marks M W, Argenta L C, Dingman R O. Traumatic arteriovenous malformation of the external carotid arterial system. *Head Neck Surg* 1984; 6: 1054–1058
70. Berenstein A, Scott J, Choi I S *et al.* Percutaneous embolization of arteriovenous fistulas of the external carotid artery. *Am J Neuroradiol* 1986; 7: 937–942
71. Albernaz V S, Tomsick T A. Embolization of arteriovenous fistulae of the maxillary artery after Le Fort I osteotomy: a report of two cases. *J Oral Maxillofac Surg* 1995; 53: 208–210
72. Debrun G, Lacour P, Vinuela F *et al.* Treatment of 54 traumatic carotid–cavernous fistulas. *J Neurosurg* 1981; 55: 678–692
73. Gronski H, Creely J J. Carotid–cavernous fistula; a complication of maxillofacial trauma. *South Med J* 1975; 68: 1096–1102
74. Jones G M, Ross J W. Carotid–cavernous fistula presenting as a late complication of facial trauma. *Br J Oral Maxillofac Surg* 1989; 27: 481–486

75. Rogers R, Epker B N. Carotid–cavernous sinus fistula accompanying mid-facial fractures: report of case. *J Oral Surg* 1972; 30: 429–432
76. Hes J, de Man K. Carotid–cavernous sinus fistula following maxillofacial trauma and orthognathic surgery. *Int J Oral Maxillofac Surg* 1988; 17: 295–297
77. Habal M. A carotid–cavernous sinus fistula after maxillary osteotomy. *Plast Reconstr Surg* 1986; 77: 981–985
78. Little J M, May J, Vanderfield G K *et al.* Traumatic thrombosis of the internal carotid artery. *Lancet* 1969; 2: 926–930
79. Brady S, Courtemanche A, Steinbok P. Carotid artery thrombosis after elective mandibular and maxillary osteotomies. *Ann Plast Surg* 1980: 6: 121–126
80. Walsh F B, Hoyt W F. *Clinical neuro-ophthalmology.* Volume I (3rd edn). Baltimore: Williams & Wilkins Co., 1969, 537
81. Tomasetti B J, Broutsas M, Gormley M *et al.* Lack of tearing after Le Fort osteotomy. *J Oral Surg* 1976; 34: 1095–1097
82. Lanigan D T, Romanchuk K, Olson C K. Ophthalmic complications associated with orthognathic surgery. *J Oral Maxillofac Surg* 1993; 51: 480–494
83. Marais J, Brookes G B. Secretomotor rhinopathy after Le Fort I maxillary osteotomy. Case report. *Int J Oral Maxillofac Surg* 1993; 22: 17–19
84. Holt J E, Holt G R, Blodgett J M. Ocular injuries sustained during blunt facial trauma. *Ophthalmology* 1983; 90: 14–18
85. Al-Qurainy I A, Stassen L F, Dutton G N *et al.* The characteristics of midfacial fractures and the association with ocular injury: a prospective study. *Br J Oral Maxillofac Surg* 1991; 29: 291–301
86. Ketchum L D, Ferris B, Masters F W. Blindness without direct injury to the globe: a complication of facial fractures. *Plast Reconstr Surg* 1976; 58: 187–191
87. Ioannides C, Treffers W, Rutten M *et al.* Ocular injuries associated with fractures involving the orbit. *J Cranio-Max-Fac Surg* 1988; 16: 157–159
88. Jayamanne D G, Gillie R F. Do patients with facial injuries to the orbito-zygomatic region also sustain significant ocular injuries? *J R Coll Surg Edinb* 1996; 41: 200–203
89. Babajews A, Williams J L. Blindness after trauma insufficient to cause bony injury: case report and review. *Br J Oral Maxillofac Surg* 1986; 24: 7–11
90. Whitaker L A, Munro I R, Salyer K E *et al.* Combined report of problems and complications in 793 craniofacial operations. *Plast Reconstr Surg* 1979; 64: 198–203
91. Gonzalez M G, Santos-Oller J M, De Vicente Rodriguez J C *et al.* Optic nerve blindness following a malar fracture. *J Cranio-Max-Fac Surg* 1990; 18: 319–321
92. Lipkin AF. Visual loss due to orbital fracture: the role of early reduction. *Arch Otolaryngol Head Neck Surg* 1987; 113: 81–83
93. Ord R A, El Attar A. Acute retrobulbar hemorrhage complicating a malar fracture. *J Oral Maxillofac Surg* 1982; 40: 234–239
94. Morris T A, Ward Booth R P. Delayed spontaneous retrobulbar hemorrhage. A case report. *J Max-Fac Surg* 1985; 13: 129–130
95. Funk G F, Stanley R B, Becker T S. Reversible visual loss due to impacted lateral orbital wall fractures. *Head Neck* 1989; 11: 295–300
96. Fujii K, Chambers S, Rhoton A. Neurovascular relationships and the sphenoidal sinus. *J Neurosurg* 1979; 50: 31–39
97. Seiff S R, Berger M S, Guyon J *et al.* Computed tomographic evaluation of the optic canal in sudden traumatic blindness. *Am J Ophthalmol* 1984; 98: 751–755
98. Anderson R L, Panje W R, Gross C E. Optic nerve blindness following blunt forehead trauma. *Ophthalmology* 1982; 89: 445–455
99. Hiranuma Y, Yamamoto Y, Iizuka T. Strain distribution during separation of the pterygomaxillary suture by osteotomes. *J Cranio-Max-Fac Surg* 1988; 16: 13–17
100. Ord R A. Post-operative retrobulbar haemorrhage and blindness complicating trauma surgery. *Br J Oral Surg* 1981; 19: 202–207
101. Ord R A, Awty M D, Pour S. Bilateral retrobulbar haemorrhage: a short case report. *Br J Oral Maxillofac Surg* 1986; 24: 1–6
102. Li K K, Meara J G, Rubin P A. Orbital compartment syndrome following orthognathic surgery. *J Oral Maxillofac Surg* 1995; 53: 964–968
103. Giammanco P, Binns P M. Temporary blindness and ophthalmoplegia from nasal packing. *J Laryngol Otol* 1970; 84: 631–635
104. Johnson L P, Parkin J L. Blindness and total ophthalmoplegia. A complication of transantral ligation of the internal maxillary artery for epistaxis. *Arch Otolaryngol* 1976; 102: 501–504
105. Beall J, Scholl P, Jafek B. Total ophthalmoplegia after internal maxillary artery ligation. *Arch Otolaryngol* 1985; 111: 696–698
106. Antoniades K, Karakasis D, Taskos N. Abducent nerve palsy following transverse fracture of the middle cranial fossa. *J Cranio-Max-Fac Surg* 1993; 21: 172–175
107. Watts P G. Unilateral abducent nerve palsy: a rare complication following a Le Fort I maxillary osteotomy. *Br J Oral Maxillofac Surg* 1984; 22: 212–215
108. Carr R J, Gilbert P. Isolated partial third nerve palsy following Le Fort I maxillary osteotomy in a patient with cleft lip and palate. *Br J Oral Maxillofac Surg* 1986; 24: 206–211
109. Al-Qurainy I A, Stassen L F, Dutton G N *et al.* Diplopia following midfacial fractures. *Br J Oral Maxillofac Surg* 1991; 29: 302–307
110. Rucker C W. Paralysis of the third, fourth, and sixth cranial nerves. *Am J Ophthalmol* 1958; 46: 787–794
111. Bowerman J E. The superior orbital fissure syndrome complicating fractures of the facial skeleton. *Br J Oral Surg* 1969; 7: 1–6

112. Zachariades N, Vairaktaris E, Papavassiliou I *et al.* The superior orbital fissure syndrome. *J Max-Fac Surg* 1985; 13: 125–128
113. Law I M. Two unusual complications of fractured condyles. *Br J Oral Surg* 1967; 5: 51–59
114. Kinnebrew M C, Dzyak W R. Modifications in the Le Fort III osteotomy. *J Oral Maxillofac Surg* 1985; 43: 995–998
115. Juniper R P, Stajcic Z. Pterygoid plate separation using an oscillating saw in Le Fort I osteotomy. *J Craniomaxfac Surg* 1991; 19: 153–154
116. Lanigan D, Loewy J. Postoperative computed tomography scan study of the pterygomaxillary separation during the Le Fort I osteotomy using a micro-oscillating saw. *J Oral Maxillofac Surg* 1995; 53: 1161–1166
117. Marsh J L, Gado M. Surgical anatomy of the craniofacial dysostosis: insights from CT scans. *Cleft Palate J* 1982; 19: 212–221
118. Rogers S N, Patel M, Beirne J C *et al.* Traumatic aneurysm of the maxillary artery: the role of interventional radiology. A report of two cases. *Int J Oral Maxillofac Surg* 1995; 24: 336–339
119. Pearson B W, MacKenzie R G, Goodman W S. The anatomical basis of transantral ligation of the maxillary artery in severe epistaxis. *Laryngoscope* 1969; 79: 969–984
120. Lanigan D T. The inherited risks of the pterygomaxillary downfracture. *Jap J Jaw Deformities* 2000; 10: 158–162

Chapter 6

Tumours and tumour-like disorders of the infratemporal fossa

R.M. TIWARI

INTRODUCTION AND HISTORICAL BACKGROUND

Although the development of surgical approaches to the infratemporal fossa and their inclusion into head and neck surgery has progressed in the last few decades, attempts to diagnose and treat lesions of this area have been made since the beginning of the nineteenth century. One of the earliest references to the anatomy of the region is to be found in the German literature.[1] In 1909 Sluder described the anatomy and relations of the sphenopalatine (pterygopalatine) ganglion.[2] His article on the aetiology, diagnosis, prognosis and treatment of sphenopalatine neuralgia in 1913 was a landmark in clinical medicine.[3] Surgical intervention in such obscure areas was restricted to the relief of pain. Carnochan in 1858 described a surgical technique for the removal of the sphenopalatine ganglion via an external incision of the cheek.[4] His approach did not gain popularity because of the associated morbidity, but the technique was improved upon by Sewall in 1926.[5] Sewall approached the area via the maxillary antrum. This was a refinement and a technique that is still used today. The following three decades saw intense development in anaesthesia, blood transfusion techniques and development of antibiotics. Conley's description in 1956 of a surgical approach to the pterygoid area was probably the first report of an operation designed to treat extension of carcinoma to the infratemporal fossa.[6] The incisions used were slightly unconventional from present-day standards and were in part abandoned by Conley later.[7]

Many of Conley's patients were recurrences or failures after initial radiotherapy. There was growing dissatisfaction amongst surgeons with the results in the treatment of carcinoma of the maxillary antrum. Barbosa stated 'it was unfortunate that in the conventional Weber–Fergusson approach to the surgical excision of maxillary sinus carcinoma, the pterygoid area, where the tumour often extended beyond the boundaries of the antrum, was the last to be tackled'. He therefore put forward a transmandibular approach comprising a segmental mandibular resection, section of the pterygoid muscles and the pterygoid plates as well as a complete resection of the maxilla.[8] In Britain Crockett described a transfacial approach to the back of the maxilla for similar indications.[9]

A systematic classification of tumours of the infratemporal fossa was first put forward by Conley.[10] Successful attempts were also under way to resect ethmoid cancers *en bloc* with the craniofacial approach and Terz *et al.* utilised this approach with segmental mandibulotomy to obtain *en bloc* resections of tumours of the infratemporal region.[11,12] These early attempts at major resections were marred by morbidity and mortality. Fisch's description of an infratemporal approach was a milestone in the development of surgery of this area. He combined his expertise in otological surgery with head and neck surgery and microsurgery. His attempts were directed to glomus tumours since this group of tumours were feared both by the otologists and head and neck surgeons alike and surgical results were not satisfactory.[13] His approaches have become legendary in his lifetime. He advocated control of the internal carotid artery in the temporal bone and rerouting of the facial nerve. He subsequently applied his technique to excision of nasopharyngeal carcinoma and carcinomas of the sphenoid and ethmoid extending to the infratemporal region as well as to extensive nasopharyngeal angiofibromas. Fisch's results were influenced by the advances in neuroradiological techniques of imaging, embolisation and balloon occlusion of the internal carotid artery.[14] There are, however, certain drawbacks to this approach. The hearing in the operated ear is permanently lost and, in spite of careful surgery, in the hands of others some degree of facial weakness persists. The postauricular incision is away from the infratemporal region and mastoid surgery is essential. These observations have led to continued search for improvement and perfection. Several modifications of Fisch's original technique have since been put forward to bypass these objections.[15] The last two decades have seen renewed interest in the surgery of this area. Improved imaging techniques, combined efforts by various surgical disciplines and improved bony and soft tissue reconstruction have introduced many more approaches.

SURGICAL LANDMARKS

In spite of the developments in technical expertise, a thorough understanding of the surgical landmarks remains an essential ingredient of surgery of this area. The surgeon involved in this work has to be familiar with the anatomy as well as the variations therefrom to achieve a successful execution of the operative procedure with efficiency. A clear concept is therefore essential. From the surgeon's point of view the infratemporal fossa is a space closed between the lateral wall of the nasopharynx and the medial surface of the mandible. The superior constrictor muscle of the pharynx which forms the lateral wall of the nasopharynx is attached to the medial pterygoid plate but ends about 2 cm below the base of the skull. The rest of the lateral wall cranial to the superior constrictor of the pharynx is formed by the pharyngobasilar fascia which is attached to the posterior border of the medial pterygoid plate. Laterally the zygomatic arch is further lateral to the superior-most part of the ascending ramus of the mandible. In a study of adult skulls carried out by this author the mean distance of the medial pterygoid plate to the zygomatic arch was 4.78 cm.[16] The posterior wall of the maxillary antrum forms the anterior boundary of the space. The pterygoid process is closely adherent to the lower part of this wall of the maxilla. In its upper part it is separated from the maxilla and forms the pterygomaxillary fissure, which leads medially to the yet limited pterygomaxillary space and to the nasal cavity (Fig. 1.1). In its uppermost part the fossa communicates anteriorly with the orbit via the inferior orbital fissure. The transverse process of the second cervical vertebra is the posterior boundary of this space and it is here posterolaterally that the internal carotid artery and the internal jugular vein are related to it.

The superior wall of the fossa is formed by the greater wing of the sphenoid and the squamous portion of the temporal bone. The bony Eustachian tube is situated posteromedially in this area. The foramen ovale and foramen spinosum are located in this area, the former being medial and just posterior to the attachment of the lateral pterygoid plate. In an adult the average distance between the root of the lateral pterygoid plate, i.e. the foramen ovale, and the zygomatic arch is 3.82 cm.[16] This information is important to the surgeon, although there are bound to be variations with age, sex and race. The mandibular nerve emerges from the foramen ovale and after a short distance of about 1 cm it divides into its lingual and inferior dental branches. The middle meningeal artery enters the middle cranial fossa through the foramen spinosum. Inferiorly the infratemporal fossa communicates with the parapharyngeal space. This 5 cm wide space between the nasopharynx and the zygomatic arch is traversed by four muscles: from lateral to medial these are the temporalis, the lateral pterygoid, the medial pterygoid and the superior constrictor of the pharynx. The temporalis originates from the squamous portion of the temporal bone and is attached to the coronoid process of the mandible. The lateral pterygoid originates from the lateral pterygoid plate and the undersurface of the body of the sphenoid and is inserted to the anterior surface of the neck of the mandible (Fig. 6.1). Between the lateral pterygoid and the temporalis muscles is the vascular space, occupied by the maxillary artery and its branches and the pterygoid venous plexus. Haemangiomas can be expected to be found in this region. Between the lateral and medial pterygoid muscles is the neural space occupied by the mandibular nerve and its branches. The otic gangion is also located here. These nerves may be the seat of benign or malignant schwannomas or neurofibromas. These tumours may arise from other nerves as well in the area such as the branches of the maxillary nerves. The cranial nerves carry a sleeve of the meninges during their passage to the infratemporal space and this may become a seat of neoplastic change leading to an extracranial meningioma.[17] More frequently

Figure 6.1 (a) Coronal section through the infratemporal and middle cranial fossa (diagrammatic). The spaces between the zygoma and the temporalis, the temporalis and the lateral pterygoid, the lateral and medial pterygoid and the medial pterygoid and the nasopharynx are depicted. (b) The situation after removal of the contents of the infratemporal fossa. The zygomatic has been reflected downwards along with the temporalis muscle.

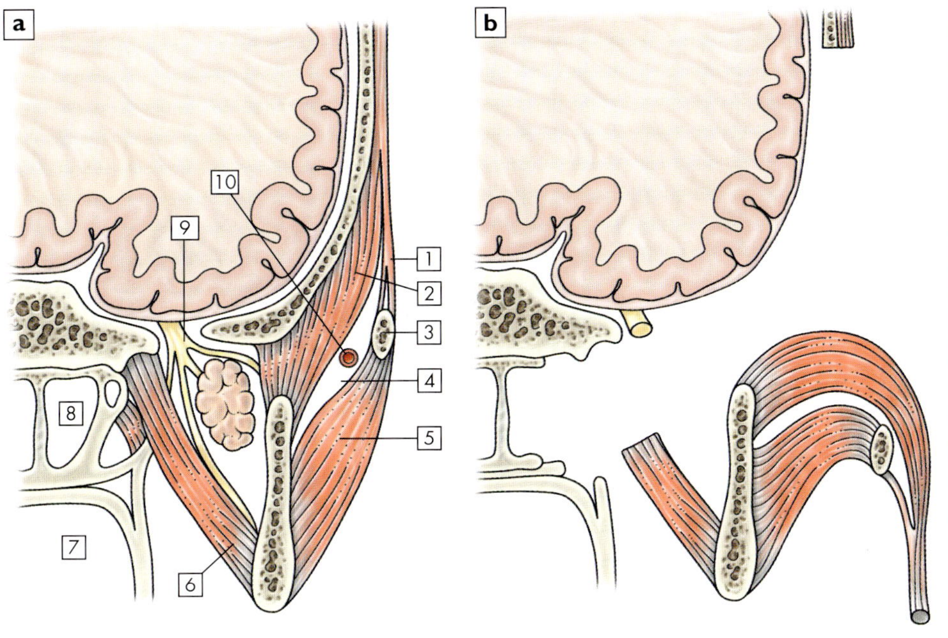

1 Temporalis fascia
2 Temporalis muscle
3 Zygomatic arch
4 Vascular space with maxillary artery
5 Masseter muscle
6 Medial pterygoid muscle
7 Nasopharynx
8 Sphenoid sinus
9 Mandibular nerve
10 Maxillary artery

meningiomas originating in the middle cranial fossa extend to the infratemporal fossa via the foramen ovale. Similarly, tumours originating in the infratemporal space or extending to it from adjoining areas may extend intracranially through these foramina. The space between the medial pterygoid muscle and the superior constrictor or the lateral wall of the pharynx is often the seat of extension of tumours originating in the nasopharynx and sphenoid sinuses. Benign lesions, such as juvenile angiofibroma, may extend through this path of least resistance. Carcinomas often extend in this way. These tumours directly or indirectly involve the Eustachian tube which is superomedial to the medial pterygoid plate, leading to conductive hearing loss, a symptom which may mislead the clinician to a wrong diagnosis. Tumours originating in the pterygopalatine fossa or extending to it may spread to the infratemporal fossa. They may also extend superiorly and anteriorly pressing on the infraorbital nerve and causing symptoms of pressure behind the eye, exophthalmos and diminution of vision.

CLASSIFICATION

Tumours of the infratemporal fossa are classified into the following groups:[10]

1. *Primary*. Tumours are rare and arise from the structures within the infratemporal fossa. They are mesenchymal in origin and comprise 25–30% of all the tumours seen in the area.
2. *Contiguous*. More common tumours and involve the infratemporal fossa by extension from adjoining areas such as the maxilla, nasal cavity, nasopharynx, sphenoid, mandible, parotid gland, external acoustic meatus and the cranial cavity.
3. *Metastatic*. Spread of malignant tumours to the infratemporal fossa by the haematogenous route is not common. Metastases from lung, ovary and breast carcinoma have been reported. Rarely the infratemporal fossa may be the seat of unusual lesions, and fibrous dysplasia and hydatid cyst have been reported.[18]

DIAGNOSIS

In principle a histopathological diagnosis must be obtained prior to definitive therapy of tumours. Imaging techniques are extremely useful in visualising the localisation and extent of primary tumours and in all those cases where the infratemporal fossa is secondarily involved by tumour extension. Primary and metastatic tumours may sometimes produce a typical picture that may be diagnostic. However, in view of the fact that the surgical treatment has consequences, histopathological diagnosis is deemed essential. Should the diagnosis show metastases, then the primary tumour has to be searched for, and the treatment is dependent on the nature of the primary tumour.

Aspiration cytology is least traumatic, but occasionally painful; CT-guided aspiration cytology is often used.[19,20] The procedure should preferably be performed by an experienced radiologist. A formal biopsy may be the only means to establish a diagnosis in some cases. This is not always easy and involves the risk of contamination of non-involved anatomical regions. A transantral approach is often used, but may not be suitable if the lesion is laterally placed. A biopsy via the lateral wall of the nasopharynx can be obtained if the lesion is close to it. In such cases a thorough radiological examination and discussion with the radiologist prior to the biopsy is essential to assess the distance of the internal carotid artery from the pharynx and the exact site of the biopsy. Radiological appearances may sometimes mimic a neoplasm and the clinician must use his/her judgement (Figs 6.2, 6.3), which are considered in more detail in Chapter 7.

SURGICAL APPROACHES

Several surgical approaches to the infratemporal fossa have been described over the years and some of them have been improved and modified. Basically the various approaches can be grouped under the following categories, which are condsidered in more

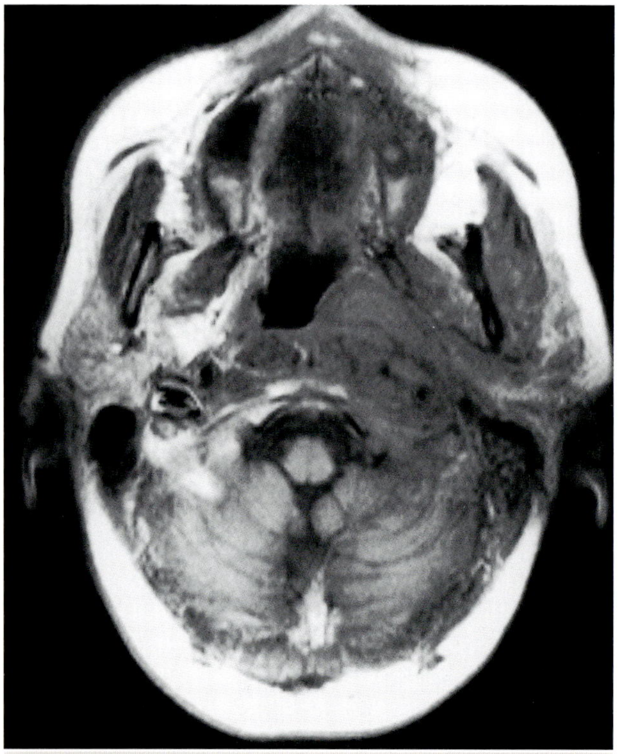

Figure 6.2 Axial MR image of a space-occupying lesion of the infratemporal fossa which appeared like a tumour, but was caused by an inflammatory process.

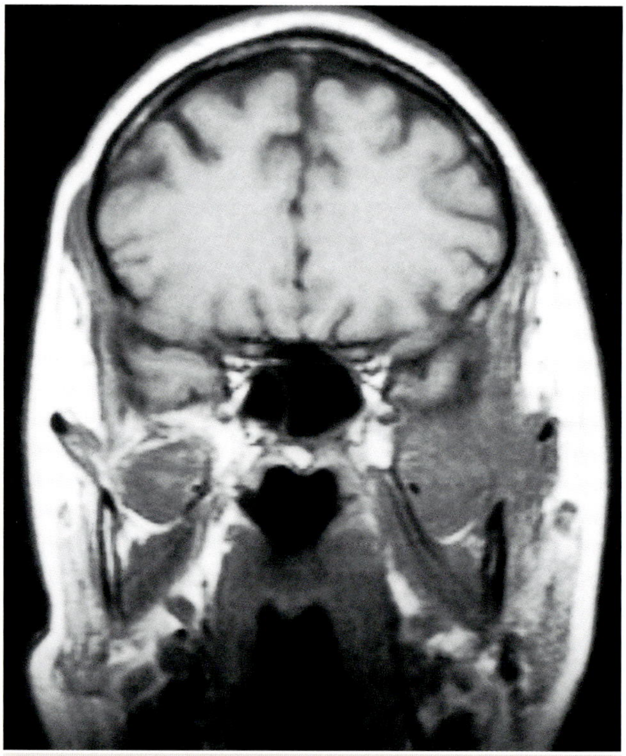

Figure 6.3 Coronal MR image of a space-occupying lesion of the infratemporal fossa which appeared like a tumour, but was caused by an inflammatory process.

detail in Chapter 7:

1. Transoral
2. Transantral
3. Transpalatal
4. Transmaxillary
5. Extended maxillotomy
6. Maxillary swing
7. Transmandibular
8. Transzygomatic
9. Facial translocation
10. Transcranial
11. Combined

The choice of approach depends upon the histopathology, localisation and mode of presentation of the tumour. Metastatic tumours, if isolated, may need other forms of therapy such as radiotherapy. Contiguous tumours would have to be excised *en bloc* with the area of primary origin of the tumour. Primary tumours require excision directed primarily to the infratemporal fossa.

TRANSORAL APPROACH

The superior gingivolabial sulcus posteriorly is close to the tuberosity of the maxilla and provides access to the lower part of the infratemporal fossa. An approach through this area does not provide enough exposure for removal of tumours, the view is obstructed by fatty tissue and there is no vascular control. However, the recess provides access for biopsy purposes especially if the lesion is located low in the infratemporal fossa. Occasionally a benign tumour may be removed through this approach.

TRANSANTRAL APPROACH

The antral cavity is entered through a sublabial incision, extending from the level of the canine to the first molar tooth and the mucoperiosteal flap is elevated until the infraorbital foramen, so as to preserve the infraorbital vessels and nerve. A window is made into the anterolateral wall of the antrum large enough to provide good exposure of the complete posterior wall of the maxillary sinus. The roots of the canine and premolars are preserved. The antral mucosa on the posterior wall is incised at its junction with the medial, lateral and superior walls, and the mucoperiosteal flap is reflected down. The bony posterior wall is incised with a fine cutting burr and the bone is fractured down. The periosteum on the outer surface of the posterior wall is incised along its medial, lateral and superior border and reflected downwards. Care is essential at the superomedial angle for the maxillary artery is close to it. It is usually embedded in a layer of fat and runs across from lateral to medial through the retromaxillary space. At the end of the procedure the bony posterior wall and the mucoperiosteal flap are replaced. This approach is not suitable for tumour excision by itself, but may be combined with other approaches. It is invariably employed for the purpose of obtaining a biopsy.

TRANSPALATAL APPROACH

The authors Kornfehl *et al.* have basically described a transpharyngeal approach via the palate.[21] The nasopharynx is reached via an 'S'-shaped incision running vertically on the soft palate and on to the anterior pharyngeal arch towards the side of the lesion (Fig. 6.4). The mucosa of the lateral wall of the nasopharynx is incised vertically,

Figure 6.4 Incision for transpalatal–transpharyngeal approach to infratemporal fossa.

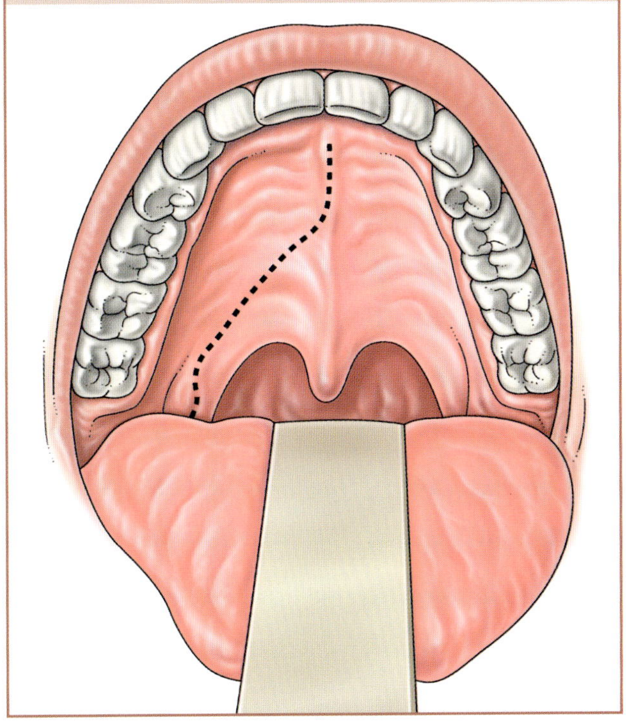

avoiding injury to the Eustachian tube, and the superior constrictor muscle of the pharynx is split to enter the most medial part of the infratemporal fossa. Kornfehl *et al.* employed this approach to extirpate a cavernous haemangioma close to the lateral pterygoid muscle which had been shown not to have any feeding vessels. This is not a safe approach for tumour excision. The internal carotid artery is close to the pharyngeal wall and it is not possible to obtain any control on the vessel. The exposure obtained is limited. Such an approach opens pathways for infection and is actually the route taken by infection and tumours spreading to the area.

Transmaxillary approach

This is the same approach as Le Fort I. It was originally described by Langenbeek in 1859 as an osteoplastic technique for tumours of the pterygopalatine fossa.[22] Le Fort's application of the technique to fractures of the maxilla found its application in orthognathic surgery where it has been used for years.[23] In recent years Stell and Belmont described its application to tumour surgery of the retromaxillary area.[24,25]

An incision is placed in the buccal sulcus above the attached gingivae between the maxillary second premolars. An orthopantogram is made prior to surgery to visualise the apices of the teeth and the incision is placed half a centimetre above this level to ensure the viability of the teeth. A mucoperiosteal flap is raised. The nasal septum is separated from the anterior nasal spine and the maxillary crest and the facial soft tissue are retracted cranially. Alternatively a facial degloving procedure is performed. An osteotomy incision is placed, using an electric burr from one maxillary tuberosity to the other. The incision passes just under the zygomatic buttress and divides the anterior nasal aperture (Fig. 6.5). An osteotomy of the medial wall of the maxilla is performed through the inferior meatus to the palatine canal. At this stage the palate and the inferior portion of the maxilla remain attached by the pterygomaxillary suture, the thin posterior wall of the maxillary sinus and the bone forming the canal of the palatine vessels. Using a curved osteotome the maxilla is separated from the pterygoid process and disimpacted downwards. The buttress of bone anterolaterally and at the piriform nasal aperture are preserved so that they can be approximated at closure. The bone from the walls of the maxillary sinus can be resected to obtain optimal exposure.

Figure 6.5 Transmaxillary approach (Le Fort I osteotomy) to infratemporal fossa (diagrammatic).

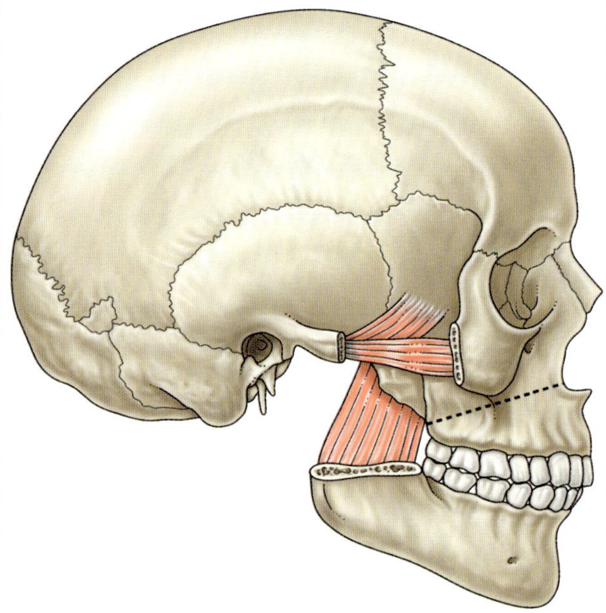

The blood supply of the inferior maxillary segment is maintained by branches from the ascending pharyngeal and facial arteries (Fig. 6.6). Intermaxillary fixation is essential at the end of the procedure and is kept for 6 weeks. This procedure is suited for benign tumours, such as nasopharyngeal angiofibroma, in appropriately selected cases.

Extended maxillotomy approach

This is essentially a transantral approach with an extended sublabial incision taken from the midline to the maxillary tuberosity and carried down to the periosteum. The posterior wall of the maxillary sinus is widely excised allowing access to the pterygomaxillary portion of the tumour. The maxillary artery is ligated to facilitate tumour removal. The medial wall of the maxillary sinus and the nasopharynx is removed. The ethmoid cells can be exposed. Lateral extension of the tumour can be exposed by removing the lateral wall of the antrum. This approach also provides a good view of the inferior orbital fissure, which is essential to visualise tumour extension (Fig. 6.7). Cranial extensions can be

Figure 6.6 Vascular supply of the maxilla. The descending palatine artery is not depicted (diagrammatic).

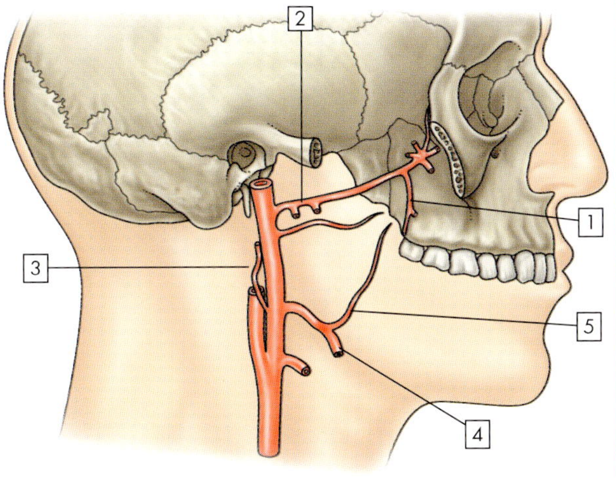

1 Greater and lesser palatine arteries
2 Maxillary artery
3 Ascending pharyngeal artery
4 Facial artery
5 Ascending palatine artery

dissected and gently pulled away. It can also be combined with a transpalatal approach. It was described by Krause and Baker who used it mainly for surgical treatment of nasopharyngeal angiofibroma.[26] These tumours arise from the pharyngobasilar fascia around puberty almost exclusively in males and are very vascular. Preoperative embolisation is helpful.

EXTENDED OSTEOPLASTIC MAXILLOTOMY

Transfacial approaches to the retromaxillary area had been described by Hernandes and Brown.[27,28] In 1993, Catalano and Biller described osteoplastic maxillotomy with three different versions to adapt to medially, centrally or laterally located tumours.[29] It is the lateral version of this approach that is suitable for approach to the infratemporal fossa.

A preliminary tarsorrhaphy is performed. Tracheotomy is optional. A Weber–Fergusson incision is used. The incision is full-thickness to expose the facial skeleton (Fig. 6.8). The periosteum should be

Figure 6.7 Extended transantral approach. Lateral and posterior walls of maxilla have been removed to provide access to retromaxillary space.

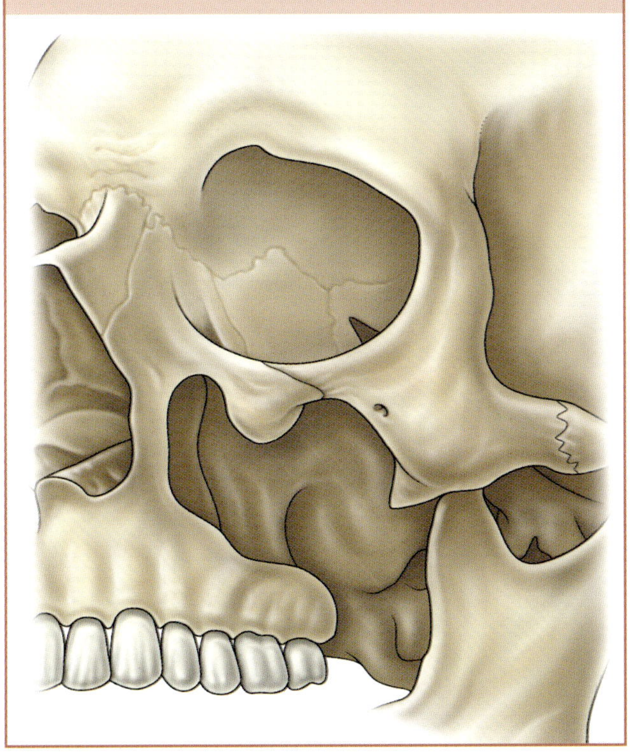

Figure 6.8 Osteotomies of the maxilla and zygomatic arch (dotted line) in extended osteoplastic maxillotomy approach to infratemporal fossa.

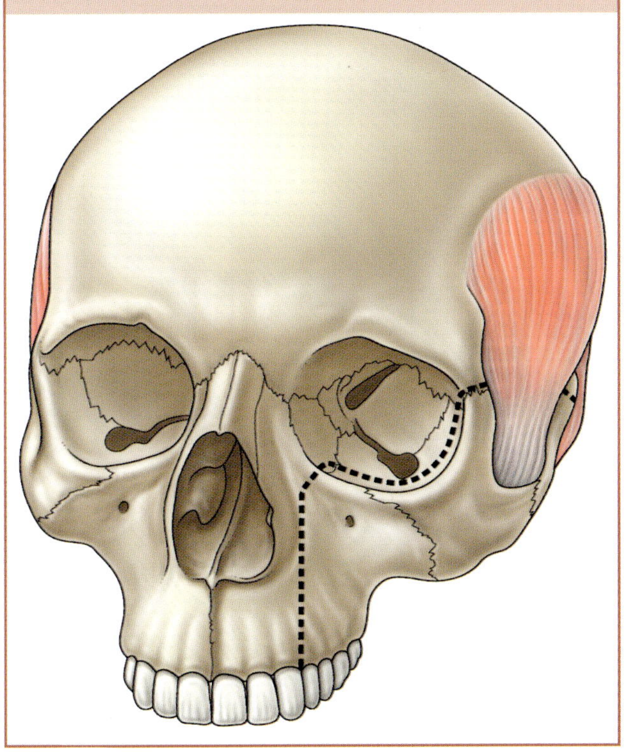

exposed but not yet incised. Special attention is paid to the vermilion borders, base of columella, alar crease and medial canthal septum. A subciliary incision is made 3 mm inferior and parallel to the ciliary margin. A skin flap is raised inferiorly to expose the orbicularis oculi muscle. Sharp dissection through the inferior aspect of this muscle allows identification and exposure of the infraorbital rim (Fig. 6.9). A transverse incision from the lateral canthus is placed horizontally to the helix (Fig. 6.10). The incision is deepened to reveal the temporalis fascia, exposing the lateral orbital rim and temporalis muscle. The muscle and the fascia are detached from the zygomatic arch and lateral orbital rim. The temporal and zygomatic divisions of the facial nerve are identified. A nerve monitor may be used. The nerve branches are cut and tagged for future identification and resuturing at the end of the procedure.

Osteotomies are placed through the anterior face of the maxilla just lateral to the medial wall; the orbital floor and the inferior orbital rim are cut. The lateral orbital wall is cut below the bend of the lateral canthus and the zygomatic arch is divided in its posterior part. Should it be necessary to visualise the area more medially then the maxillary cut is placed appropriately. The palate is cut 1.5 cm lateral to the midline and the cut passes through the interdental space. Alternatively the tooth may be extracted. A transverse incision 5 mm anterior to the junction of the hard and soft palate is placed to reduce the risk of a fistula (Fig. 6.11). The osteoplastic unit consisting of skin, subcutaneous tissue, periosteum, bone and related neurovascular structures is mobilised laterally and inferiorly (Fig. 6.12). The temporalis muscle can be exposed by adding a frontotemporal incision behind the hairline and the temporalis muscle is detached. This can include the coronoid process and part of the ascending ramus of the mandible. Catalano and Biller advocated detaching the muscle superiorly for a combined neurosurgical approach.

It is imperative during mobilisation of the osteoplastic unit to keep the posterior wall of the maxilla intact since in many instances this forms the anterior boundary of the tumour.

Further dissection is dictated by the tumour extent, histology and proximity to neurovascular structures.

Figure 6.9 Exposure of the orbital floor and osteotomy sites for frontal and zygomatic processes as well as the orbital floor (dotted line) in extended osteoplastic maxillotomy approach to infratemporal fossa.

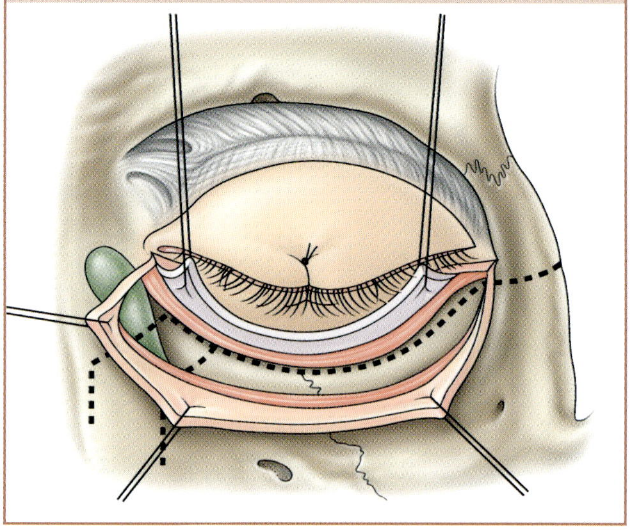

Figure 6.10 Skin incision for extended osteoplastic maxillotomy approach to infratemporal fossa.

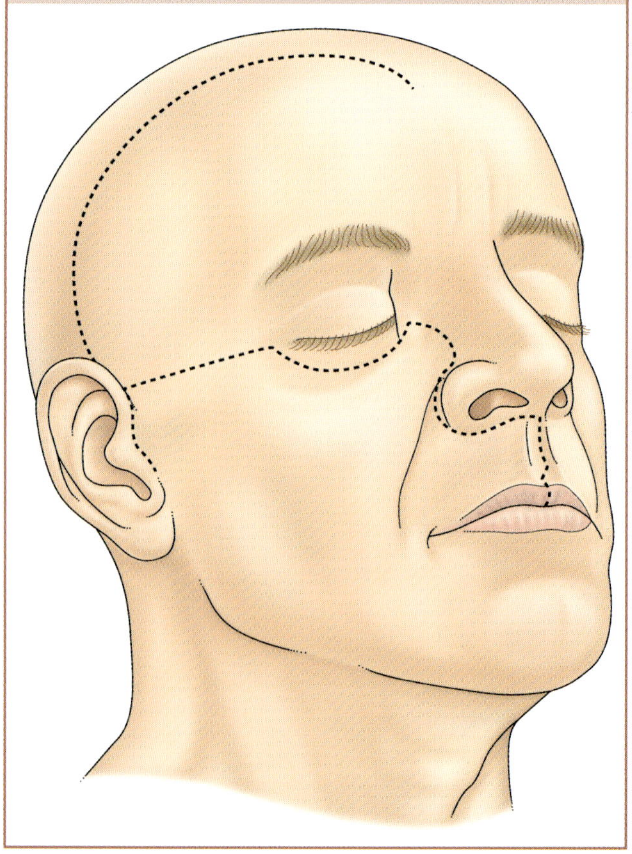

Figure 6.11 Palatal incisions and osteotomy for extended osteoplastic maxillotomy.

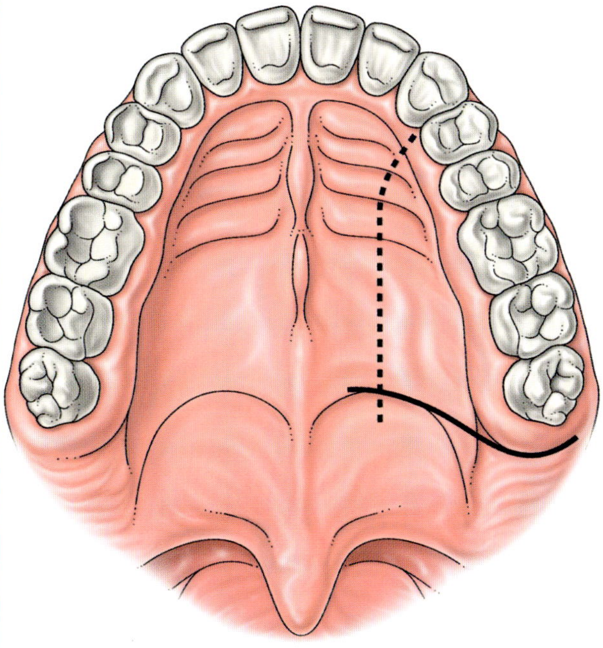

Figure 6.12 Exposure of infratemporal fossa through an extended osteoplastic maxillotomy approach. The zygoma is also resected. Posterior wall of maxilla is kept intact to ensure *en bloc* excision of tumour.

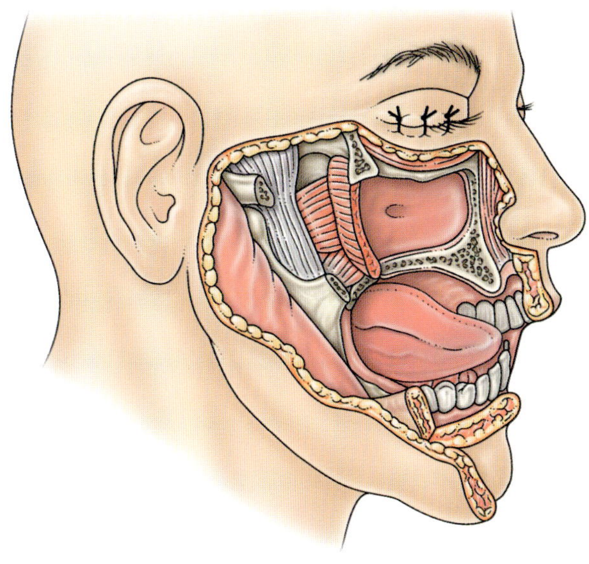

Maxillary swing

In 1991 Wei *et al.* described an osteoplastic approach to the nasopharynx similar to the one described above. The maxilla is separated from the facial skeleton, but remains attached anteriorly to the cheek flap. It is suitable only for medially located lesions.[30]

Transmandibular approach

The concept of approaching the retromaxillary area through a mandibulotomy is not new and has been advocated by Conley and Barbosa.[7,8] The infratemporal fossa communicates inferiorly with the neck. If the mandible is laterally retracted and the medial pterygoid muscle is detached from its mandibular attachment the infratemporal space can be reached. This approach provides good control of the vessels and nerves and *en bloc* resection of nasopharynx, posterior maxilla, infratemporal fossa structures, mandibular ramus and parotid gland can be performed.[31,32] The procedure has been modified by Attia *et al.* to obtain wide field exposure without sacrifice of either mandibular function or the sensory supply of the face and oral cavity. The facial nerve is preserved.[33] The ramus of the mandible is exposed via a suprahyoid incision and the marginal mandibular branch of the facial nerve is preserved. The facial vessels are ligated and divided. The mandibular osteotomies are arranged to spare the inferior alveolar nerve and vessels and are positioned under the intercondylar notch above the opening of the mandibular canal and just medial to the mental foramen (Fig. 6.13). Detachment of the medial and lateral pterygoid muscles and the sphenomandibular ligament allows the mandibular segment to be reflected superiorly (Fig. 6.14). This provides direct access to the infratemporal fossa; osteosynthesis of the mandible and intermaxillary fixation is performed. The procedure preserves function, exposure is good and is cosmetically acceptable.

Transzygomatic approach

The concept of access to the infratemporal fossa via an osteotomy of the zygomatic arch has been described in the past by Barbosa, Crockett, Conley, and Samy and Girgis.[8–10,34] The basic techniques since have remained the same. There has been renewed interest in this approach by various surgical disciplines to use this as

Figure 6.13 Sites for mandibular osteotomies in transmandibular approach to infratemporal fossa (after Attia *et al.*[33]).

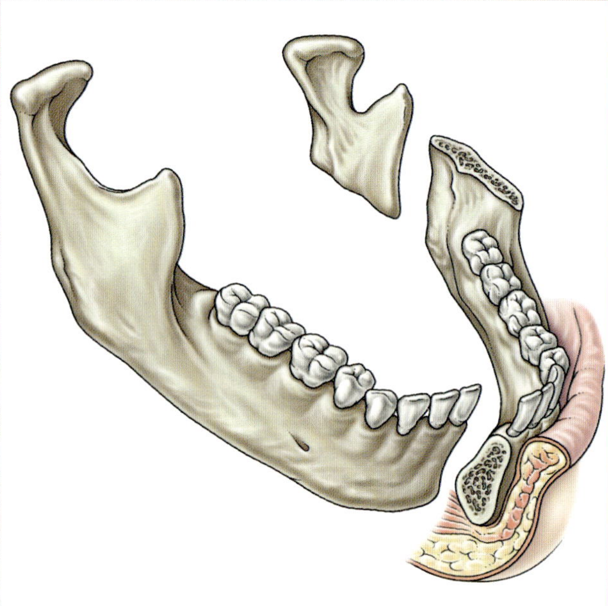

Figure 6.14 Transmandibular approach to infratemporal fossa (after Attia *et al.*[33]). Inferior alveolar vessels are preserved. Exposure is excellent. Temporary intermaxillary fixation is required.

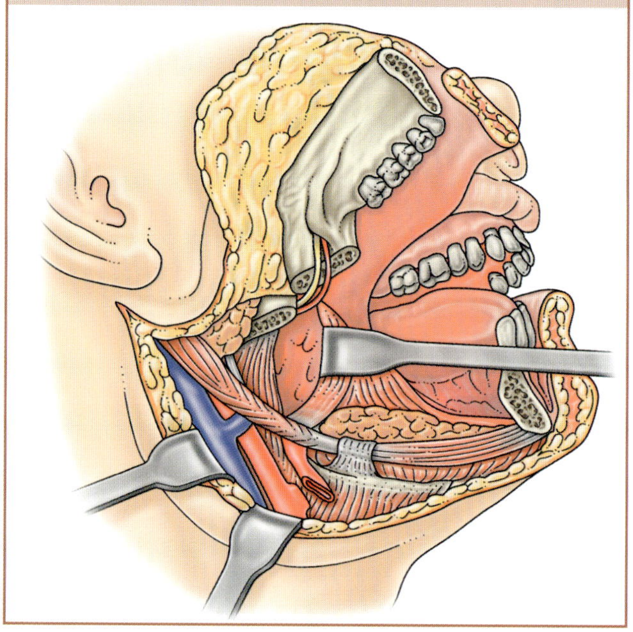

a combined procedure in order to obtain optimal access and visualisation to achieve *en bloc* excision.[35] The approach has been further improved upon by the anatomical studies to prevent inadvertent injury to the frontal and zygomatic branches of the facial nerves (Fig. 6.15).[36–38] The incision begins on the scalp behind the hairline and runs preauricularly then under the ear lobule behind the ascending ramus of the mandible to the hyoid bone (Fig. 6.16). The skin flap anteriorly on the scalp above the zygoma is not prepared, but the temporalis fascia is opened and dissection proceeds deep to it until the zygoma where the fascia is attached (Fig. 6.17). The frontal branch and the rami of the zygomatic branch of the facial nerve are located in the areolar tissue superficial to the temporalis fascia. Wetmore *et al.* advocated to displace the areolar tissue inferiorly to the lower border of the zygomatic arch and anteriorly to include the orbicularis oculi muscle while exposing the periorbita of the lateral orbit (Fig. 6.18).[36] Alternatively, if the main trunk of the facial nerve has been dissected, as would be the case where a tumour is originating from the parotid gland, the frontal and zygomatic branches are then exposed and gently retracted. In order to obtain maximum exposure the zygomatic arch needs to be divided as far back and as far forwards as possible. The superior border is dissected free while the masseter muscle is left attached to the inferior border. Appropriate drill holes are made prior to the osteotomies to ensure precise approximation after surgery. As the zygoma is reflected downwards, a few fibres of the masseter muscle that take their origin from the temporalis tendon need to be released.[9] The coronoid process along with a part of the ascending ramus of the mandible anterior to the mandibular canal is exposed. Drill holes are made with the miniplates in position and osteotomy is performed. The temporalis muscle with the coronoid process is reflected superiorly. This manoeuvre exposes the infratemporal space. Further steps depend on the pathology of the tumour. In the event that the pathology is deep seated in the nasopharynx and further dissection is needed, the maxillary artery is at this stage ligated and divided as it traverses over the lateral pterygoid muscle. The attachment of the lateral pterygoid to the anterior surface of the mandibular condyle is released. The pterygoid venous plexus is carefully avoided or coagulated. The medial pterygoid muscle is now exposed. Its attachment to the mandible is cut.

Figure 6.15 Relationship of superficial temporal artery and facial nerve branches to the zygoma. The preauricular incision is designed to preserve the nerves and vessels.

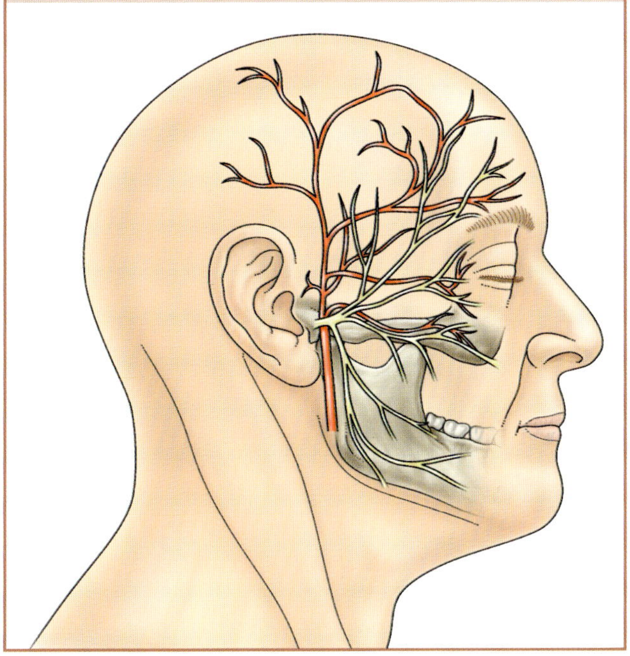

Figure 6.16 Incision for preauricular transzygomatic approach.

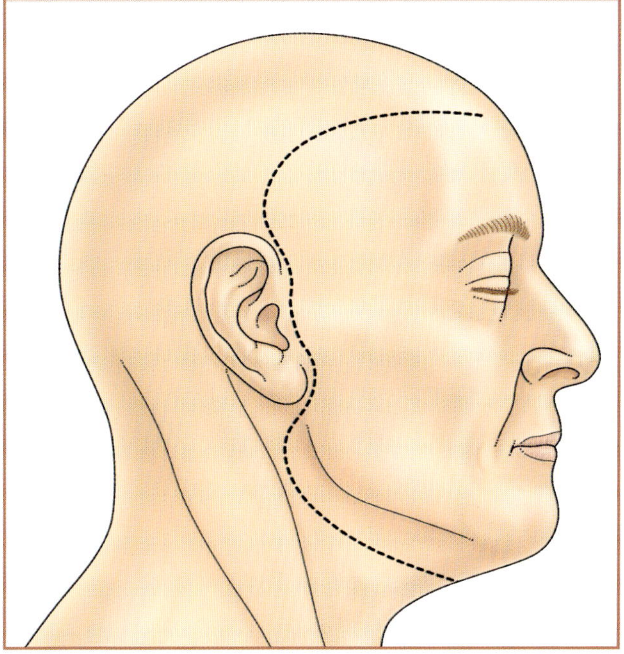

Figure 6.17 Preauricular transzygomatic approach. Dotted line indicates incision on the temporal fascia. Dissection deep to the fascia prevents inadvertent injury to frontal branch of facial nerve.

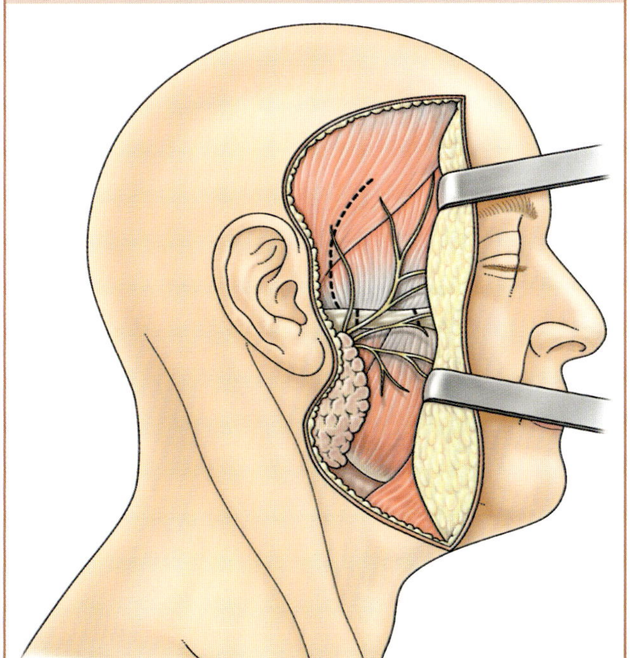

Figure 6.18 Preauricular transzygomatic approach. Mobilisation of the anterior flap and lateral orbital margin provides better exposure.

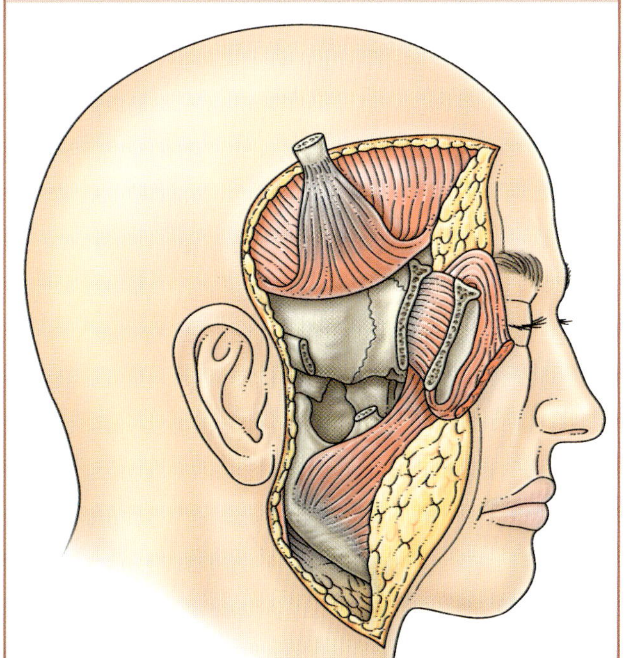

The mandible is retracted laterally. It helps to mobilise the mandible further if the temporomandibular joint is disarticulated. The lateral pterygoid plate is visualised. If the pterygoid muscles need to be removed *en bloc* (for instance with the maxilla) then an osteotomy is placed close to the skull base. Before doing so, the mandibular nerve, which runs on a posterior plane, is retracted posteriorly. The middle meningeal artery is just posterior to the nerve and is ligated and divided. Most pathologies are encountered in this area, which lies at a depth of approximately 4.5 cm from the zygoma in an adult. The superior constrictor muscle of the pharynx and the lateral pharyngeal wall are close by and can be opened. In the event of the tumour originating from the nasopharynx, it is now exposed. The posterior wall of the maxilla and the pterygomaxillary fissure are anterior to this plane and the inferior orbital foramen is anterosuperior.

Transfacial approaches

An interesting new addition to this armamentarium was from Janecka in 1990.[39] Janecka described an approach which combined facial soft tissue translocation and craniofacial osteotomies. The facial incision resembles a Weber–Fergusson incision, which is extended through the medial canthus of the eye, includes the lower lid, passes through the lateral canthus and is extended to the preauricular area at the superior attachment of the helix (Fig. 6.10). A superior frontotemporal incision over the scalp is combined and may be extended inferiorly to the hyoid bone. Craniofacial osteotomies include the maxilla with floor of the orbit and complete zygoma as separate units. The coronoid process is removed (Fig. 6.19). Exposure is excellent (Fig. 6.20). The drawback is that the zygomatic and frontal branches of the facial nerve are sectioned. However, microanastomoses are performed at the end and a tarsorrhaphy is needed. Removal of the floor of middle cranial fossa and the use of the temporalis muscle with the deep temporal artery for reconstructive purposes is the same as in most other approaches. At the end of the procedure the removed bones are replaced and fixed by miniplates.

Tumours in the infratemporal fossa

The pathologies encountered in the infratemporal fossa are varied. Extension of tumours from adjoining anatomical areas such as the nasopharynx, paranasal sinuses, oropharynx, parotid gland and cranium is the

Figure 6.19 Osteotomies in facial translocation approach. Anterior maxilla and zygoma are temporarily removed (shaded area). The coronoid process is resected.

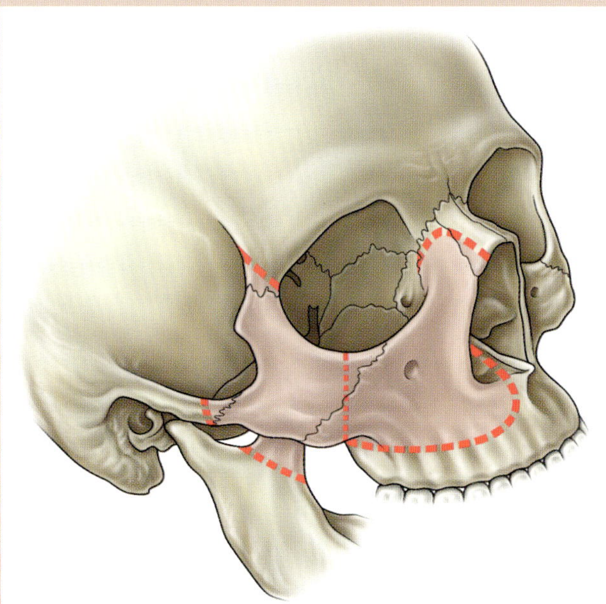

Figure 6.20 Exposure of the infratemporal fossa through facial translocation approach. Posterior antral wall is intact.

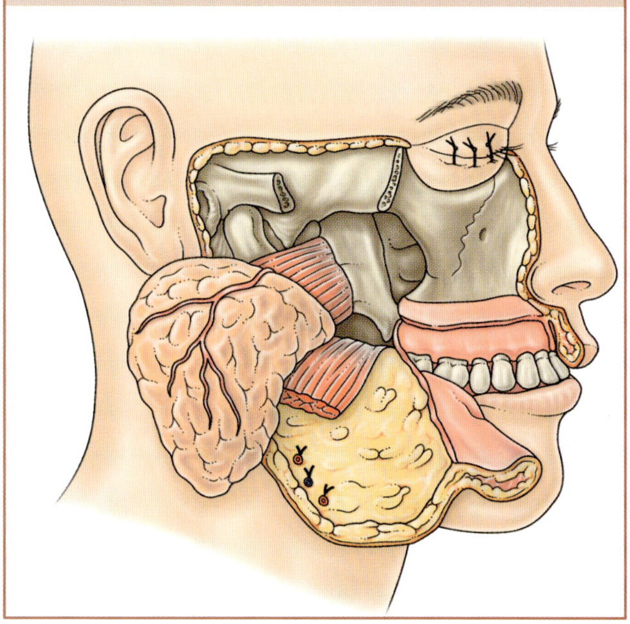

largest group and comprises both benign and malignant tumours. Approach to the infratemporal fossa is then combined with the approach to the location of the primary pathology to obtain *en bloc* excision. This group of tumours is diagnosed relatively early because they usually present with symptoms pertaining to the site of their origin.

Primary tumours of the infratemporal fossa may remain undiagnosed for long periods of time. They are mesodermal in origin and both benign and malignant tumours are encountered. Symptoms are few and insidious and often referred to other areas. Conductive hearing loss, trismus and sensory disturbances of the face are often present, but their onset may be so gradual that the patient may not have taken them seriously for a long time before seeking medical attention.

Space-occupying lesions originating in the infratemporal space tend to push the thin wall of the maxilla forwards, while tumours growing in the sinus push the posterior wall of the maxilla into the space. Table 6.1 shows the primary tumours encountered by the author. Table 6.2 lists the tumours infrequently seen, and reported in the literature. Sites of tumours extending to the infratemporal fossa are listed in Table 6.3. In Table 6.4 tumour-like lesions encountered in the infratemporal fossa and reported in the literature are listed.

Table 6.1 Primary tumours and tumour-like lesions of the infratemporal fossa

Pathology	
Primary tumours	Meningioma
	Schwannoma
	Neurofibroma
	Malignant lymphoma
	Haemangioma
	Rhabdomyosarcoma
	Fibrosarcoma
	Histocytosis X

Table 6.2 Rare primary tumours of the infratemporal fossa

Angioma
Chondrosarcoma
Chondroblastoma
Leiomyosarcoma
Liposarcoma
Solitary fibrous tumour (pleural tumour)
Osteosarcoma
Synovial sarcoma
Giant cell tumour
Histocytoma
Chordoma
Lymphangioma

Table 6.3 Sites of tumours extending to the infratemporal fossa

Maxilla
Mandible
Nasopharynx
Sphenoid sinus
Ethmoid sinus
Parotid gland
Temporal fossa
Temporomandibular joint
External auditory meatus
Cranium

Table 6.4 Tumour-like lesions of the infratemporal fossa

Fibrous dysplasia
Desmoplastic fibroma
Myositis ossificans
Neurofibromatosis
Cystic hygroma
Hydatid cyst

COMBINED APPROACHES

Efforts of surgical oncologists to achieve whenever possible *en bloc* radical excision of tumours and the relatively limited access obtained by any single approach have made combined approaches necessary. It offers the patients the maximum benefit of the technical 'know-how' of the surgical team and the best opportunity for surgical excision. A combination of median mandibulotomy and maxillectomy by Weber–Fergusson incision has been reported by Lawson *et al.* in recent years for carcinoma of the maxilla extending posteriorly (Fig. 6.21).[40] The mandibulotomy approach along with a maxillectomy approach via the Weber–Fergusson incision has been further combined with temporary excision of the malar complex and resection of the coronoid process to provide better access for excision of large juvenile nasopharyngeal angiofibroma.[41]

Combinations of transmandibular and transzygomatic approaches have been used for a variety of pathologies originating from the paranasal sinuses, parotid gland, orbit, parapharynx and cranial cavity.[42,43] It has long been realised that in order to obtain control of the superior extents of tumours in this area, removal of the floor of the middle cranial fossa and retraction of the dura in combination with the other approaches is likely to provide superior access. This concept has been highlighted in several publications.[44,45] In the experience of this author, a combination of preauricular and paramedian transmandibular approach combined with a facial degloving approach gives excellent results and avoids a facial scar.

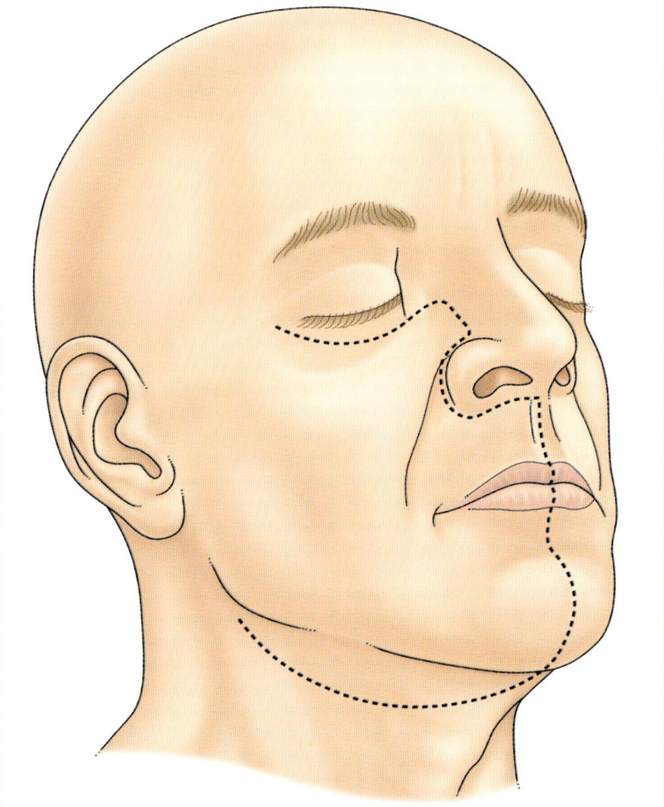

Figure 6.21 Incision for combined transmaxillary and transmandibular approach. This incision combines the Weber–Fergusson approach with a suprahyoid lip splitting incision with median mandibulotomy. It can be extended to include a preauricular incision.

REFERENCES

1. Krause W. *Anatomy des Menschen*. Leipzig: Verlag. S. Hirzel, 1905; 547
2. Sluder G. The anatomic and clinical relations of the sphenopalatine (Meckel's) ganglion to the nose and its accessory sinuses. *NY Med J* 1900; 293–298
3. Sluder G. Etiology diagnosis. Prognosis and treatment of sphenopalatine ganglion neuralgia. *JAMA* 1913; 61: 1202–1205
4. Carnochan. Excision of the trunk of the second branch of the fifth nerve. *Am J Med Soc* 1858; 1: 134
5. Sewall E C. An operation for the removal of the sphenopalatine ganglion. *Ann Otol Rhinolaryngol* 1926; 35: 1
6. Conley J J. The surgical approach to the pterygoid area. *Ann Surg* 1956; 144: 139
7. Dingman D L, Conley J J. Lateral approach to the pterygomaxillary region. *Ann Otol Rhinol Laryngol* 1970; 79: 967–969
8. Barbosa J F. Surgery of extensive cancer of paranasal sinuses. *Arch Otolaryngol* 1961; 73: 129–133
9. Crockett D J. Surgical approach to the back of the maxilla. *Br J Surg* 1962; 50: 819–821
10. Conley J J. Tumours of the infratemporal fossa. *Arch Otolaryngol* 1964; 79: 498–504
11. Ketcham A S, Wilkins R H, Van Buren J M. A combined intracranial facial approach to paranasal sinuses. *Am J Surg* 1966; 106: 698–703
12. Terz J J, Alkone F J, Lawrence W. Craniofacial resection for tumours invading the pterygoid fossa. *Am J Surg* 1969; 118: 732–740
13. Fisch V. Infratemporal fossa approach to tumours of the temporal bone and base of skull. *J Laryngol Otol* 1978; 92: 949–967
14. Valavanis A. Preoperative embolisation of head and neck. Indications, patient selection, goals and precautions. *Am J Neuroradiol* 1987; 7: 943–952
15. Holliday M J, Nachlas N, Kennedy D W. Uses and modifications of the infratemporal fossa approach to skull base tumours. *Ear Nose Throat J* 1986; 65: 101–106
16. Tiwari R M. Surgical landmarks of the infratemporal fossa. *J Craniomaxillofac Surg* 1998; 26: 84–86
17. Farr H W, Gray G F Jr, Vrana M, Panio M. Extracranial meningioma. *J Surg Oncol* 1973; 5: 411–420
18. Johnson A T, Maran A G D. Extracranial tumours of the infratemporal fossa. *J Laryngol Otol* 1982; 96: 1017–1026
19. Shapsay S M, McCain C F, Ukmakli A *et al.* Diagnosis of infratemporal fossa tumours using percutaneous core needle biopsy. *Head Neck Surg* 1979; 2: 35–41
20. Spearman M, Curtin H, Dusenberg D *et al.* Computer tomography directed fine needle aspiration of skull base, parapharyngeal and infratemporal masses. *Skull Base Surgery* 1995; 5: 199–205
21. Kornfehl J, Gstottner W, Kontrus M, Sediny R. Transpalatine excision of a cavernous haemangioma of the infratemporal fossa. *Eur Arch Otorhinolaryngol* 1996; 253: 172–175
22. Von Langenbeek. Beitrage zur Osteoplastik. In: Goschen A (ed) *Die osteoplastische Resektion des Oberkiefers*. Berlin: Reimer, 1859
23. Le Fort R. Etude experimentale sur les fractures de la machoire superieure. *Rev Chir* 1901; 23: 208
24. Stell P M. Le Fort I osteotomy for approach to nasopharynx. *Clin Otolaryngol* 1983; 9: 59–61
25. Belmont J R. The Le Fort I osteotomy approach for nasopharyngeal and nasal fossa tumours. *Arch Otolaryngol Head Neck Surg* 1988; 114: 751–754
26. Krause C J, Baker S R. Extended transantral approach to pterygomaxillary tumours. *Ann Otol Rhinol Laryngol* 1982; 91: 395–398
27. Hernandez A F. Transfacial access to the retromaxillary area. *J Maxillofacial Surg* 1986; 14: 165
28. Brown A M S, Lavery K M, Millar B G. The transfacial approach to the postnasal space and retromaxillary structures. *Br J Maxillofac Surg* 1991; 29: 230–236
29. Catalano P J, Biller H F. Extended osteoplastic maxillotomy. *Arch Otolaryngol Head Neck Surg* 1993; 119: 394–400
30. Wei W I, Lam K H, Sham J S T. New approach to the nasopharynx, the maxillary swing approach. *Head Neck* 1991; 13: 200–207
31. Krespi Y P, Sisson G A. Transmandibular exposure of the skull base. *Am J Surg* 1984; 148: 534–538
32. Krespi Y P. Lateral skull base surgery for cancer. *Laryngoscope* 1989; 99: 514–524
33. Attia E L, Bentely K C, Head T, Mulder D. A new external approach to the pterygomaxillary fossa and parapharyngeal space. *Head Neck Surg* 1984; 6: 884–891
34. Samy L L, Girgis J H. Transzygomatic approach for naospharyngeal fibromata with extrapharyngeal extension. *J Laryngol Otol* 1965; 79: 782–795
35. Dingman D L, Conley J. Lateral approach to the pterygomaxillary region. *Ann Otol Rhinol Laryngol* 1970; 79: 967–969
36. Wetmore S J, Suen J Y, Synderman N L. Preauricular approach to infratemporal fossa. *Head Neck Surg* 1986; 9: 93–103
37. Al Kayat A, Bramley P. A modified preauricular approach to the temporomandibular joint and malar arch. *Br J Oral Surg* 1980; 17: 91–103
38. Amirati M, Spallone A, Cheatham M, Becker D. An anatomical study of the temporal branch of the facial nerve. *Neurosurgery* 1993; 33: 1039–1045
39. Janecka I P, Sen C H, Sekhar L N, Arriaga M. Facial transdislocation. A new approach to the cranial base. *Otolaryngol Head Neck Surg* 1990; 103: 413–419
40. Lawson W, Naidu R K, Le Benger J, Biller H. Combined median mandibulotomy and Weber–Fergusson maxillectomy. *Arch Otolaryngol Head Neck Surg* 1990; 116: 596–599

41. Mann W J, Gilsbach J, Seeger W, Floel H. Use of a malar bone graft to augment skull base access. *Arch Otolaryngol* 1985; 111: 30–33
42. Friedman W H, Katsantonis G P, Cooper M H, Lee J M, Strelzow V V. Stylomandibular dissection. A new method for *en bloc* resection of malignancies of infratemporal fossa. *Laryngoscope* 1981; 91: 1869–1879
43. Rosenblum B N, Katsantonis G P, Cooper M H *et al.* Infratemporal fossa and lateral skull base dissection. Long term results. *Otolaryngol Head Neck Surg* 1990; 102: 106–110
44. Sekhar L N, Schramm V L, Jones N F. Subtemporal preauricular infratemporal fossa approach. *J Neurosurg* 1987; 67: 488–499
45. Mickye B, Close L, Schaefer S, Samson D. A combined frontolateral and lateral infratemporal fossa approach to the skull base. *J Neurosurg* 1988; 68: 678–683

Chapter 7

Surgical approaches to the infratemporal fossa

B.T. EVANS, D. WIESENFELD, L. CLAUSER AND C. CURIONI

INTRODUCTION

Surgical approaches to pathology in the infratemporal fossa have developed over a relatively short period of time. As recently as the 1960s malignant tumours involving the infratemporal fossa, and the adjacent middle cranial fossa, were considered by some to be inoperable due to the difficulty in achieving adequate tumour clearance and excessive blood loss. Early attempts at improving access frequently resulted in both severe cosmetic deformity and disturbance of function – usually the result of the resection of structures uninvolved by the pathology in order to increase exposure.

PATHOLOGY

Tumours of the infratemporal fossa are uncommon and may be benign or malignant. Rarely, they may arise from within the infratemporal fossa itself or, more commonly, involve it by direct spread from adjacent areas such as the maxilla, oral cavity, nasopharynx, or skull base (Chapter 6).

Primary tumours arising from the structures within the infratemporal fossa include meningiomas, schwannomas, neurofibromas, angiomas and chondro-, fibro- and rhabdomyosarcomas. Oral cavity and maxillary lesions may include oral, antral and nasal squamous cell carcinomas, minor salivary gland tumours – particularly adenoid cystic carcinomas of the palatal glands, lymphomas, malignant fibrous histiocytomas, giant cell tumours and odontogenic tumours such as ameloblastomas. Skull base lesions include meningiomas of the sphenoid wing and chordomas. Nasopharyngeal angiofibromas may bulge into the infratemporal fossa. Involvement by deep lobe tumours of the parotid gland is well recognised.

APPLIED ANATOMY

The infratemporal fossa has been described in detail in Chapter 1. Briefly, it lies beneath the base of the skull between the side wall of the pharynx and the ascending ramus of the mandible. It is a quadrangular space. The *roof* is formed by the infratemporal surface of the greater wing of the sphenoid and a small portion of the squamous temporal bone immediately anterior to the articular eminence of the temporomandibular joint. The roof of the fossa therefore forms part of the floor of the middle cranial fossa. The *anterior boundary* is formed by the posterior wall of the maxilla. The *lateral boundary* is formed by the zygomatic arch, masseter and temporalis muscles and the ascending ramus of the mandible, completed posteriorly by the stylomandibular ligament, parotid gland and the carotid sheath. The *medial boundary* is formed by the lateral pterygoid plate anteriorly and the superior constrictor, tensor and levator veli palatini posteriorly (Fig. 7.1). The infratemporal fossa has no anatomical floor. It is limited anteriorly by the attachment of the medial pterygoid muscle to the medial aspect of the mandible and communicates with the parapharyngeal space anterior to the internal carotid artery.

The aim in the treatment of malignant tumours is *en-bloc* clearance, i.e. three-dimensional tumour clearance with an intact anatomical barrier between the lesion and the resection margin. The medial boundary of the dissection conceptually poses the most difficulty for surgeons visualising an *en-bloc* resection of the infratemporal fossa. Fortunately, as described by Friedman,[1] there exists a natural plane of cleavage deep to the medial pterygoid muscle, between it and the superior constrictor muscle. These muscles are separated by a layer of loose areolar tissue – an extension of the buccal pad of fat. This natural plane of

Figure 7.1 Medial boundary of the infratemporal fossa.

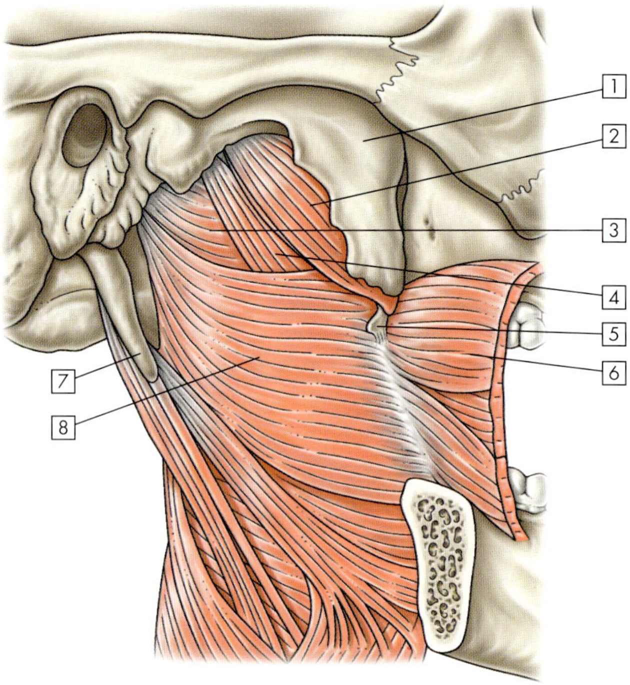

1 Lateral pterygoid plate
2 Tensor veli palatini muscle
3 Pharyngobasilar fascia
4 Levator veli palatini muscle
5 Pterygoid hamulus
6 Buccinator muscle
7 Styloid process
8 Superior constrictor muscle

cleavage forms the medial boundary of the resection for malignant tumours *confined* to the infratemporal fossa. Medial (deep) to this plane, from anterior to posterior, is the buccinator, pterygomandibular raphe and the superior constrictor. The tensor palati, levator palati and cartilaginous Eustachian tube are exposed in the dissection. Two prominent reference points are available for the identification of this critical medial plane of dissection: the styloid process and the pterygoid hamulus (Fig. 7.2). The foramen lacerum, carotid canal and jugular foramen are situated medial to this plane. The 'surgical block' is wedge shaped with the base lateral (Fig.7.3).[1,2]

Medially, the pterygopalatine fossa, containing the maxillary nerve and artery and the pterygopalatine ganglion, communicates with the infratemporal fossa via the pterygomaxillary fissure. The infratemporal fossa and the pterygopalatine fossa combine to form the so-called 'retromaxillary space'. Tumours do not respect anatomical boundaries. Posteriorly penetrating malignant maxillary tumours, for example, frequently invade both spaces, i.e. the tumour is not confined to the infratemporal fossa. Tumour clearance medially in these circumstances requires the resection of the pterygoid plates and the attached musculature. The maxillary division of the trigeminal nerve may be biopsied at the skull base nerve, i.e. foramen rotundum, for prognostic purposes.

The lateral pterygoid muscle arising from the infratemporal surface of the skull, and the lateral aspect of the lateral pterygoid plate inserting into the mandibular condyle and meniscus, crosses the entire infratemporal fossa in an anterior–posterior/medio–lateral direction. This muscle effectively divides the infratemporal fossa into *superior* and *inferior* compartments (Fig. 7.4).

Resection of the floor of the middle cranial fossa is usually required for adequate clearance of malignant tumours involving the superior compartment. The limit of the medial resection margin in such cases is the line running medial to the foramen rotundum, foramen ovale, foramen spinosum, and lateral to the internal carotid artery (foramen lacerum). Traced posteriorly in the floor of the middle fossa this line passes through the junction of the squamous and petrous temporal bones (see Fig. 7.66).

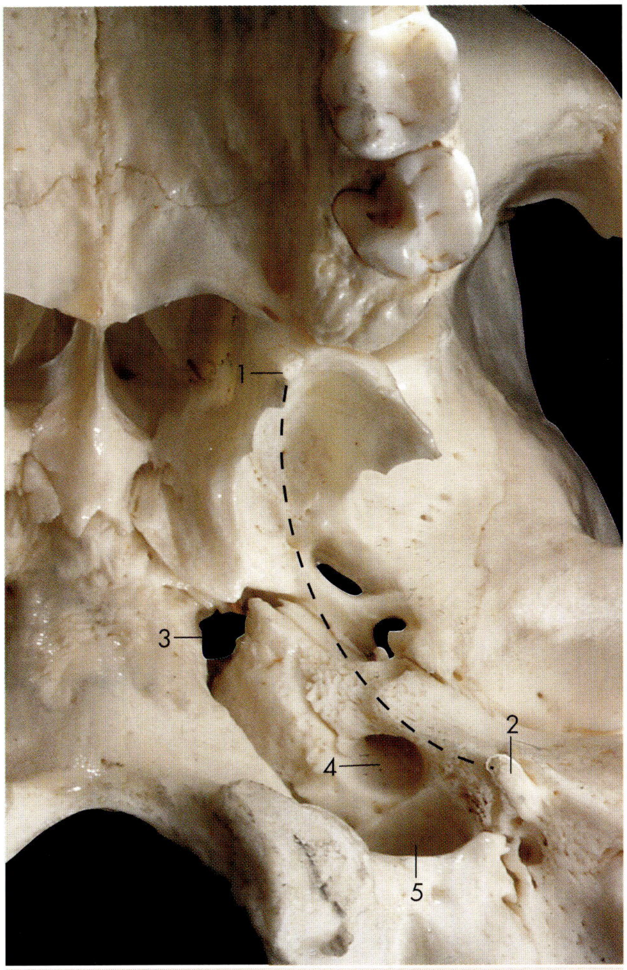

1 Pterygoid hamulus
2 Styloid process
3 Foramen lacerum
4 Carotid canal
5 Jugular foramen

Figure 7.2 Styloid process and the pterygoid hamulus – reference points for the medial plane of dissection in the infratemporal fossa. The stylohamular dissection. Foramen lacerum, carotid canal and the jugular foramen are medial to this plane.

Figure 7.3 Surgical block of the infratemporal fossa – Friedman wedge.

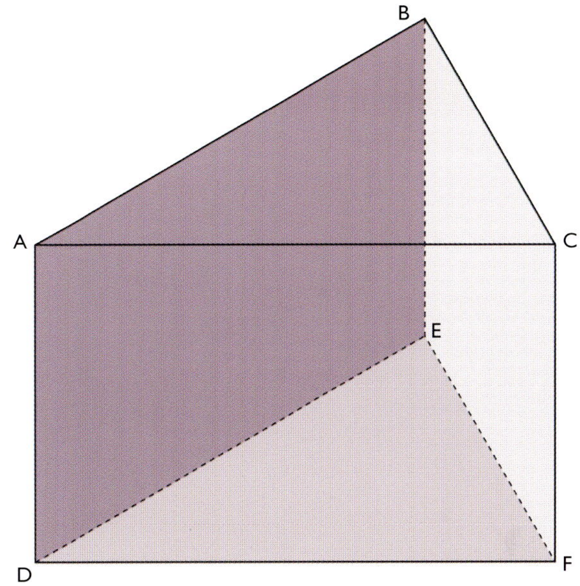

ACFD Lateral plane (mandibular ramus and zygomatic arch)
BCFE Anterior plane (posterior wall of maxilla)
ABED Medial plane (pterygoid hamulus and styloid process)
ABC Superior plane (sphenoid–middle cranial fossa)
AD Posterior plane (styloid process and mastoid process)

Figure 7.4 The infratemporal fossa is effectively divided into superior and inferior compartments by the lateral pterygoid muscle.

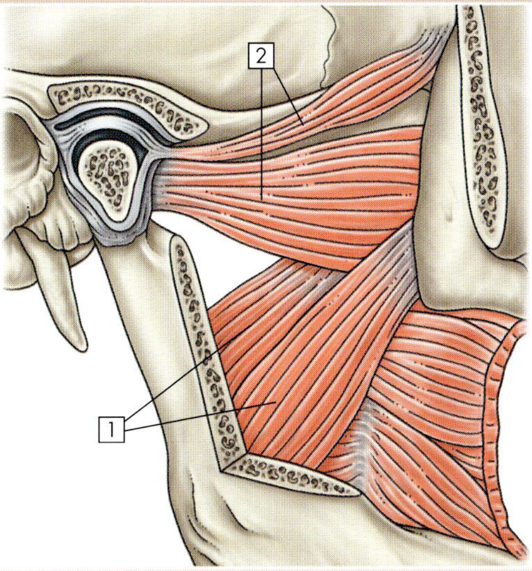

1 Medial pterygoid muscle
2 Lateral pterygoid muscle

SURGICAL APPROACHES

These may be anterior, lateral or combinations.

- *Anterior*
  1. Transmandibular
  2. Transfacial
  3. Intraoral
- *Lateral*
  1. Transmandibular
  2. Transzygomatic
- *Combinations*

These approaches may be combined with a (fronto)temporal craniotomy as necessary.

The ideal surgical approach to the infratemporal fossa should:

- Provide increased and more direct exposure of the pathology and the adjacent neurovasculature with:
  - a short straight line between the surgeon and the pathology, and
  - a wide arc of exposure in three dimensions.
- Be extensile, i.e. capable of being extended peroperatively.
- Minimise brain retraction where exposure of the intracranial contents is required.
- Have minimal morbidity functionally or cosmetically.
- Result in minimal increase in overall operating time.
- Avoid facial skin incisions.

The different approaches to the infratemporal fossa fulfil these ideals to varying extents. The approach adopted will depend upon the site, nature and extent of the pathology. There is no one correct/ideal approach to the infratemporal fossa; the only requirements are that the approach adopted provides adequate access for the lesion and the morbidity of the approach is considered acceptable.

The potential morbidity of the approach must be weighed against the morbidity of inadequate access with possible incomplete removal of the pathology and avoidable damage to adjacent structures. This applies particularly with benign pathology such as nasopharyngeal angiofibromas. In view of its proximity to critical areas such as the middle cranial fossa, orbit and nasopharynx, *the infratemporal fossa must be regarded as a malignant site.*

The various surgical approaches to the region have been made possible by advances in imaging, surgical techniques developed in the fields of head and neck, craniofacial, orthognathic and skull base surgery, and an improved understanding of the anatomy of the skull base. The availability of rigid fixation systems specifically designed for the craniofacial region ensures accurate replacement of mobilised facial and skull bone segments.

ANTERIOR APPROACHES

Transmandibular approach: extended mandibular swing

The mandibular swing approach for improving access to pathology in the oral cavity and the pharynx is a technique used frequently in head and neck surgery. The lower lip is divided in the midline, the mandible sectioned anteriorly and the hemimandible swung laterally, providing increased and more direct exposure to the oral cavity and adjacent areas; the exposure achieved being comparable to jaw resection. For exposure of the infratemporal fossa and the adjacent skull base the dissection is taken further laterally. A second osteotomy of the mandibular ramus above the lingula may be added if necessary, allowing further superior and outward mandibular retraction.

There are three separate elements to the approach:

- an incision to divide the lower lip and chin
- division of the mandible anterior to the mental foramen – preserving ipsilateral lower lip sensation
- dissection of the tissues in the floor of the mouth, submandibular region and neck.[3,4]

Surgical technique

A full thickness vertical incision is made through the midline of the lower lip; a v-shaped notch may be incorporated in the midline lip incision and the vermilion incision for accurate realignment on closure and to prevent scar contracture. The incision is continued downwards to the base of the hollow between the lip and the chin and curved around the chin button to the submental fold. The concavity of the incision is towards the side of the lesion (and the neck dissection if required), reducing the possibility of ischaemic necrosis of the chin prominence.

The submandibular incision curves downwards

from the midpoint of the submental fold (linking with the lip-splitting incision) to the level of the hyoid bone and curves upwards to the mastoid process (Fig. 7.5). A simple skin/platysmal flap preserving the marginal mandibular branch of the facial nerve may suffice, with a limited submandibular dissection. Alternatively a formal neck dissection, either radical or functional, may be required. The skin incision is modified according to the requirements of the case, as for example, in deep lobe parotid tumours, a superficial parotidectomy is performed initially and the parotidectomy incision linked naturally with the submandibular incision inferiorly. The parotid incision superiorly links readily with a bicoronal incision if required (Fig. 7.6).

Exposure of the major neurovascular structures is straightforward. The skin flap is raised in the subplatysmal plane from the menton to the mastoid process. The sternocleidomastoid muscle is retracted laterally, the posterior belly of the digastric retracted medially, and the carotid sheath exposed at the level of the carotid bifurcation. The internal jugular vein, carotid artery and vagus nerve are identified and traced superiorly. If a neck dissection is not required the submandibular gland is removed to improve the exposure of both these vessels and the lingual and hypoglossal nerves (Fig. 7.7).[5]

Figure 7.5 Incision notched at vermilion border and lower lip curving around chin point, linking with submandibular incision.

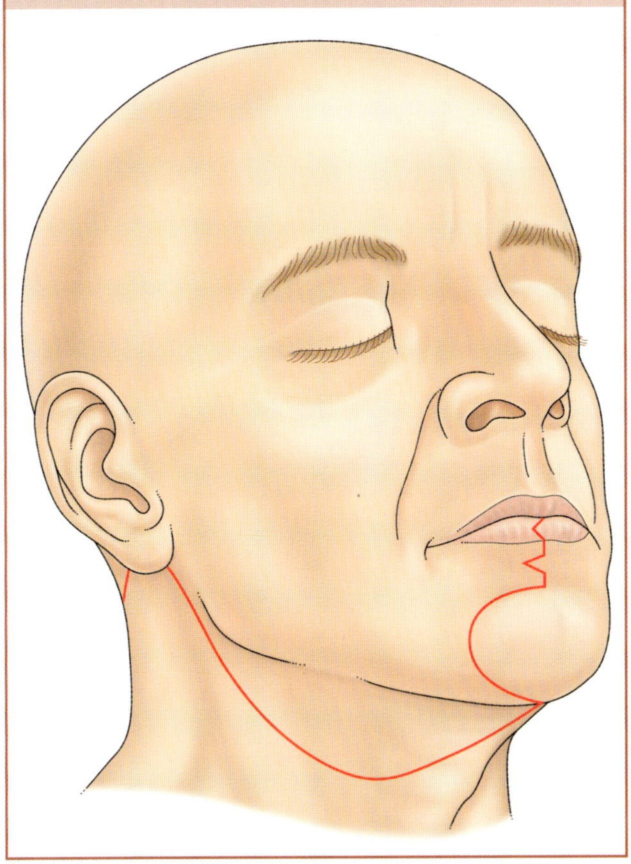

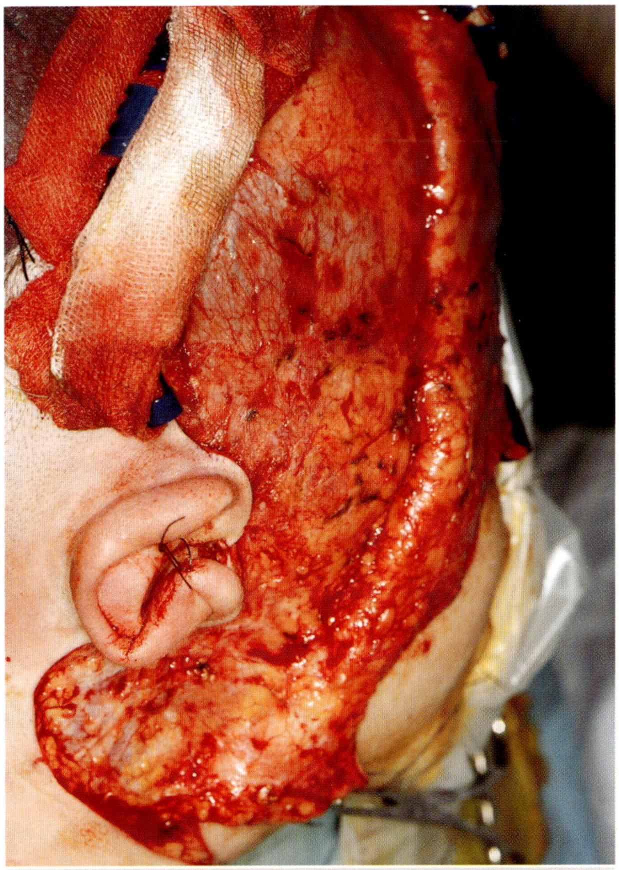

Figure 7.6 Scalp and parotidectomy flaps elevated – in the superficial muscular aponeurotic system (SMAS) layer.

Intraorally the incision through the labial mucosa and the attached gingiva is stepped so as not to lie directly over the subsequent osteotomy bone cut. Full thickness mucoperiosteal flaps are raised labially. Periosteal stripping is kept to a minimum – ideally no more than the width of one tooth on either side of the osteotomy cut. The lingual mucoperiosteum is elevated away from the osteotomy site and protected.

The mandible is divided with a fine saw blade between the roots of the standing teeth. A simple stepped osteotomy is employed (Fig. 7.8). In children a straight osteotomy bone cut may be used to reduce the risk of damage to the developing dentition (Fig. 7.9). Despite the availability of fine saws and osteotomes, sectioning the mandible between the roots of the teeth without damaging them can be difficult. If in doubt it is preferable to remove a single tooth and place the osteotomy cut through the socket. Prior to completing the division of the mandible, bone plates are placed across the osteotomy bone cut and subsequently removed to aid accurate reapproximation of the bone ends. In dentate patients an eyelet wire may be placed in each quadrant to allow temporary intermaxillary fixation when reapproximating the mandible. This may be supplemented with a wire ligature to the teeth on either side of the osteotomy. These simple additional measures aid accurate realignment of the mandible and restoration of the occlusion.

The osteotomy is placed most commonly either in the midline, or in the premolar region anterior to the mental foramen. For exposure of the infratemporal fossa the premolar site is usual as it is uncommon in such cases for the tumour or the radiation field to involve this area. Sectioning the mandible in the premolar region requires minimal periosteal stripping to expose the mandible for both sectioning and fixation and also avoids the need to divide the genial muscles and the anterior belly of the digastric.[3]

When exposing the infratemporal fossa a *modification of the usual intraoral dissection on the lingual aspect of the mandible is helpful.* Instead of the dissection extending along the floor of the mouth lateral to the submandibular duct (leaving a cuff of mucosa attached to the lingual aspect of the mandible, as is the

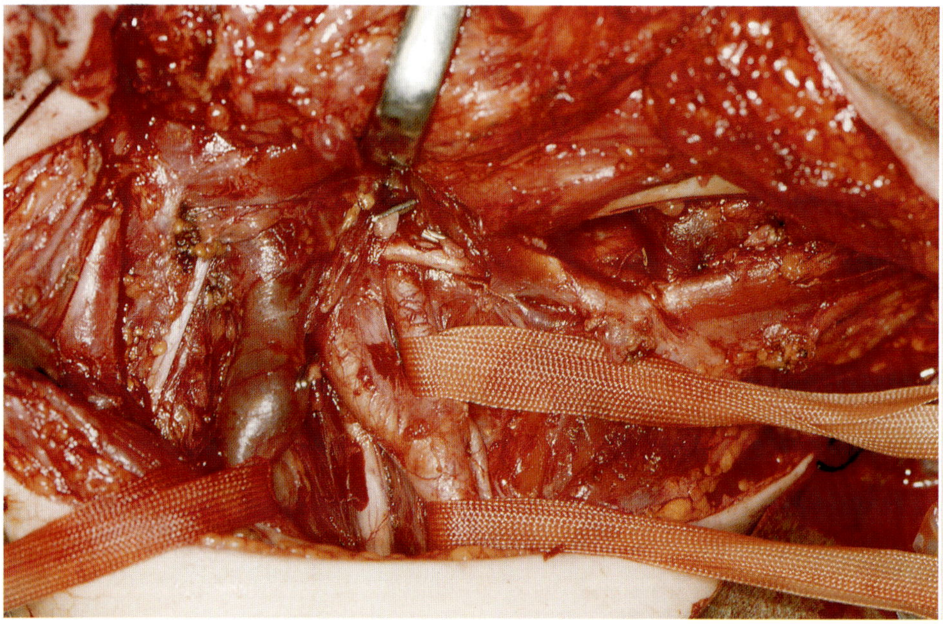

Figure 7.7 Neck contents following removal of the submandibular gland. Tapes around the internal jugular vein, common and external carotid arteries. The spinal accessory, vagus and hypoglossal nerves are readily identifiable.

case when this approach is used for access to the oral cavity/pharynx), the lingual incision is taken through the gingival margins of the teeth around the last molar tooth and up the anterior border of the ascending ramus. A full thickness lingual mucoperiosteal flap is therefore elevated off the lingual aspect of the mandible. The viability of a cuff of soft tissue left attached to the lingual aspect of the mandible in this more extensive dissection would be doubtful. With lateral retraction of the mandible the mylohyoid and medial pterygoid attachments are dissected free; the lingual and hypoglossal nerves are kept under direct vision. The inferior alveolar neurovascular bundle is identified and protected at the level of the lingula. The sphenomandibular ligament is dissected off the lingula and the stylomandibular ligament is dissected

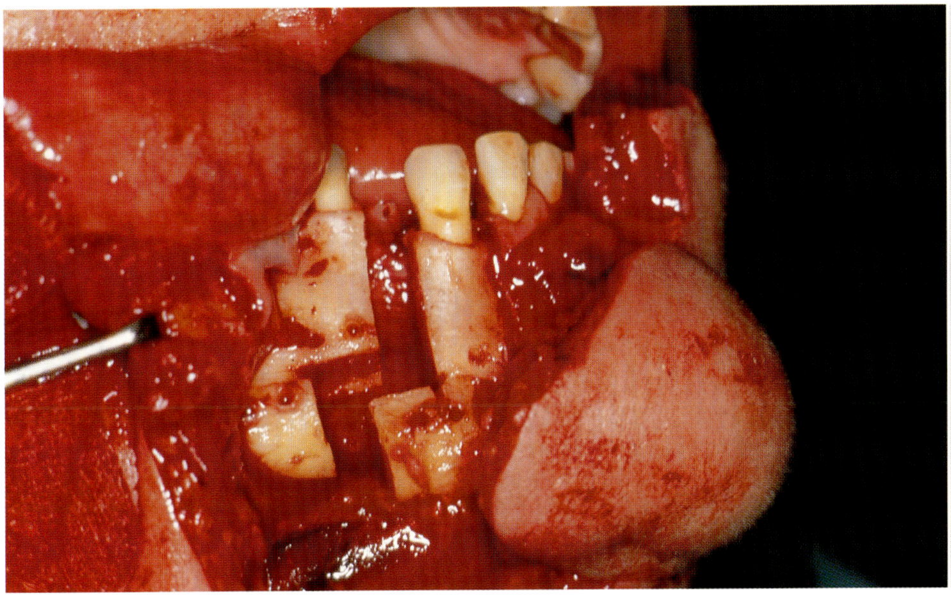

Figure 7.8 Stepped osteotomy of the mandibular body – used in adults.

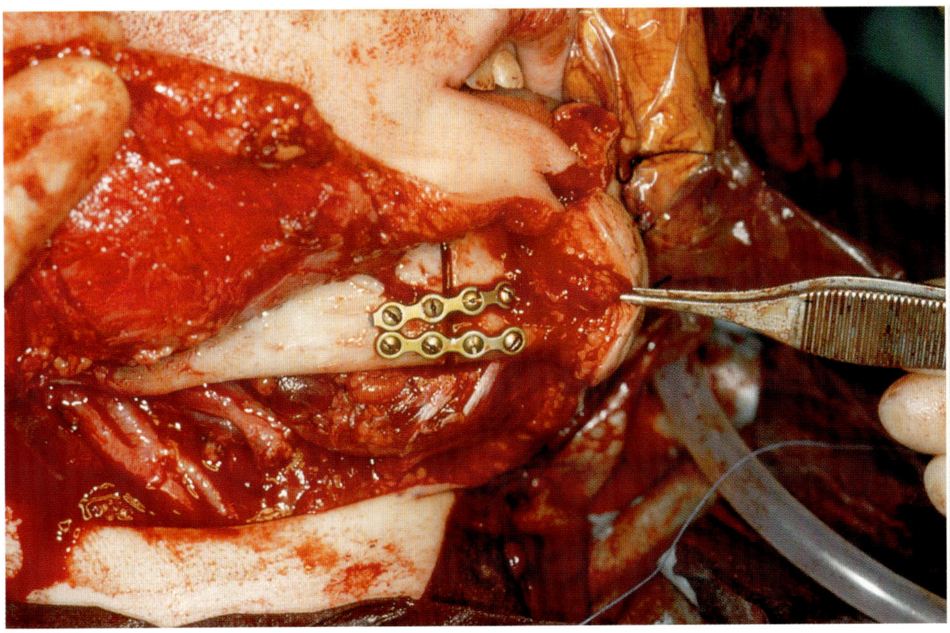

Figure 7.9 Straight mandibular osteotomy is preferred in children to reduce the risk of injury to the developing dentition.

off the posterior border of the mandible. Wide superior and lateral retraction of the mandible is now possible (Fig. 7.10).[6]

For *en-bloc* clearance of tumours *confined* to the infratemporal fossa, the medial border of the resection extends from the styloid process to the pterygoid hamulus which corresponds to the plane between the medial pterygoid and the superior pharyngeal constrictor. This plane lies lateral to the foramen lacerum and the jugular foramen and therefore allows the surgeon to spare the internal carotid artery and internal jugular vein. The pterygoid musculature and the other contents of the infratemporal fossa are incorporated in the surgical block (see Fig. 7.3).[1]

The natural plane of cleavage between the medial pterygoid and superior constrictor muscles can be developed by gentle blunt dissection medial to the styloid process. This dissection is made easier by the resection of the styloid process and the division of the muscles and ligaments attached to it. This manoeuvre also facilitates the dissection of the internal carotid artery and internal jugular vein superiorly to the skull base. *To avoid accidental damage to both the internal carotid artery and the internal jugular vein, these structures should be formally identified and traced superiorly rather than relying on blunt dissection in the plane between the medial pterygoid and the superior constrictor for their protection.*

Attia *et al.*[5] described an approach to the pterygomaxillary fossa and the parapharyngeal space in which a horizontal osteotomy of the mandibular ramus above the lingula was added to the usual anterior mandibular osteotomy to allow the ascending ramus to be retracted further laterally and superiorly for improved access.

Before performing this second osteotomy however, a *trial of retraction* is carried out. It is frequently possible to achieve the necessary exposure, without the need for the second osteotomy, by gentle retraction of the ramus in a lateral and superior direction. In adults the detachment of the lateral pterygoid insertion from the neck of the mandibular condyle may permit a few extra degrees of elevation of the ramus. This manoeuvre is usually not required in children.

The authors avoid the second osteotomy, if possible, in view of the excellent exposure achieved in most cases with the single anterior osteotomy bone cut and soft tissue dissection, as described above – the horizontal ramus bone cut providing only a further 5–10° elevation of the mandibular ramus. In addition, the ramus bone cut is sited where the area of bone contact is minimal. As limitation of mouth opening is usual following the extensive soft tissue dissection, early and vigorous jaw exer-

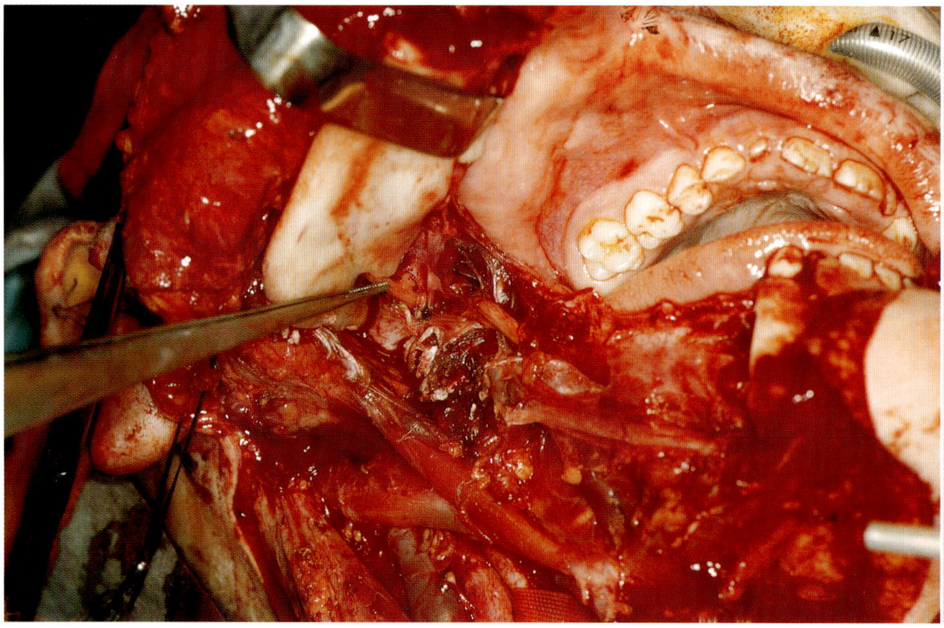

Figure 7.10 Superior and lateral retraction of the mandible providing wide exposure of the infratemporal fossa. Exposure of the superior compartment may be sufficient for the resection of benign pathology. Additional exposure is usually required for the resection of malignant tumours approaching the skull base. Forceps identify the inferior alveolar neurovascular bundle.

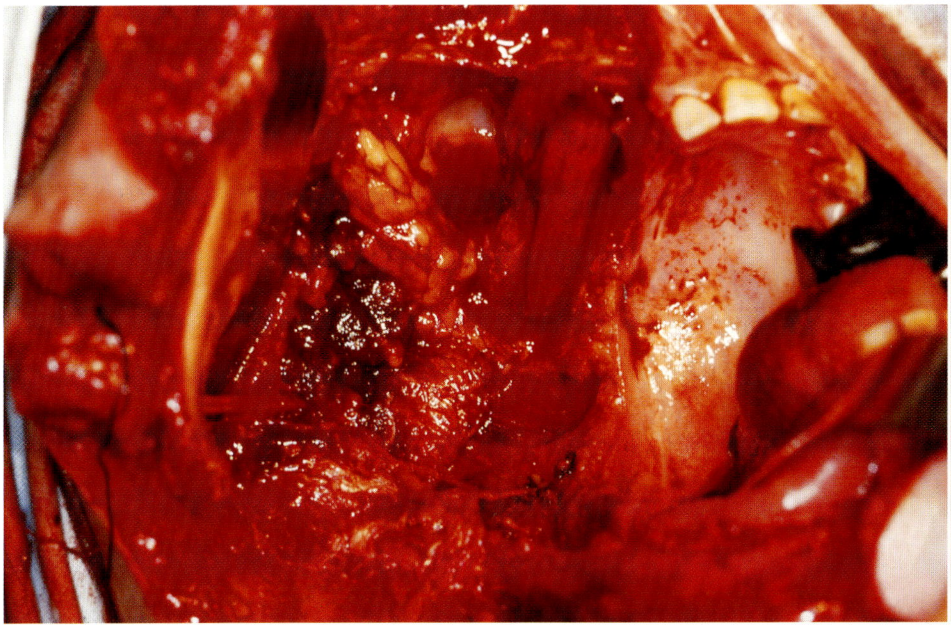

Figure 7.11 Inferior maxillectomy with clearance of the inferior compartment of the infratemporal fossa for a chondrosarcoma of the maxilla in the molar/tuberosity region – access via an extended mandibular swing approach.

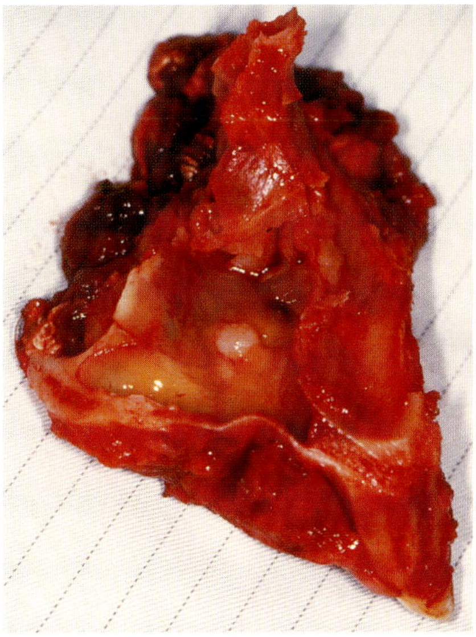

Figure 7.12 Superior view of excised specimen in Fig. 7.11 – the pterygoid plates sectioned at the level of the skull base.

cises are necessary to re-establish the original mouth opening. These exercises may be delayed when the second osteotomy is employed, to avoid displacement of the ramus bone fragments. The second osteotomy in the mandibular ramus is avoided if the extended mandibular swing is used for the resection of maxillary tumours with extension into the infratemporal fossa, as the osteotomy cut is likely to be in communication with the surgical defect (Figs 7.11, 7.12).

A comparison of the exposure achieved with and without the second (ramus) osteotomy is evident from examination of the clinical photographs in Figures 7.10 and 7.13. Although the case illustrated in Figure 7.10 is from a child aged 9 years, the authors are able to achieve essentially the same degree of exposure in adults and avoid the second osteotomy.

If required, the osteotomy in the vertical ramus is performed with a fine saw or drill above the lingula. The masseter is dissected off the lateral aspect of the mandible over a limited area. A bone plate is adapted to the lateral ramus, fixed with screws, removed, and the osteotomy completed.

The extended mandibular swing approach provides excellent three-dimensional exposure of the infratemporal fossa from the skull base to the hyoid bone with good control of the important neurovasculature, i.e. the internal and external carotid arteries, internal jugular vein, and the lingual, inferior alveolar, vagus, and hypoglossal nerves. The elevation of the ascending ramus with the attached masseter and temporalis muscles effectively removes the lateral boundary of the infratemporal fossa out of the operative field. This provides a short straight line between the surgeon and the pathology with a wide arc of exposure.

This approach may be used as the sole route of access or can be combined readily with either a transfacial or lateral transzygomatic approach. Used alone, it is particularly useful for extensive benign pathology, such as juvenile angiofibromas, and for malignant lesions extending posteriorly from the maxilla and involving the inferior compartment of the infratemporal fossa. For malignant tumours involving the superior compartment for which resection of the roof of the infratemporal fossa/floor of the middle fossa and an intracranial dissection is necessary, it is combined with the lateral–transzygomatic approach (Figs 7.6, 7.7, 7.9, 7.10, 7.14–7.18).

The case illustrated in Figs 7.6, 7.7, 7.9, 7.10, 7.14–7.18 demonstrates the resection of an epithelioid

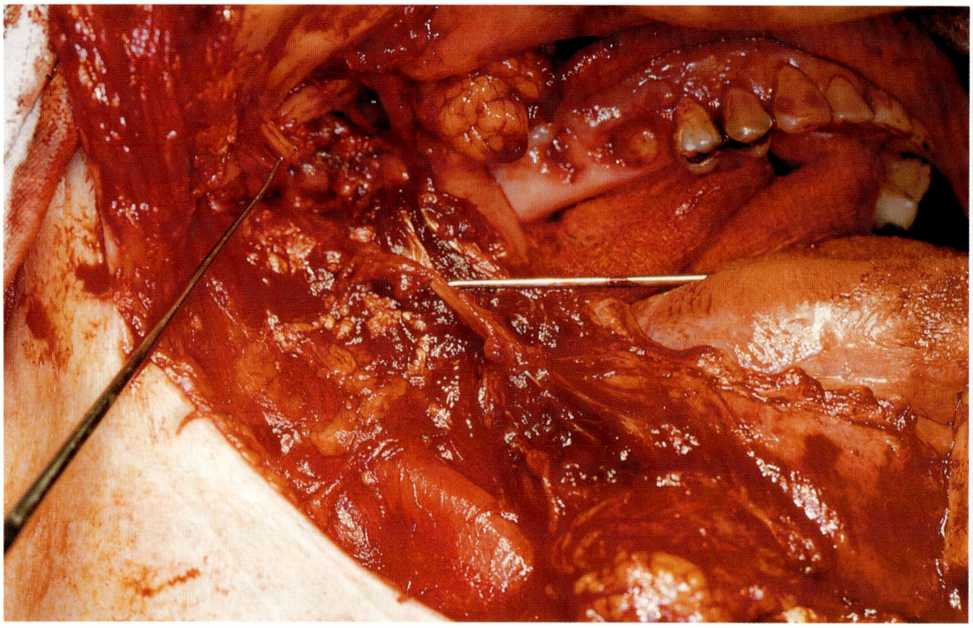

Figure 7.13 Exposure achieved with a second osteotomy of the mandibular ramus as advocated by Attia *et al.* in 1984.[6] Inferior alveolar nerve identified with hook. Lingual nerve identified with metal instrument anteriorly.

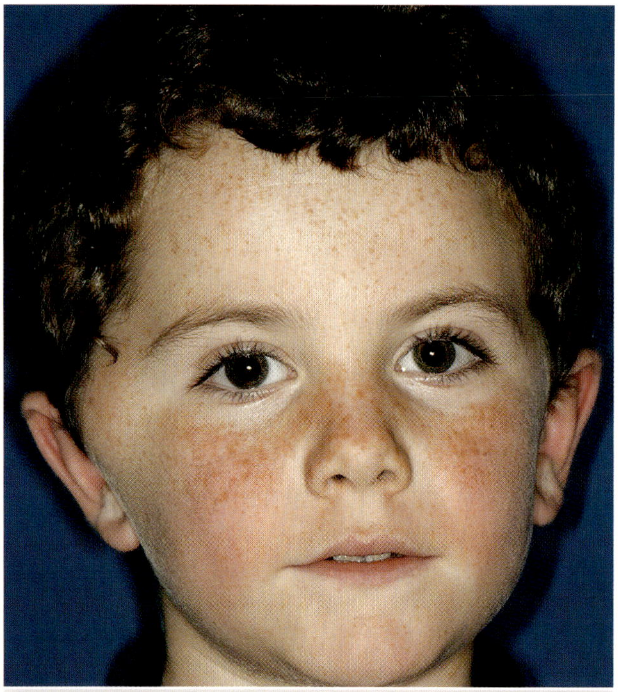

Figure 7.14 Preoperative view of boy of 9 years with an epithelioid sarcoma involving the temporal and infratemporal fossae.

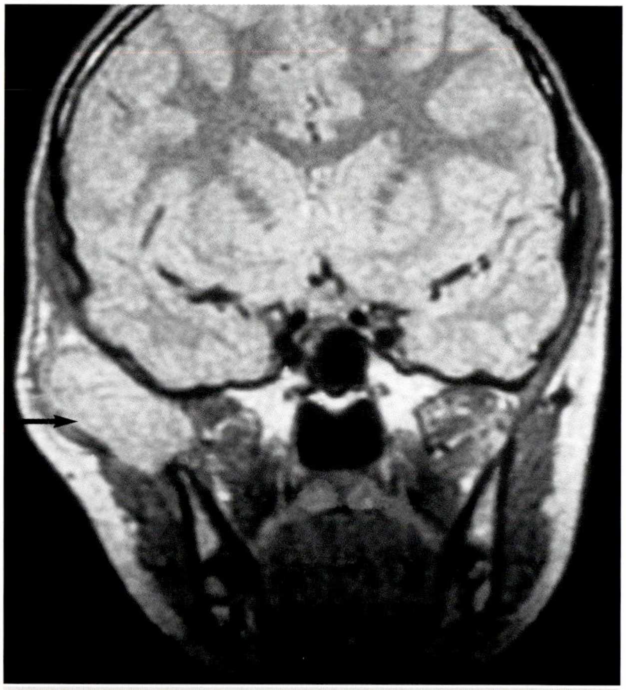

Figure 7.15 Coronal CT scan demonstrating tumour.

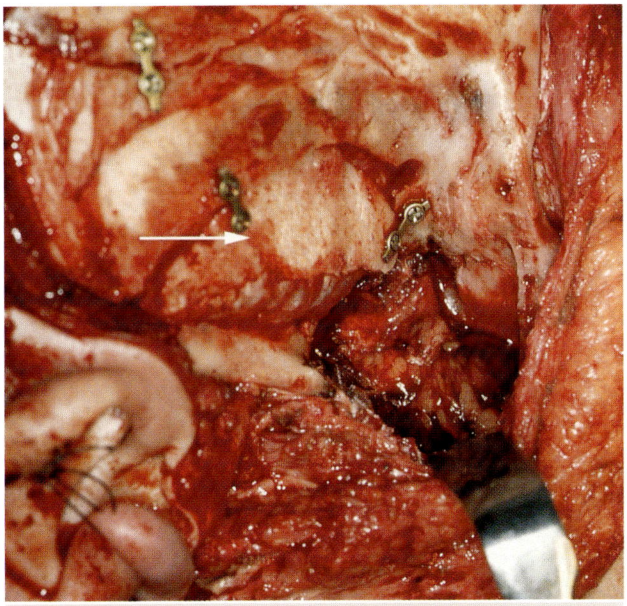

Figure 7.16 Exposure of the superior compartment of the infratemporal fossa. In this case the body of the zygoma and arch were resected for tumour clearance. Equal exposure can be achieved with an osteotomy – see lateral transzygomatic approaches below. Note the bone graft reconstructing the squamous temporal bone (arrow).

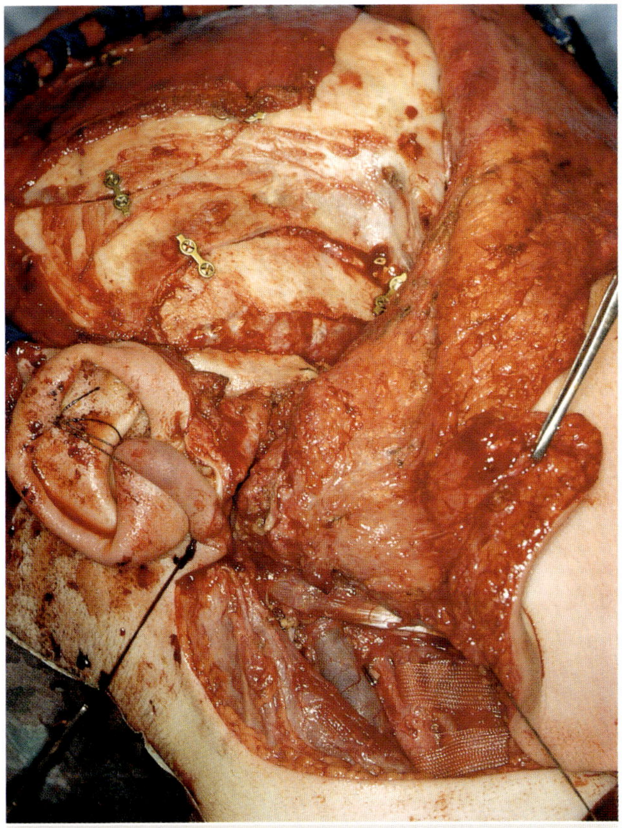

Figure 7.17 Lateral view at completion of the resection – the parotid gland and facial nerve are in the soft tissue bridge.

sarcoma involving the temporal and infratemporal fossae in a 9-year-old boy. An extended mandibular swing was combined with a lateral transzygomatic approach for exposure of the infratemporal fossa. In this case the zygoma was resected for tumour clearance rather than mobilised and replaced. An initial partial superficial parotidectomy was performed to isolate and protect the frontal branch of the facial nerve. The temporalis fascia and the body/arch of the zygoma, a portion of the squamous temporal bone and the lateral aspect of the middle fossa floor were resected with the tumour specimen. The squamous temporal bone was replaced with an inner table calvarial bone graft. The patient is alive and well 7 years after surgery with no evidence of recurrent disease. The lingual and inferior alveolar nerves were isolated and preserved (case operated on jointly with Miss D. Lang, Consultant Neurosurgeon, Southampton University Hospitals).

Figure 7.18 Postoperative appearance at 6 months.

The disadvantages of this approach are the need for a facial incision to avoid sectioning the inferior alveolar nerve and the need to enter the oral cavity. The latter is a disadvantage if the pathology itself does not directly involve the mouth. If the inferior alveolar nerve is to be sacrificed in the resection, then the lip split may be avoided and the mandible divided at the junction of the body and ascending ramus. This divides the inferior alveolar nerve in its canal, which both avoids the lip incision and the need to enter the mouth. Access to the infratemporal fossa is similar to that achieved with the extended mandibular swing.[1–4,6–13]

TRANSFACIAL APPROACHES FOR ACCESS TO TUMOURS OF THE INFRATEMPORAL FOSSA

The basic concept of the transfacial approaches to deep-seated pathology is the mobilisation of pedicled bone flaps in which the midfacial bone segment remains pedicled to the soft tissues of the cheek thereby retaining their blood supply. Although this concept is not new, the development and refinement of these techniques took place in Europe, particularly in Italy and Spain, during the 1980s.[14–17]

As with other approaches, the planning of the facial dismantling procedure is made in concert with the radiologist, pathologist and, when necessary, the neurosurgeon. If more than one approach is possible, the chosen technique should have the least morbidity – the objective being the preservation of function with minimal scarring and deformity, bearing in mind that the main goal is adequate tumour resection.

In tumour surgery three points are considered:

- the best surgical route to the tumour
- the extent of the resection
- the type of reconstruction and/or reassembling.

Based upon the authors' clinical experience with transfacial approaches it has been found helpful to divide these into four levels. This classification system was proposed by Clauser[18] with the aim of simplifying surgical planning. The approach adopted is determined by the anatomical site of the pathology. The appropriate pedicled bone cheek flaps (PBCF) may therefore be selected to provide the required exposure.

Level 1: Lesions of the ethmoid, sphenoid, upper nasopharynx and anterior cranial base. *Suggested PBCF*: Nasal cheek flap + Le Fort I – optional nasal maxillo cheek flap (NMCF)

Level 2: Lesions involving the retropharynx and clivus.
Suggested PBCF: NMCF ± contralateral maxillo cheek flap (MCF)

Level 3: Lesions of the retromaxilla or pterygomaxillary space.
Suggested PBCF: MCF.

Level 4: Lesions of the parapharyngeal space, infratemporal fossa.
Suggested PBCF: MCF or mandibular cheek flap or combination of the two.

Lesions of the infratemporal fossa are therefore usually approached using the MCF or mandibular cheek flap or a combination of the two approaches. In some situations the NMCF may be considered.

With these approaches the patient usually has an oral endotracheal tube. Oral intubation rarely gives rise to restricted surgical access, while nasal intubation would be inappropriate with the NMCF. Tracheostomy is usually unnecessary. However, if extensive postoperative oedema is expected or if the tube hinders surgical access, a tracheostomy is performed electively.

As a rule most of the dismantling procedures heal very well with minimal morbidity. In the classic MCF, the maxilla is severed from its bony connections, remaining attached only by the soft tissues of the cheek, and swung laterally to provide access to deeper areas. The volume of the oral cavity is increased temporarily, permitting improved instrumentation. This also enables the use of the microscope, thereby aiding three-dimensional tumour clearance. The advantages of improved access are felt to far outweigh the surprisingly small increased surgical morbidity.

Notes on surgical anatomy

Starting from experience in orthognathic surgery, in which the osteotomised bone fragments preserve their blood supply through the covering soft tissue pedicle, we applied the concept of performing three-dimensional pedicled osteotomies in midfacial dismantling procedures. In these cases the pedicle is changed in that the blood supply is vascular through the cheek components.

The MCF can be defined as a composite flap of soft tissues, maxillary skeleton and blood supply. It can be considered an osteomyocutaneous flap.

The skeleton and the cartilaginous structure of the nose are included when performing an NMCF. In the classic MCF the pedicle is made up of the cheek's soft tissues: skin, subcutaneous tissue, muscles associated with the buccal fat pad, vestibular mucosa and periosteum. The vascular supply is derived from the external carotid artery system (facial artery, superior labial artery, maxillary artery and its branches) and from the jugular venous system.

In order to perform a pedicled bone cheek flap, an understanding of the basic extraoral and intraoral incisions and osteotomies is necessary. The classic techniques proposed by Curioni[15,16] for the MCF and NMCF are therefore described.

Extraoral incisions

Maxillo cheek flap

The basic extraoral incision follows a Weber–Fergusson type line (Fig. 7.19). A subciliary incision is made through the palpebral skin. The underlying orbicularis muscle is detached from the skin until the inferior orbital rim is reached. The periosteum is incised 2–3 mm below the inferior orbital rim and reflected to expose the anterior maxillary wall where an upper horizontal osteotomy will be performed. The paranasal incision is made down to the periosteum and the periosteum reflected over a limited area to create the necessary space for performing the paranasal osteotomy. At the lateral limit of the subciliary incision a subperiosteal tunnel is created below the zygoma back to the junction of the posterior maxilla/pterygoid plates.

Naso maxillo cheek flap

For this flap the previously described paranasal incision is made on the contralateral side and continues upwards on the nasal skin medial to the medial canthus, to the level of the root of the nose. It is then taken transversely across the nasal root, again inferior to the medial canthus, to the opposite side extending laterally into the subciliary incision and is continued as described above (Fig. 7.20). In the NMCF the cutaneous flap therefore remains attached to nasal skeleton thereby allowing the dismantling of the nasomaxillary unit as a pedicled flap. (Where a combination MCF and NMCF is required the incisions described are bilateral. The combination of these two flaps provides wide exposure of the clivus.)

Figure 7.19 Facial incision for the MCF.

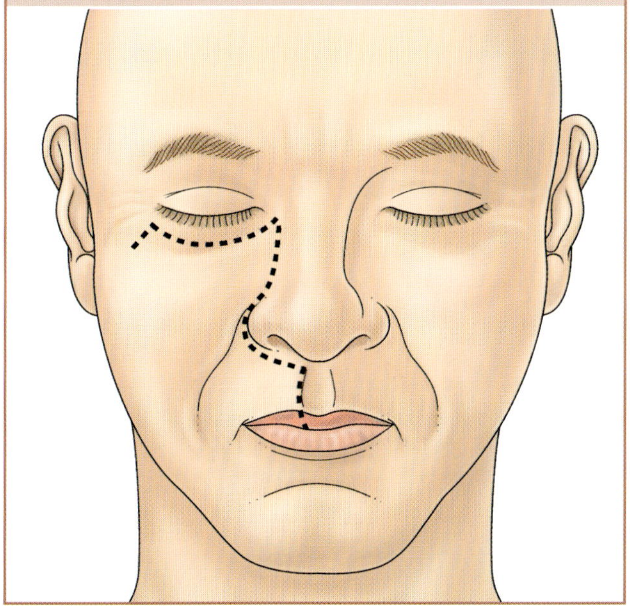

Figure 7.20 Facial incision for the NMCF.

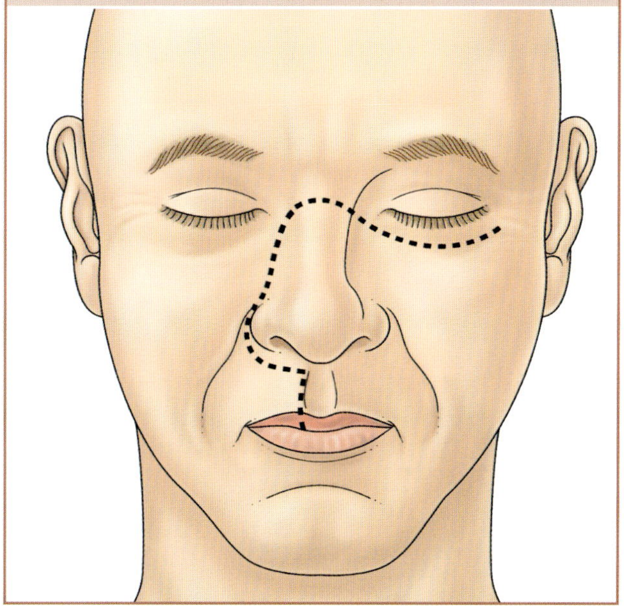

Intraoral incisions

It is important that all the mucosal incisions and the underlying osteotomy bone cuts are stepped to ensure that all soft tissue margins are replaced over sound bone.

Maxillo cheek flap

On the labial aspect of the alveolus the site of the incision is determined by the site of the underlying osteotomy bone cut. It may be either a few millimetres lateral to the midline for a midline osteotomy between the upper central incisors (to ensure the soft tissue incision and the bone cut are not coincident), or in the region of the lateral incisor/canine. The palatal incision may be either median or paramedian in the sagittal direction, and is turned transversely at the level of the maxillary tuberosity, at the junction of the hard and soft palates (Fig. 7.21).

Naso maxillo cheek flap

The incision through the labial mucosa/attached gingiva is in the lateral incisor canine region – again stepped in relation to the underlying bone cut. The transverse palatal incision performed at the margin of the hard and soft palate is modified from the MCF in that it starts from the *contralateral* paramedian sagittal incision (Fig. 7.22).

Osteotomies

As already mentioned, the osteotomies create bone blocks that remain pedicled to the cheek. The bone blocks mobilised can include: a hemimaxilla; a hemimaxilla with the nose; a hemimaxilla with the zygoma; a hemimaxilla with the orbital floor. The extent of the bone block that is disarticulated is dependent on the requirements of the individual case. Exposure may be increased peroperatively if necessary by mobilising further adjacent bone segments, i.e. these approaches fulfil the requirement of being *extensile*. In practical terms with appropriate preoperative planning, further extension of the exposure is required infrequently.

Maxillo cheek flap

Sagittal osteotomy (Fig. 7.23)

There are two possible sites for the osteotomy – both involve a full-thickness osteotomy of the alveolar process.

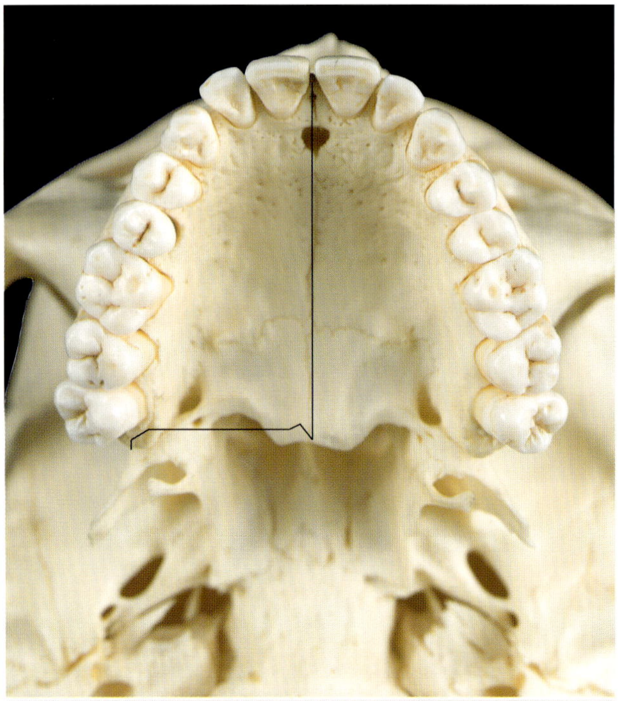

Figure 7.21 Intraoral incisions for the MCF.

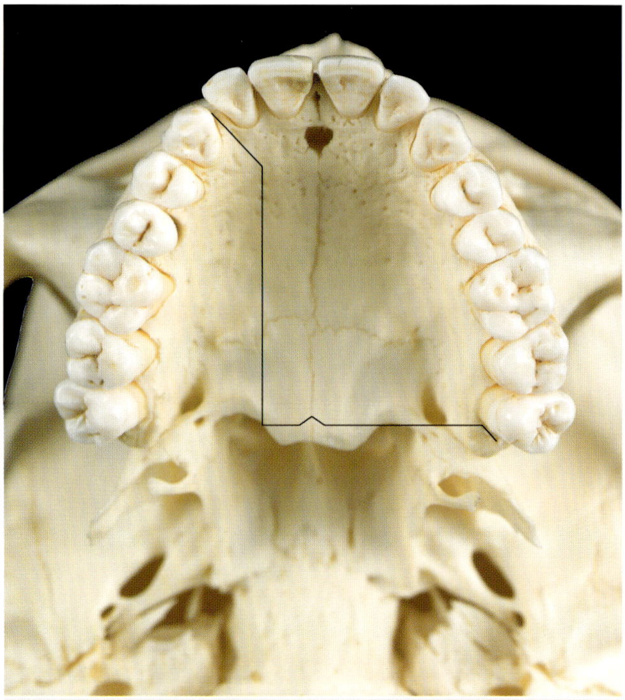

Figure 7.22 Intraoral incisions for the NMCF.

1. Median type: The osteotomy is performed through the alveolar process between the upper central incisor teeth (median suture of the palate), the bone cut extending up to the nasal floor. Prior to making the bone cut, the nasal septum is detached from the nasal floor and the mucosa of the floor of the nose elevated – the same technique used in the Le Fort I osteotomy.
2. Paramedian type: The osteotomy is between the lateral incisor and the canine teeth. The bone cut does not extend into the nasal floor but is taken lateral to the piriform fossa, thus becoming both paramedian and paranasal.

The osteotomy (either median or paramedian) is continued posteriorly from the alveolar process to the posterior ridge of the palatine bone.

Vertical and transverse osteotomies (Fig. 7.24)

In the median type, the osteotomy extends laterally from the piriform fossa, continues laterally beneath the zygoma and is angled downwards to reach the junction of the posterior maxilla and the pterygoid plates.

In the paramedian type, the osteotomy extends superiorly through the anterior wall of the maxilla, lateral to the piriform fossa, until it reaches the inferior orbital rim where it becomes transverse. It then follows a similar path to that of the median osteotomy bone cut, continuing laterally beneath the zygoma to the posterior maxilla/pterygoid plates. Where necessary the osteotomy can be taken into the orbit to include part of the orbital floor. The orbital floor is included if later anastomosis of the infraorbital nerve is planned – see below. (The paramedian approach is used when the MCF is used with the NMCF. See Fig. 7.27.)

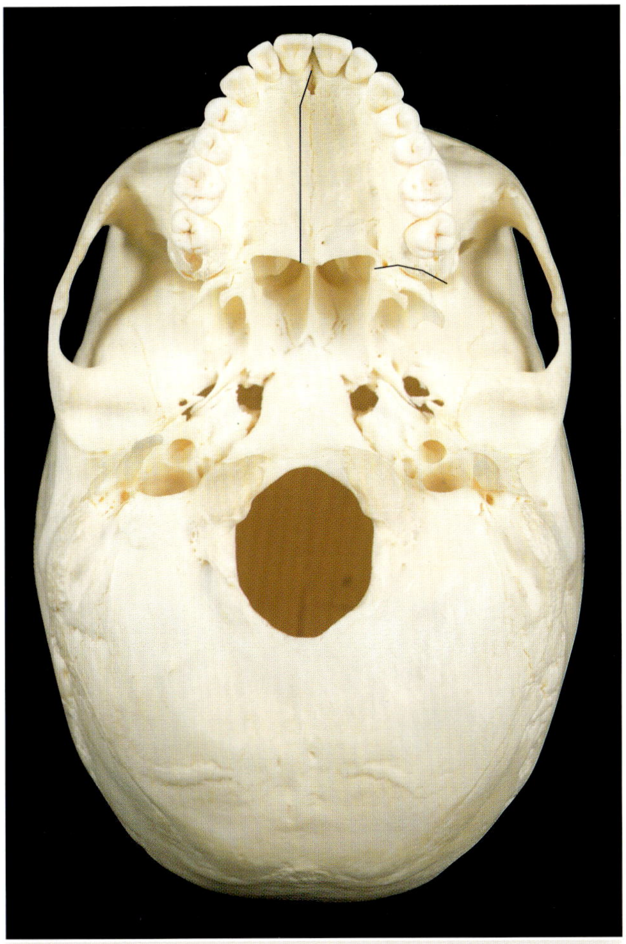

Figure 7.23 Sagittal osteotomy MCF.

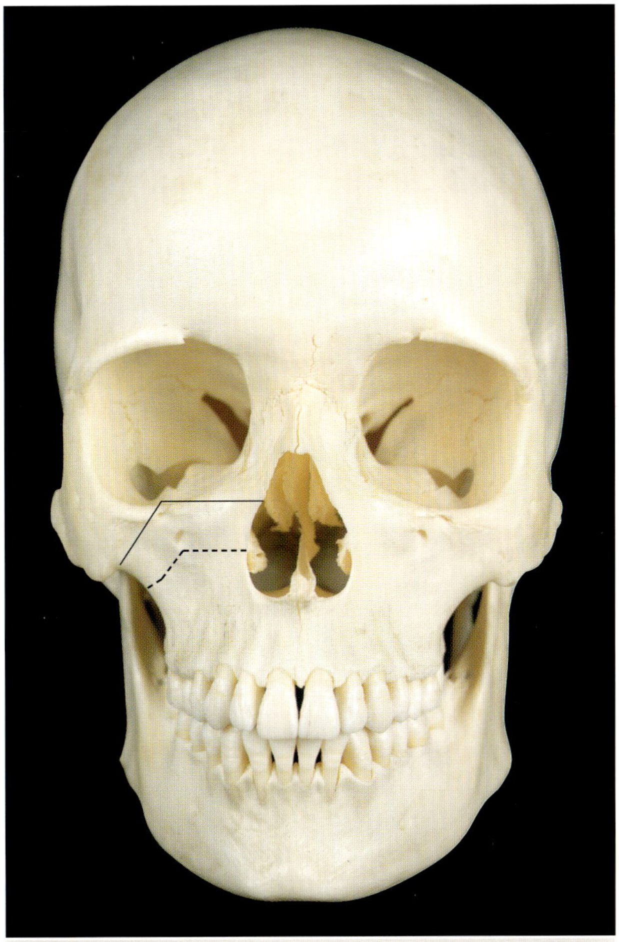

Figure 7.24 Vertical and transverse osteotomies MCF.

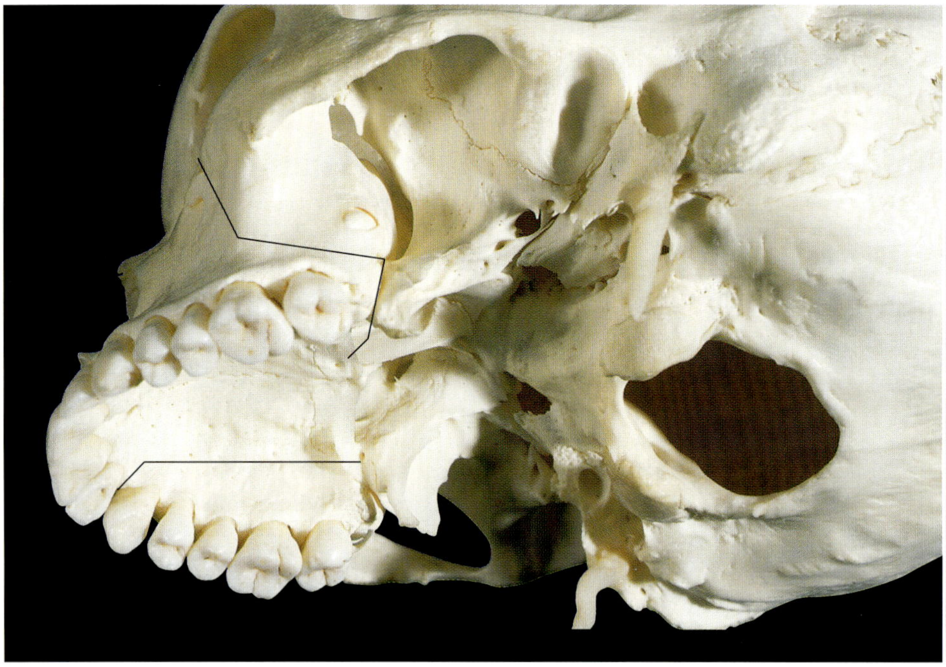

Figure 7.25 Maxillary tuberosity and pterygoid plates separated with curved chisel.

Maxillary tuberosity/pterygoid osteotomy (Fig. 7.25)

This is performed in the classic manner with a curved chisel separating the maxilla from the pterygoid plates.

Naso maxillo cheek flap (Figs 7.26, 7.27)

The paranasal osteotomy (paramedian type) described above, is taken superiorly into the frontal process of the maxilla until it reaches the frontomaxillary and frontonasal sutures, where it continues transversely across the root of the nose. The osteotomy then extends inferiorly through the frontal process of the maxilla on the opposite side, beneath the medial canthus, to the inferior orbital rim. The transverse osteotomy is then continued as described in the MCF.

Completion of the nasal bone cuts in the NMCF requires gentle tapping with an osteotome angled downwards to section the perpendicular plate of the ethmoid (as in the Le Fort II osteotomy).

The nasal bones and the hemimaxilla are then pedicled as a single block and moved laterally.

Additional mobilisation of the zygoma

For additional lateral access to the infratemporal fossa, the zygoma may be disarticulated and retracted laterally. The authors prefer to dismantle the zygoma separately from the MCF rather than *en-bloc* with the maxilla. The zygoma is usually pedicled to the masseter muscle. The lateral

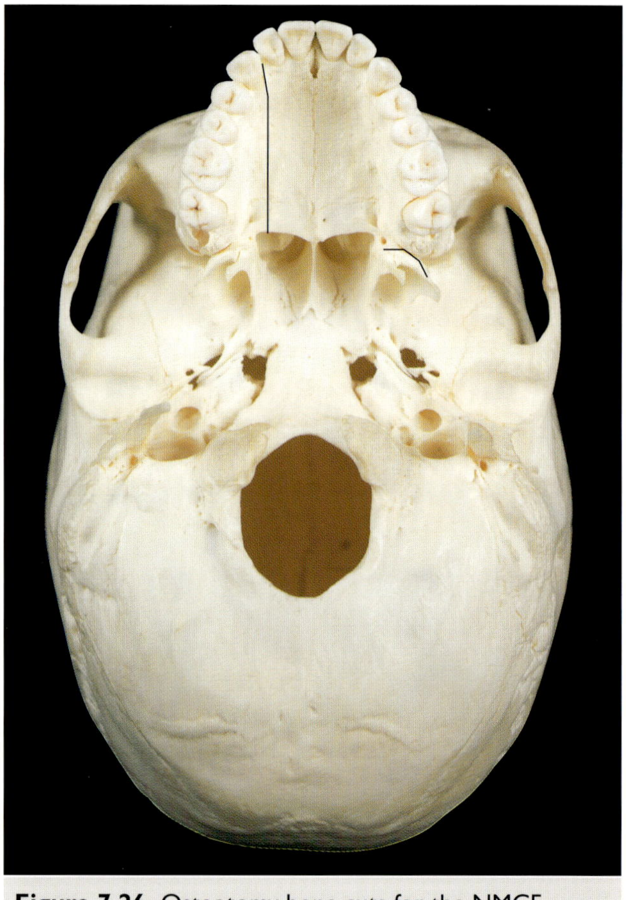

Figure 7.26 Osteotomy bone cuts for the NMCF.

orbital rim is sectioned below the frontozygomatic suture – this bone cut is made through the lower eyelid incision. The zygomatic arch is sectioned anterior to the articular eminence of the temporomandibular joint – the posterior osteotomy is made via a preauricular incision as used in temporomandibular joint surgery.

Taking the zygoma and the attached masseter laterally with the maxilla brings the temporalis tendon and the coronoid process into the operative field. A coronoidectomy is then performed and the temporalis tendon retracted superiorly. The two manoeuvres of lateral retraction of the zygoma/masseter and the superior retraction of the temporalis effectively take the lateral boundary of the infratemporal fossa out of the operative field, providing both anterior and lateral exposure of this area. There is little value in coronoidotomy, i.e. temporary mobilisation of the coronoid process with later replacement. Functionally, this is of no value and merely increases the likelihood of postoperative trismus.

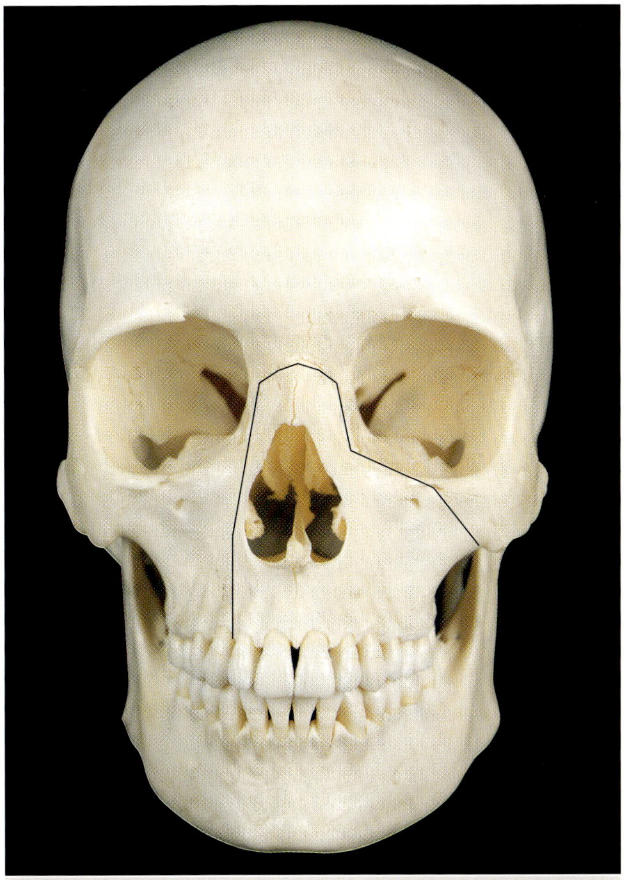

Figure 7.27 Osteotomy bone cuts for the NMCF.

Illustrative cases

Case 1

Embryonal rhabdomyosarcoma in the infratemporal fossa in a 5-year-old patient (Figs 7.28–7.30). The tumour was exposed and resected via an MCF. The pterygoid plates were included in the resection, being sectioned at the base of the skull.

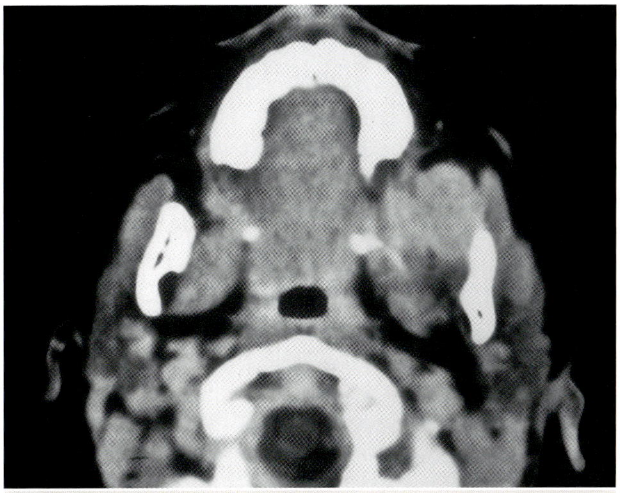

Figure 7.28 Case I – CT scan of embryonal rhabdomyosarcoma.

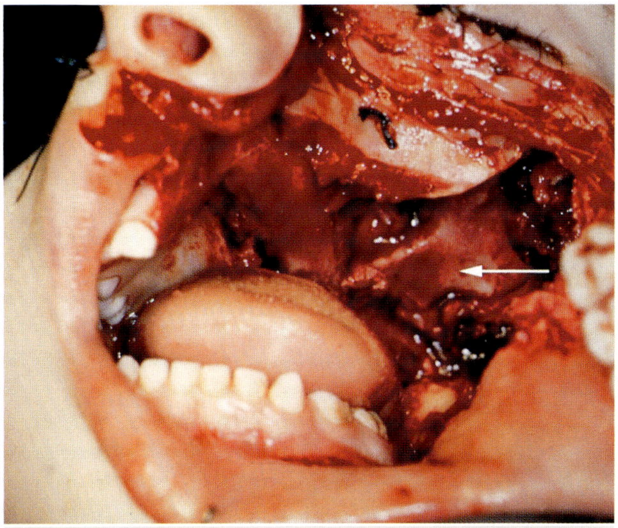

Figure 7.29 Case I – intraoperative view of the infratemporal fossa/retromaxillary space obtained after reflection of the hemimaxilla pedicled on the soft tissues. The pterygoid plates were included in the resection. The arrow points to the roof the infratemporal fossa/floor of the middle cranial fossa.

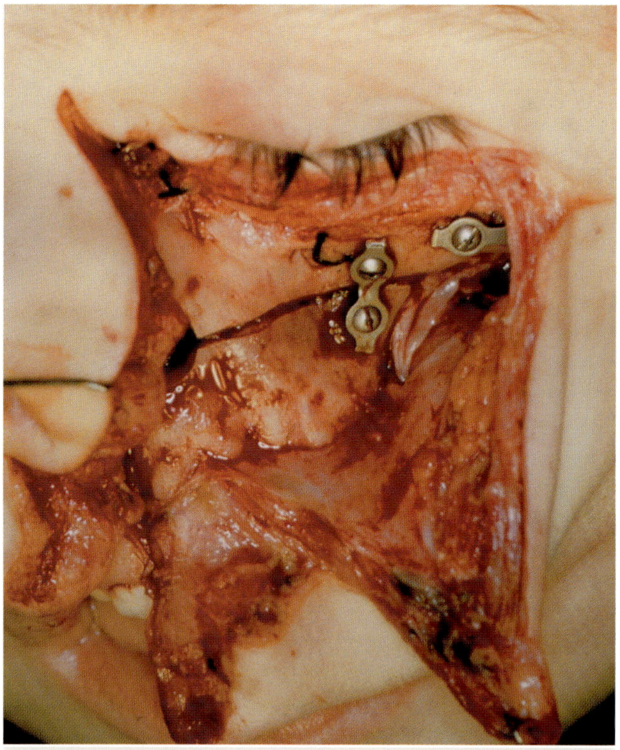

Figure 7.30 Case 1 – Fixation of the MCF, providing excellent contour and stability.

Case 2

Pleomorphic adenoma filling the infratemporal fossa and the parapharyngeal space (Figs 7.31–7.34). The tumour in this case was exposed and resected via an MCF combined with a mandibular cheek flap (extended mandibular swing). As an oral endotracheal tube was used, a maxillary splint and interocclusal acrylic wafer were used to ensure the re-establishment of the correct occlusion at the completion of the procedure.

Case 3

Angiofibroma of the infratemporal fossa extending into the middle cranial fossa (Figs 7.35–7.41). The lesion was exposed and resected via an MCF in combination with temporal craniotomy for access to the middle cranial fossa.

Discussion

The transfacial approaches of most value in providing access to the infratemporal fossa/retromaxilla/pterygomaxillary space are the MCF combined where necessary, with the mandibular cheek flap, i.e. the extended mandibular swing. The zygoma may also be dismantled for additional lateral exposure if needed.

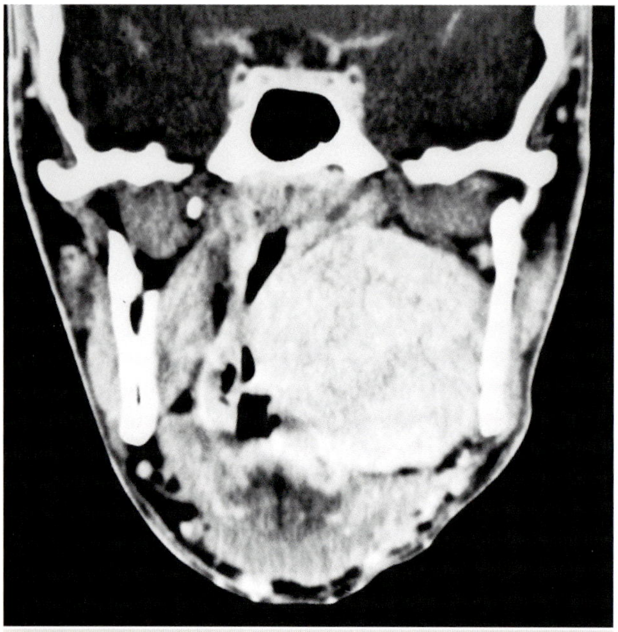

Figure 7.31 Case 2 – Coronal CT scan of the lesion with contrast.

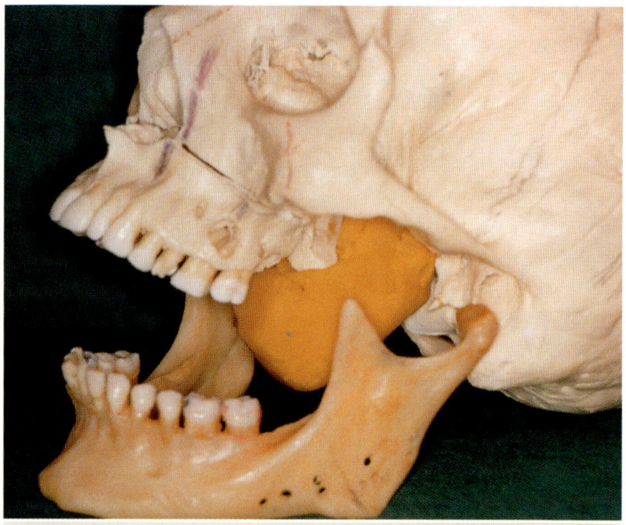

Figure 7.32 Case 2 – Localisation of the tumour on the dried skull. Temporary mobilisation of the maxilla and the mandible provides wide exposure. The tumour may be removed without traction being applied directly to the tumour itself, very significantly reducing the likelihood of tumour disruption and spillage.

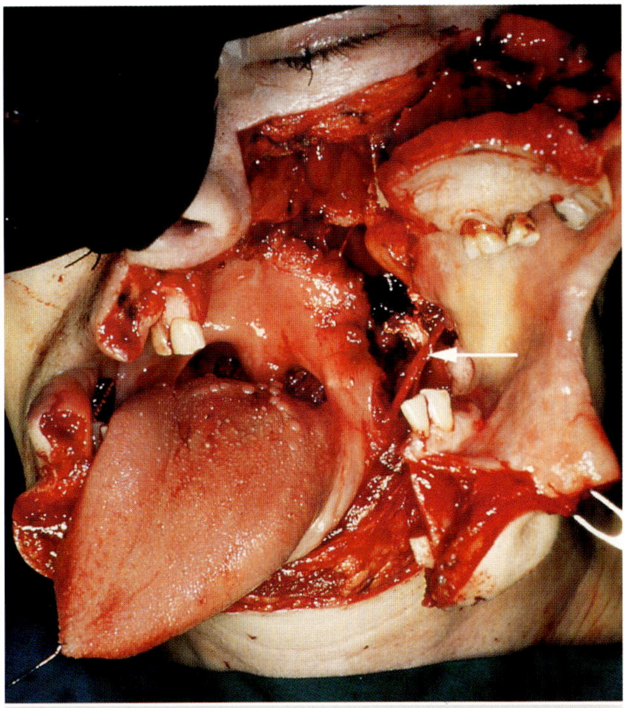

Figure 7.33 Case 2 – Dismantling of the left hemimaxilla combined with the lateral reflection of the mandible providing excellent exposure of the infratemporal fossa and parapharyngeal space for tumour resection. Lingual nerve (arrow).

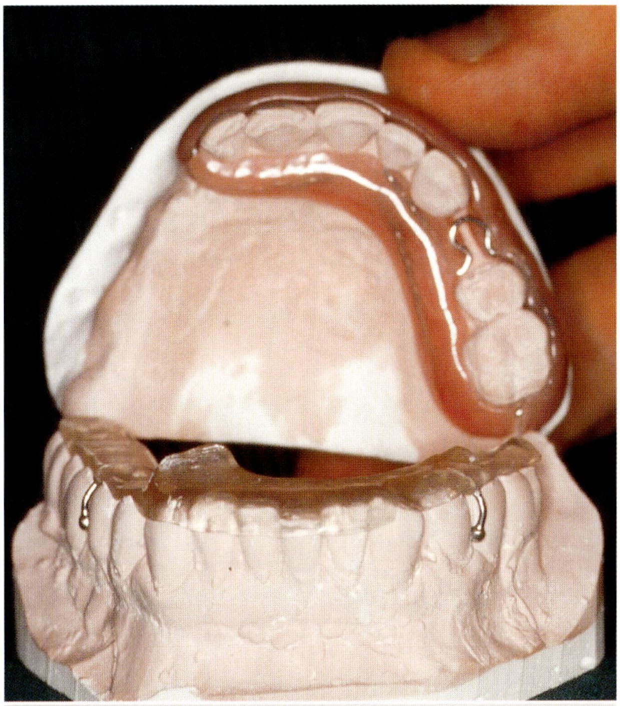

Figure 7.34 Maxillary splint and acrylic interocclusal wafer used to re-establish correct occlusion.

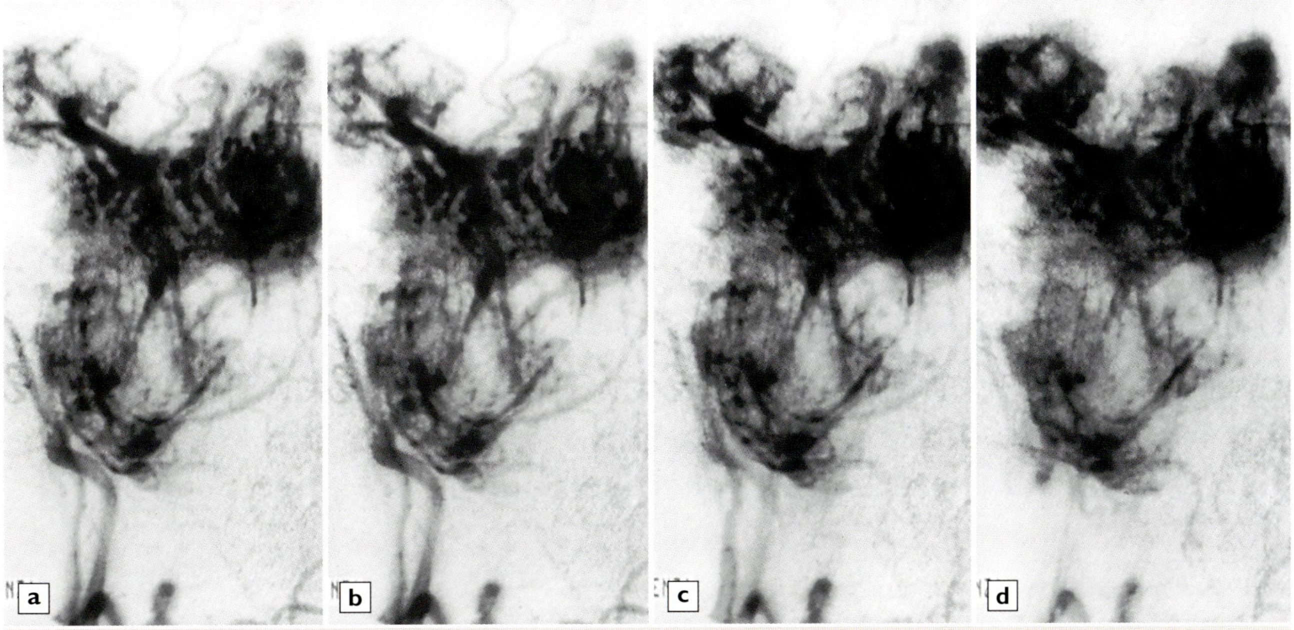

Figure 7.35 Case 3 – Pre-embolisation angiogram.

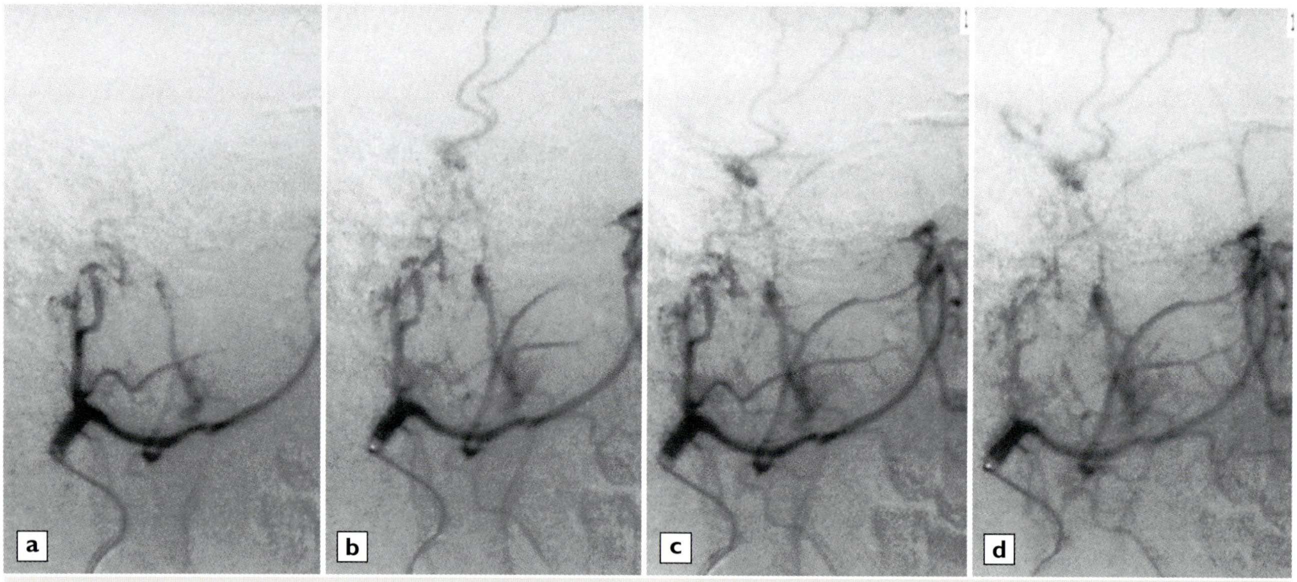

Figure 7.36 Case 3 – Postembolisation angiogram.

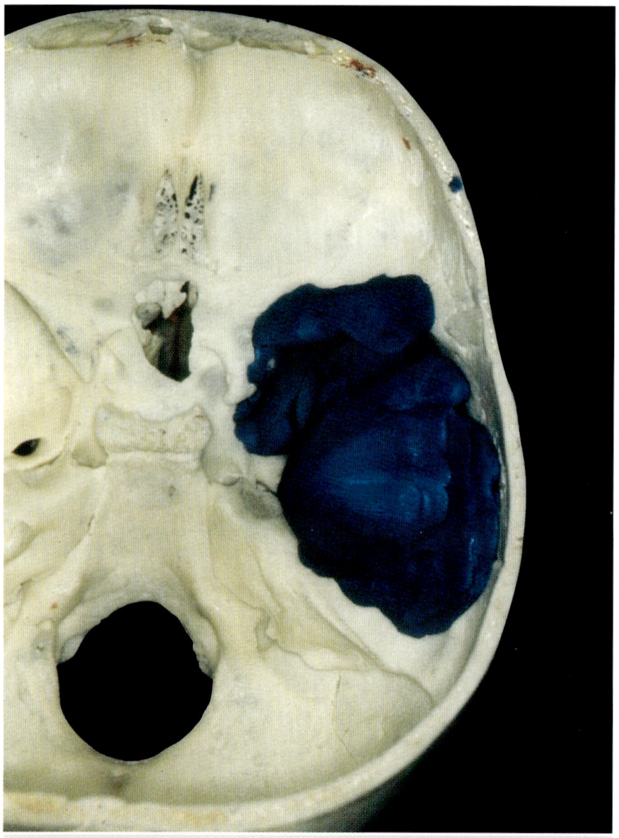

Figure 7.37 Case 3 – Localisation of the lesion in the middle cranial fossa on a dried skull.

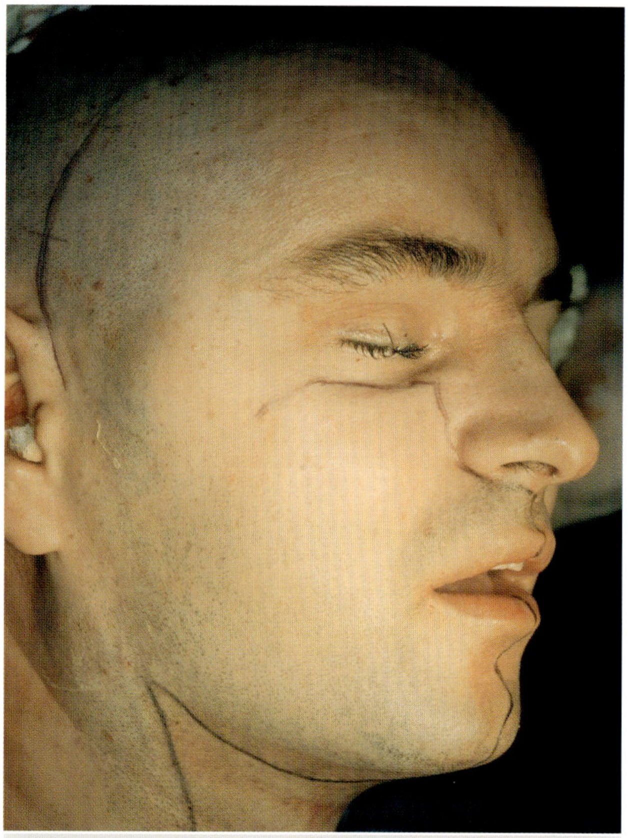

Figure 7.38 Case 3 – Planning of the incision lines. The vertical cervical incision is for the exposure of the carotid system. There was no need to divide the lip and extend the incisioin into the submandibular region as suggested by the incision lines marked.

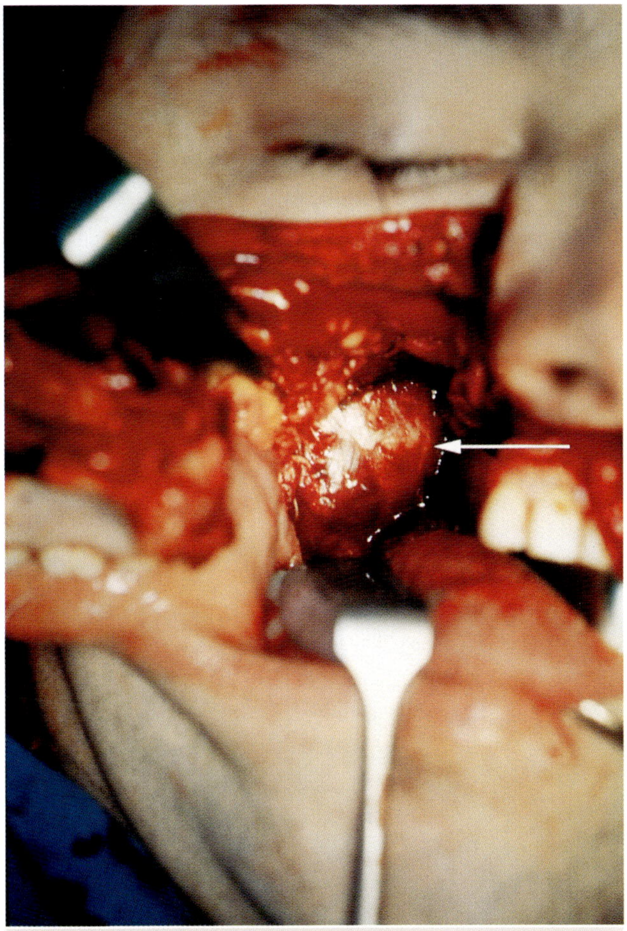

Figure 7.39 Case 3 – The MCF has been swung laterally exposing the infratemporal element of the lesion (arrow).

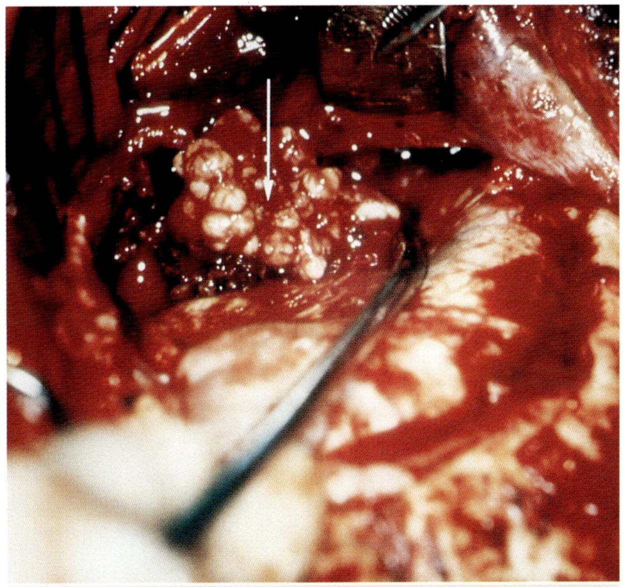

Figure 7.40 Case 3 – Intracranial dissection of the middle cranial fossa. The temporal lobe is retracted by a malleable spatula and the angiofibroma visualised. Arrow points to angiofibroma.

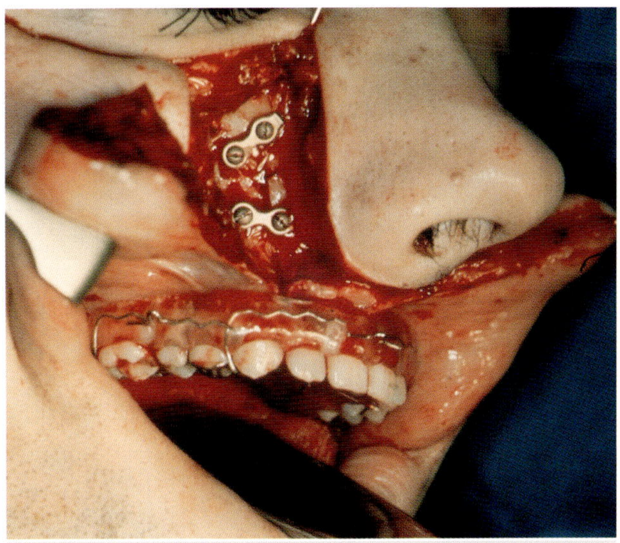

Figure 7.41 Case 3 – The maxilla reassembled on a prepared palatal splint.

By taking the maxilla laterally out of the operative field, the MCF removes the anterior boundary of the infratemporal fossa. The inclusion of the zygoma and/or the mandible from either an anterior or lateral approach then removes its lateral boundary. The addition of a (fronto)temporal craniotomy allows the roof to be removed under direct vision. The potential for extensile exposure with these various approaches is obvious.

The transfacial approaches are of value in the treatment of malignant disease and extensive benign disease in the infratemporal fossa and adjacent areas. Exposure superiorly extends to the skull base and allows resection of the pterygoid plates and the attached muscles at this level. The infraorbital nerve can be traced to the foramen rotundum and biopsied, if necessary, at this site. The medial resection margin, which is blind in the (lateral) transzygomatic approach, is under direct vision. A further advantage is that the buccal pad of fat is in the operative field and readily available for use in local reconstruction if appropriate. If the pathology involves the posterior maxilla, such as with posteriorly penetrating malignant tumours, the PBCF can be modified to incorporate only the anterior maxilla. This provides the required exposure while avoiding the resection of tissues uninvolved by tumour.

If it is necessary to detach the medial canthal ligament, the strong anterior limb of the ligament may be reattached with a suture, e.g. 3/0 braided Dexon, to a microplate at the anterior lacrimal crest. If sectioning and lateral retraction of the nasolacrimal sac is required, a silicone drainage tube is inserted on closure and left *in situ* for 4–6 months to re-establish drainage.

Sectioning of the infraorbital nerve is necessary in the MCF and the NMCF as the intact nerve hinders retraction of the flap. If the infraorbital nerve does not need to be sacrificed in the resection the osteotomy can include the inferior orbital rim; this allows the nerve to be sectioned within the orbit and the ends of the nerve to be tagged with a suture for later repair with epineural sutures (8/0 nylon). Notwithstanding the possibilities of nerve repair, the morbidity associated with the loss of sensation in the infraorbital nerve dermatome must be weighed against the potential morbidity of inadequate access.

The transfacial approaches do not expose the carotid arteries or the internal jugular vein. If the nature or the extent of the pathology requires this, then these vessels are exposed in the neck in the usual fashion, traced superiorly as necessary and isolated (Case 3).

The disarticulation of the midfacial bone segments in the transfacial techniques fulfils the requirements of providing both a short straight line between the surgeon and the pathology in the infratemporal fossa and a wide arc of exposure in three dimensions. Exposure can be extended if necessary by further osteotomies of the adjacent midfacial skeleton. In most instances with careful surgical planning additional exposure is not usually required. The transfacial approaches are readily combined with other approaches.[14–17,19–22]

INTRAORAL APPROACHES

These approaches to the infratemporal fossa all provide limited access that cannot be easily extended peroperatively. They have all been advocated as an approach to the infratemporal fossa only in *carefully selected cases in which the lesions are small, benign and well circumscribed.* Their role in all but a small number of cases is limited. *They have no role in the treatment of malignant pathology.*

Le Fort I approach

The exposure of the nasal cavity, maxillary sinuses and the nasopharynx provided by down-fracturing the maxilla at the Le Fort I level is familiar to all orthognathic surgeons. This approach has been used in the treatment of both benign and malignant lesions in these sites.

The removal of the posterolateral wall of the maxilla under direct vision when the maxilla has been downfractured provides limited exposure to the infratemporal fossa by removing its anterior boundary.

Surgical technique

The technique is the same as that used in orthognathic surgery. A labial mucoperiosteal incision is made from the first molar region on each side, 5–10 mm above the mucogingival junction, thereby preserving sufficient mucosa inferiorly for easy closure. Subperiosteal dissection of the maxilla back to the pterygoid plates is carried out – if possible avoiding perforating the periosteum at the back of the maxilla. The mucoperiosteum is elevated off the nasal floor and for a limited extent off the septum and lateral wall of the nose to allow its protection during the osteotomy. A horizontal osteotomy cut (using either a fine reciprocating saw or a drill) is made, extending from the lateral aspect of the piriform fossa (just above the dental roots) back to the pterygoid plates. Prior to mobilising the maxilla, its position is prelocalised with low profile plates and screws at the piriform and the zygomatic buttress, which are then removed (Fig. 7.42). The maxilla is freed at the pterygoid plates with a curved chisel and then downfractured pedicled to the palatal soft tissues. The posterolateral wall of the maxilla (behind the zygomatic buttress) both superiorly and inferiorly is then removed with rongeurs. If vertical exposure is restricted posteriorly, this may be improved to a limited degree by sectioning the greater palatine neurovascular bundle.

The lesion is then resected and the maxilla replaced in its original position by replacing the bone plates and screws. Temporary intermaxillary fixation by means of an eyelet wire in each dental quadrant helps to ensure the correct occlusion. The wires are removed after the maxilla has been plated in its original position.

The advantages of this approach are that it avoids a facial incision and that only the posterolateral aspect of the maxilla requires removal to improve access. The access provided, however, is limited and cannot be extended peroperatively. It is therefore only suitable for

Figure 7.42 Osteotomy at the Le Fort I level. Maxilla prelocalised with bone plates at the piriform fossa and the zygomatic buttress.

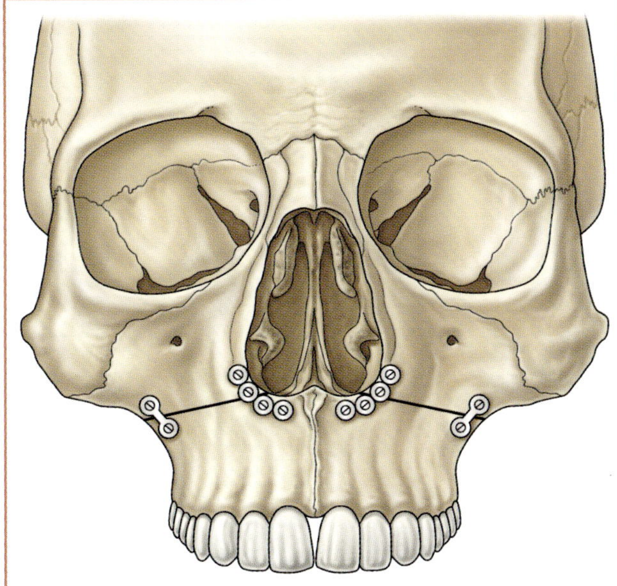

providing access to the infratemporal fossa in a small number of cases and has no role in providing access for the resection of malignant tumours in this region.

Transantral approach

This approach has been advocated for the removal of circumscribed benign masses such as angiofibromas of the nasopharynx extending into the infratemporal fossa.

Surgical technique

The maxilla is exposed subperiosteally back to the pterygoid plates via a Weber–Fergusson incision sectioning the infraorbital nerve. The anterior, lateral, medial and posterior walls of the maxilla are then removed with either a drill or rongeurs, leaving the alveolus and the orbital floor intact. The lateral nasal wall with the inferior and middle turbinates is also resected. With the access provided, the lesion is then removed from the infratemporal fossa and adjacent nasopharynx.

It is difficult to see a continued role for this approach when other approaches, particularly the transfacial approaches, provide significantly better access without the requirement for extensive resection of the maxilla. If very limited access is all that is required without the need for a facial incision and sectioning of the infraorbital nerve, the Le Fort I approach (recognising its limitations) is the preferred option.

Transpalatine approach

Limited exposure of the inferomedial aspect of the infratemporal fossa can be achieved with this approach.

Surgical technique

A 'lazy S' incision dividing the soft palate is made commencing from the junction of the hard and soft palate in the midline extending laterally to the palatoglossal arch. The pharyngeal mucosa and the superior pharyngeal constrictor muscle are divided, preserving the Eustachian tube, providing limited exposure of the infratemporal fossa.[26]

Concerns with this approach are palatal shortening/dehiscence with possible velopharyngeal incompetence. Surgeons experienced with the Le Fort I approach would use it in preference to this technique. Conversely, those less familiar with facial osteotomies may prefer this approach. Both provide limited access to the infratemporal fossa.[13,23–25]

LATERAL APPROACHES

Lateral transmandibular approaches

The lateral transmandibular approaches to the infratemporal fossa, by definition, do not involve the lip-split incision described above for the extended mandibular swing approach. They may therefore avoid entering the oral cavity.

The exposure of the infratemporal fossa obtained from a lateral approach via the mandible is variable. An osteotomy of the mandibular ramus with separation/distraction of the bone ends (inverted-L) can provide *limited exposure*, suitable for the removal of benign pathology. More extensive exposure similar to that achieved with the extended mandibular swing may be achieved by elevation of the ascending ramus and masseter muscle out of the operative field with a simple osteotomy at the junction of the body and ascending ramus, deliberately sectioning the inferior alveolar nerve. The latter technique has obvious advantages if the inferior alveolar nerve is to be sacrificed in the resection in any event.

Wide lateral exposure of the infratemporal fossa and adjacent areas may, of course, be achieved by resection of the ascending ramus of the mandible, and, where necessary, the zygoma and overlying skin. An aggressive approach such as this is occasionally required for locally advanced carcinomas of the parotid gland for example. Blanchaert and Ord[7] describe what they term 'a vertical ramus compartment resection' of the mandible in which the ascending ramus, masseter and medial pterygoid muscles are resected *en-bloc* for malignant or locally aggressive tumours arising from within the vertical ramus of the mandible and which invade the infratemporal fossa and adjacent areas. The lingual and the inferior alveolar nerves are, of course, sacrificed in the resection.

When contemplating a lateral approach via the mandible for benign lesions in the infratemporal fossa, *a trial of retraction* of the mandibular ramus after detaching the stylomandibular ligament and prior to any osteotomy bone cut, can occasionally provide sufficient access (Figs 7.43, 7.44). Additional limited access can be obtained by detaching the attachments of the masseter and medial pterygoid muscles from the mandibular angle and dislocating the mandibular condyle anteriorly. If these manoeuvres fail to provide sufficient access, the planned osteotomy bone cuts are made.

This conservative approach should *not* be adopted in the removal of deep lobe parotid tumours in this region. Wide access is essential for the safe removal of these tumours. Restricted access inevitably results in pressure being applied to the tumour during its removal, significantly increasing the risk of tumour spillage. Tumour recurrence (particularly multifocal) at this site is a surgical disaster.

The angle of the mandible/ascending ramus in the lateral approaches is reached via a cervical incision without splitting the lower lip. The incision usually extends from the tip of the mastoid process down to the level of the hyoid bone, ending anteriorly at the mandibular symphysis. This incision may be linked with a standard parotidectomy incision if necessary.

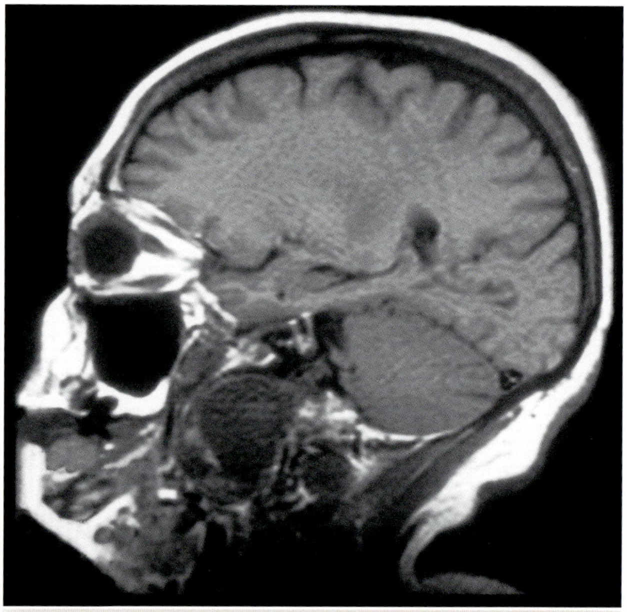

Figure 7.43 CT of schwannoma of the vagus in the infratemporal fossa.

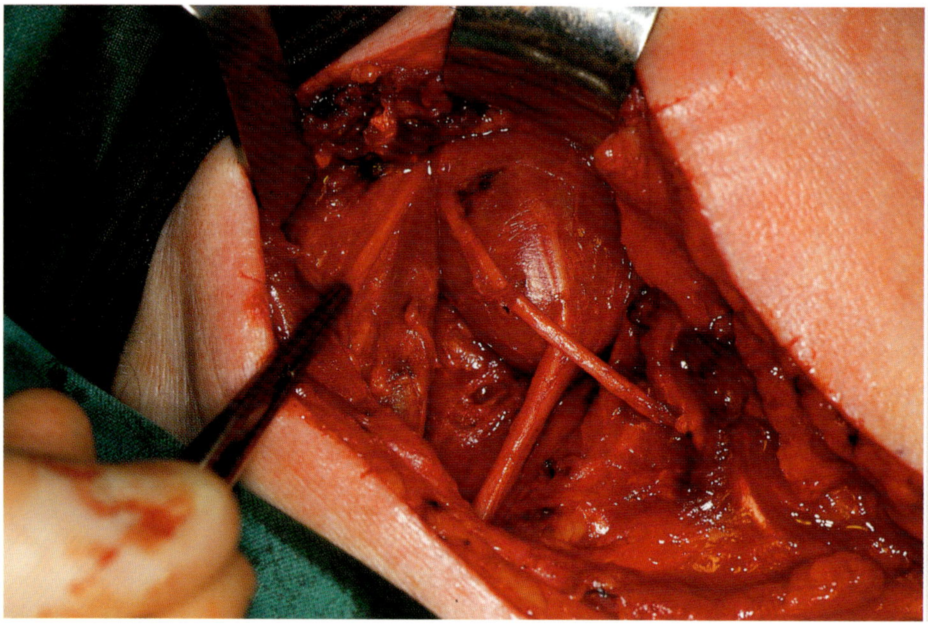

Figure 7.44 'Trial of retraction' of the mandible providing adequate access. The hypoglossal nerve may be seen crossing both the schwannoma and the vagus.

The inverted 'L'/'C' osteotomy

Although described by Flood and Hislop[10] as an approach to tumours in the parapharyngeal space, the inverted 'L' can be used as an approach to pathology in the infratemporal fossa.

The incision adopted depends on the pathology. If the tumour is related to the deep lobe of the parotid gland, then a standard parotidectomy incision with cervical extension is used. A preliminary superficial parotidectomy is performed to provide exposure to the deep lobe of the parotid gland and the lower branches of the facial nerve. If the lesion is unrelated to the parotid gland and a parotidectomy is not required, a cervical incision extending from the tip of the mastoid process to the hyoid bone is used, extended anteriorly as necessary (as described above).

The masseter muscle is elevated from the lateral aspect of the ramus and reflected superiorly to expose the sigmoid notch, anterior and posterior border of the ramus and lower border of the mandible in the region of the angle. The medial pterygoid attachment is carefully reflected off the medial aspect of the ramus, care being required superiorly in the region of the lingula. Complete detachment may not be possible until after the osteotomy has been completed.

An inverted 'L' osteotomy is then performed. The horizontal bone cut is made above the lingula and linked with the vertical bone cut. The intersection of the horizontal and vertical osteotomy cuts should be 8 mm above and 11 mm behind the midpoint of the waist of the ascending ramus. This avoids sectioning below the lingula and damaging the inferior alveolar nerve (Fig. 7.45). The osteotomy design may be modified by curving the vertical limb anteriorly to form a 'C'-shaped osteotomy. The bone cut is completed anteriorly with a short vertical bone cut (Fig. 7.46). Prior to completing the osteotomy cuts, the bone fragments are prelocalised with two mini bone plates. The plates are removed and the osteotomy cuts completed.

The proximal fragment is elevated laterally and superiorly and the distal fragment displaced both anteriorly and superiorly, as necessary – limited only by the

Figure 7.45 The intersection of the horizontal and vertical osteotomy cuts (point X) should be at a point 8 mm above and 11 mm behind the mid-point of the waist of the ascending ramus.

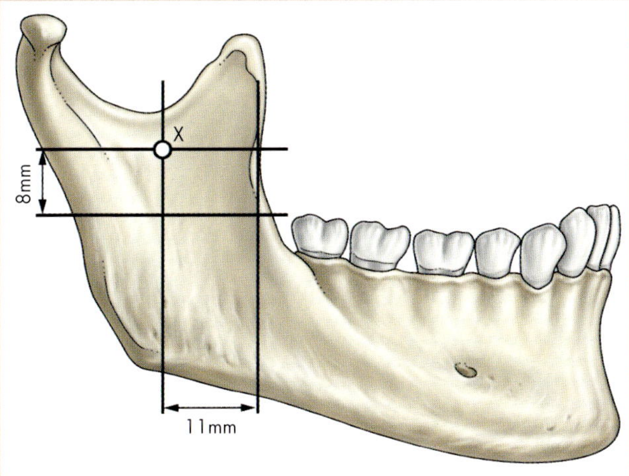

Figure 7.46 Inverted 'L' and inverted 'C' osteotomy.

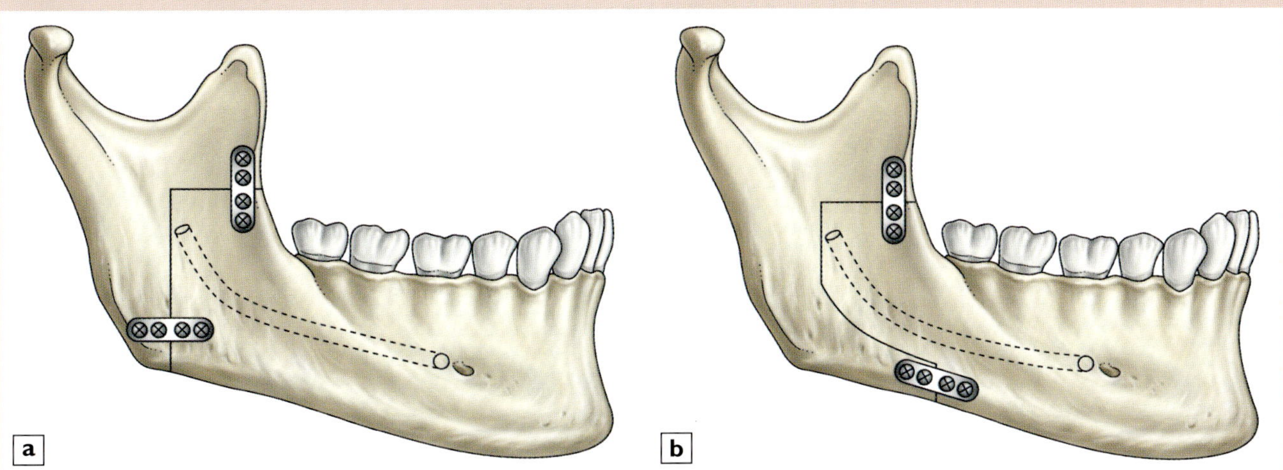

tension on the inferior alveolar nerve. A separation of 4–5 cm between the ramus bone fragments is possible.

When closing the wound, the teeth are held firmly in the correct occlusion and the mandibular fragments reduced into the correct position and the bone plates replaced. The pterygomasseteric sling is re-approximated and the neck closed in layers with vacuum drainage in the usual fashion (Figs 7.47–7.49).

The case illustrated is that of a deep lobe parotid tumour approached via an inverted L osteotomy as described above. In this instance a preliminary superficial parotidectomy was performed.

The advantages of this osteotomy design are that it:

- provides a broad area of bone contact
- elevates the coronoid process and the temporalis muscle out of the operative field
- preserves the inferior alveolar nerve
- permits easy identification of the lingual nerve up to the base of the skull.

Access superiorly with this technique can equal that provided by the mandibular swing approach but the inferior alveolar nerve remains within the operative field. As noted, the inferior alveolar nerve limits the anterior movement of the distal fragment. In this situation access may be improved peroperatively by either:

- dividing the lip and adding a further osteotomy anterior to the mental foramen to allow both lateral retraction and further elevation of the body/vertical ramus of the mandible; or
- sectioning the mandibular body posterior to the mental foramen, dividing the inferior alveolar nerve and avoiding splitting the lower lip.

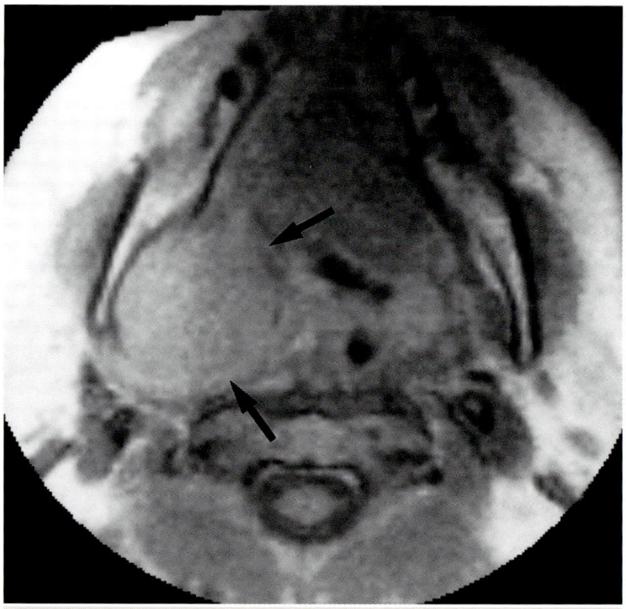

Figure 7.47 MRI scan of tumour (arrowed) on the medial aspect of the mandibular ramus. (Reproduced with permission of Mr T Flood, Consultant Oral and Maxillofacial Surgeon, Odstock Centre for Maxillofacial Surgery, Salisbury.)

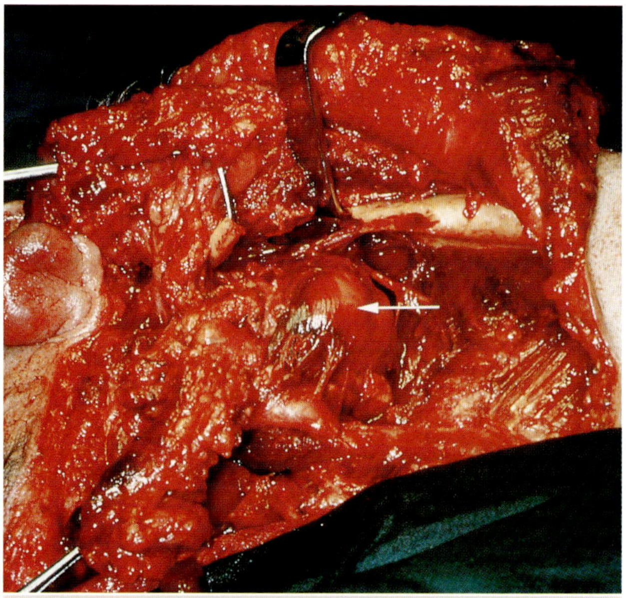

Figure 7.48 Mandible divided and proximal and distal fragments separated. The tumour is exposed and the inferior alveolar and lingual nerves are visible in the operative field. The arrow identifies the parotid mass. (Reproduced with permission of Mr T Flood, Consultant Oral and Maxillofacial Surgeon, Odstock Centre for Maxillofacial Surgery, Salisbury.)

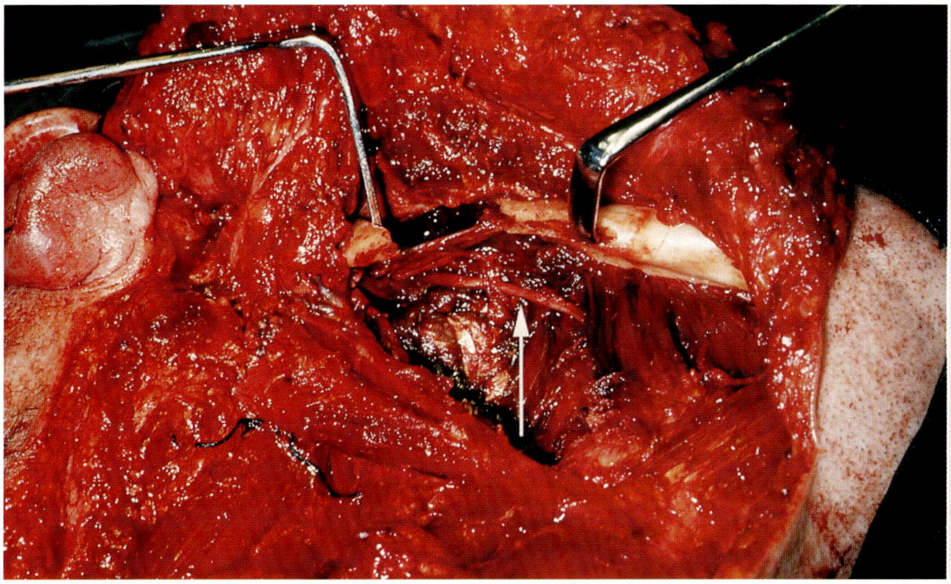

Figure 7.49 Operative site following the tumour resection. Inferior alveolar and lingual nerves preserved intact. The arrow identifies the lingual nerve. (Reproduced with permission of Mr T Flood, Consultant Oral and Maxillofacial Surgeon, Odstock Centre for Maxillofacial Surgery, Salisbury.)

Extended exposure: ramus/body osteotomies

Surgical techniques

If the pathology requires the resection of the inferior alveolar nerve, then a simple osteotomy of the mandible at the junction of the body and the ascending ramus, sectioning the inferior alveolar nerve within its bony canal, will suffice (Fig. 7.50). This is both a quick and easy technique which, with appropriate stripping of the medial pterygoid muscle, sphenomandibular and stylomandibular ligaments, can provide access similar to the extended mandibular swing while avoiding a facial incision. A further advantage with this approach is that it may also avoid entering the oral cavity, if this is not required for the resection. As with the extended mandibular swing, when this approach is used for the resection of malignant tumours involving the superior compartment of the infratemporal fossa, it may be readily combined with the lateral transzygomatic approach.

Figure 7.50 Osteotomy at the junction of the body and ascending ramus of the mandible deliberately sectioning the inferior alveolar neurovascular bundle.

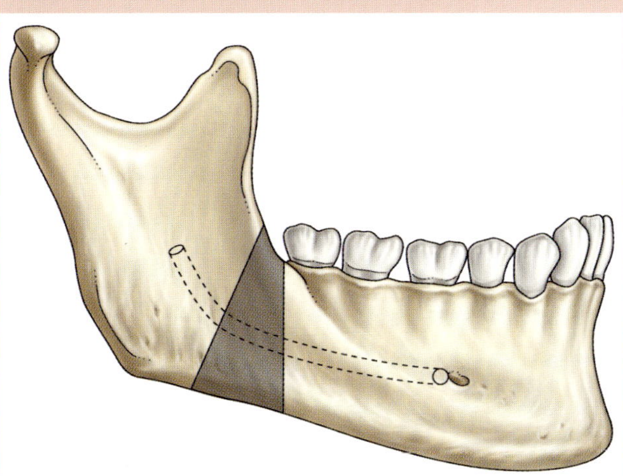

Seward[11,12] and others later[8,9] describe an approach to the parapharyngeal space in which an osteotomy of the body of the mandible anterior to the mental foramen was performed via a cervical incision *without dividing the lower lip and chin*. When used for the removal of deep lobe parotid neoplasms, a superficial parotidectomy is carried out initially with identification and preservation of the facial nerve. When a parotidectomy is not required, the lower border of the mandible is exposed via the cervical incision in the usual fashion. The periosteum of the lower border of the mandible is divided and the mental nerve identified. An osteotomy of the body of the mandible is performed in a convenient interdental space anterior to the mental foramen. Before completing the bone cuts the mandible is prelocalised by placing bone plates and screws across the site of the bone cut. The plates and screws are removed and the mandible divided. The mylohyoid and the medial pterygoid attachments to the medial aspect of the mandible are freed up to the lingula and the inferior alveolar neurovascular bundle identified. The stylomandibular and sphenomandibular ligaments are also detached. The mandible may now be elevated exposing not only the parapharyngeal space but also the infratemporal fossa, particularly the inferior compartment. An osteotomy of the mandibular condylar neck may be added if additional superior access is required.

The advantages of this approach are:

- the inferior alveolar nerve is preserved and taken out of the operative field
- the lingual nerve is readily identified and preserved if necessary
- it is easily changed into an extended mandibular swing approach as necessary by the division of the lip and chin
- the oral cavity is not necessarily entered.

The principal disadvantage is the relatively restricted mandibular retraction compared with the extended mandibular swing. The intact lower lip and chin do not permit the same degree of both elevation and outward rotation of the mandible that is possible when these structures are sectioned. This is a relatively minor criticism considering the ease with which it can be converted into the extended mandibular swing.

Lateral transzygomatic approaches

The lateral transzygomatic approaches to the infratemporal fossa and adjacent areas involve the disarticulation and inferior displacement of the zygoma, usually pedicled to the masseter muscle, as well as the displacement of the temporalis muscle. The temporalis muscle may be displaced in either a superior or inferior direction. These manoeuvres improve access to the superior compartment of the infratemporal fossa by removing its lateral boundary.[27–32]

Approach for pathology confined to the infratemporal fossa – subcranial approach

The (subcranial) lateral transzygomatic approach to the infratemporal fossa, posterior maxilla and orbits was described by Obwegeser.[32] He described it as the 'temporal' approach. This approach is suitable for pathology confined to the infratemporal fossa itself or invading the infratemporal fossa from adjacent areas (other than the base of the skull/middle cranial fossa). Orbital exposure is limited compared with the more extensive exposure possible with a craniotomy and resection of the greater wing of the sphenoid.

In view of the relatively limited exposure of the infratemporal fossa achieved with this approach, its use should be restricted to the treatment of benign lesions. Its role in the treatment of malignant pathology in this area should be that of providing *additional or supplementary exposure* if required.

Surgical technique

A coronal incision is made extending from the tragus of the ear on the affected side to the contralateral temporoparietal suture. The skin flap is elevated in the subgaleal plane. The temporalis fascia is exposed down towards the zygomatic arch identifying the layer of fat between the superficial and deep layers of the temporal fascia. A vertical incision is made through the fascia commencing at the junction of the pinna and the scalp at the helix, down to the zygomatic arch. An oblique incision is then made through the temporal fascia, again commencing at the junction of the pinna and the scalp at the helix, and angled at approximately 45° to the zygomatic arch. Anteriorly the incision through the fascia should approach no closer than 3 cm to the frontozygomatic suture (lateral aspect of the eyebrow; Fig. 7.51). The temporalis fascia is raised with the skin flap protecting the zygomatic branches of the facial nerve. The superior orbital rim is exposed by incising the pericranium approximately 3 cm above the rim, continuing the dissection inferiorly beneath the pericranium. The supraorbital neurovascular bundle is identified and, if necessary, released from its bony canal with an osteotome.

The amount of the zygoma mobilised is dependant upon the requirements of the case and may involve its complete disarticulation, including the lateral orbit, or the arch alone. The posterior cut is made just anterior to the articular eminence of the mandibular (glenoid) fossa, with the anterior cut determined by the need for reflecting either the arch alone or the entire zygoma; the latter if access to the orbit is required. The zygomatic segment is then retracted inferiorly, pedicled on the masseter muscle. Additional vertical exposure is achieved by opening the mouth as wide as possible with a mouth prop. The temporalis insertion into the coronoid process of the mandible is displayed.

A ramus osteotomy is then completed, usually an oblique coronoidotomy, allowing reflection of the coronoid process with the attached muscle insertions superiorly. (The ramus osteotomy may be varied in design to include the mandibular condyle with the coronoid process for further lateral exposure.) The coronoid process is retained attached to the temporalis insertion as it is useful for retraction purposes. In elevating the temporalis muscle above the level of the infratemporal crest of the greater wing of the sphenoid, the middle and deep temporal vessels are invariably sectioned thereby compromising the blood supply of the temporalis muscle. Some blood supply to the muscle is maintained via vessels perforating the squamous temporal bone. Inevitably, however, there will be a degree of fibrosis of the temporal muscle subsequently. It is obvious that if the temporalis muscle is now dissected off the squamous temporal bone its blood supply is lost. This must be taken into consideration if the surgeon is contemplating using the muscle for subsequent reconstruction.

Elevation of the temporalis muscle exposes the medial and lateral pterygoid muscles. The pterygoid muscles are divided/resected as appropriate, depending upon the site of tumour (Fig. 7.52). The maxillary artery will be encountered and should be ligated and divided. Frequent haemorrhage from the maxillary artery is not unusual as its various branches are encountered. The pterygoid venous plexus can also prove a troublesome source of persistent haemorrhage. Following resection of the upper (infratemporal) head of the lateral pterygoid muscle, it is possible to identify the mandibular nerve and its branches, which can be followed to foramen ovale (Fig. 7.53). The mandibular nerve may be resected and a segment sent for histological examination if required. The pterygoid plates are immediately anterior to the foramen ovale. Their junction with the base of the skull is easily palpated with an instrument such as

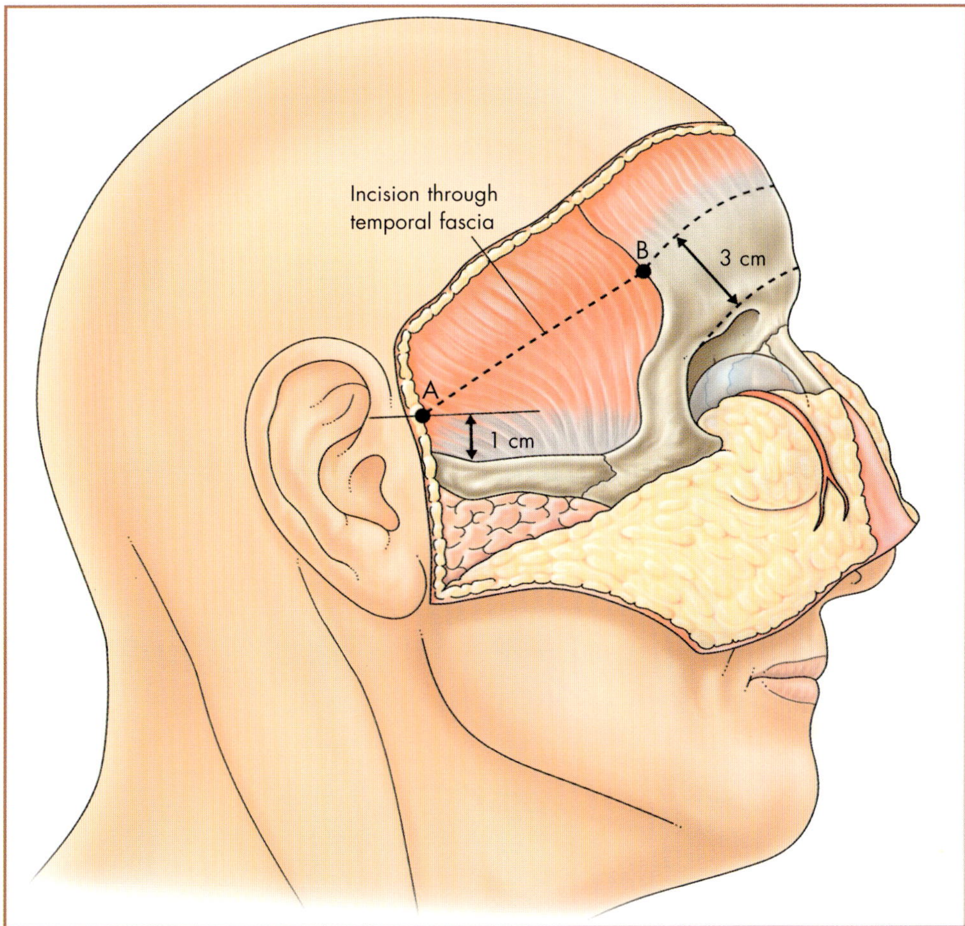

Figure 7.51 The incision through the temporal fascia is made along the line A–B to ensure the preservation of the zygomatic branches of the facial nerve.

Figure 7.52 Following the reflection of the zygomatic arch inferiorly, coronoidotomy (coronoidectomy) and the superior reflection of the temporalis muscle, the infratemporal fossa is exposed.

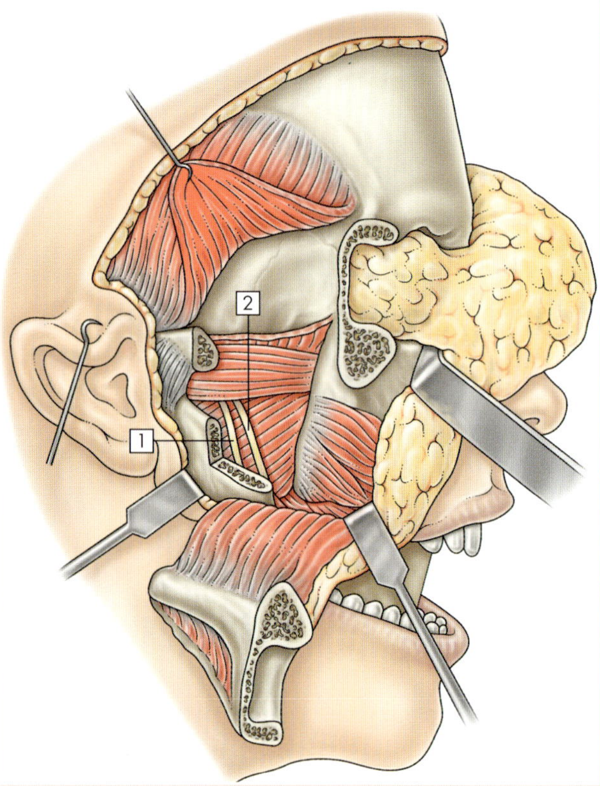

1 Inferior alveolar nerve
2 Lingual nerve

Howarth's nasal raspatory. The lateral aspect of the lateral pterygoid plate is exposed superiorly by gentle stripping of the lateral pterygoid muscle off the bony surface. The dissection is then taken anteriorly and the pterygomaxillary fissure exposed.

If exposure of the maxillary division of the trigeminal nerve at the level of foramen rotundum is required for both nerve section and for histological sampling (to assess the presence of tumour), gentle bone removal with a burr of the greater wing of the sphenoid at the root of the lateral pterygoid plate will give direct visual access to the foramen rotundum.

With the lateral transzygomatic approach, the inferior exposure reaches the level of the maxillary teeth (Fig. 7.54). Following resection of the lateral and medial pterygoid muscles, medial exposure reaches the lateral wall of the nasopharynx.

In the case of maxillary tumours with posterior extension into the infratemporal fossa, the maxillectomy – either partial or total – is completed anteriorly either with an intraoral incision alone or via a Weber–Fergusson approach as necessary with the additional exposure provided by mobilising the zygoma from a lateral approach. The pterygoid plates can be divided from the skull base and delivered *en bloc* with the operative specimen, through the mouth. Bleeding is controlled in the usual fashion with ligation, diathermy or packing with resorbable packs as necessary. The coronoid process is not reattached in order

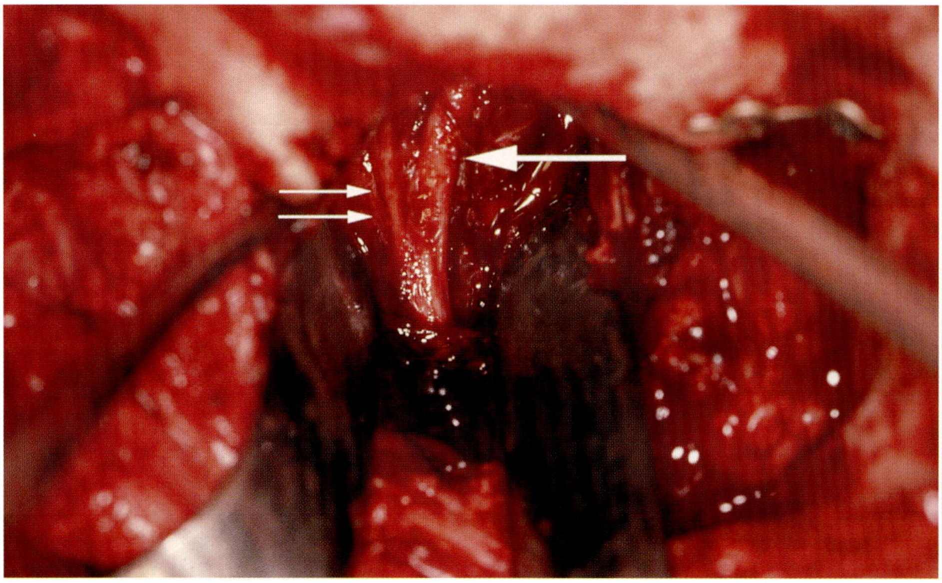

Figure 7.53 The lingual nerve (large arrow) identified adjacent to the base of the skull following resection of the upper head of the lateral pterygoid muscle. The chorda tympani nerve (double, small arrows) may be seen joining the lingual nerve. The tensor veli palatini muscle lies deep to the nerve. The depth of the dissection at this point is approximately 4 cm deep to the articular eminence of the temporomandibular joint.

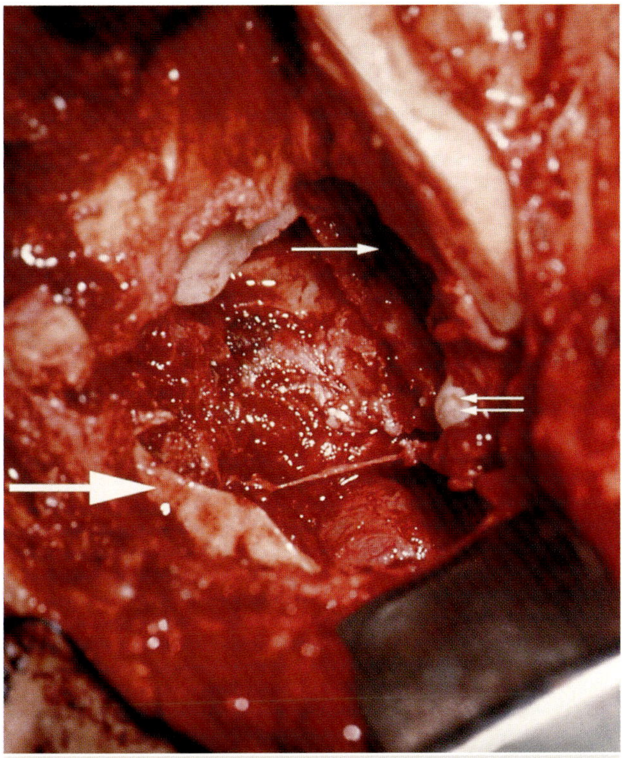

Figure 7.54 With displacement of the zygoma exposure inferiorly reaches the level of the maxillary teeth. The posterior wall of the maxilla has been resected. Note the mandibular condyle (large arrow), posterior wall maxilla/maxillary antrum (single, small arrow) and the maxillary third molar tooth (double, small arrows).

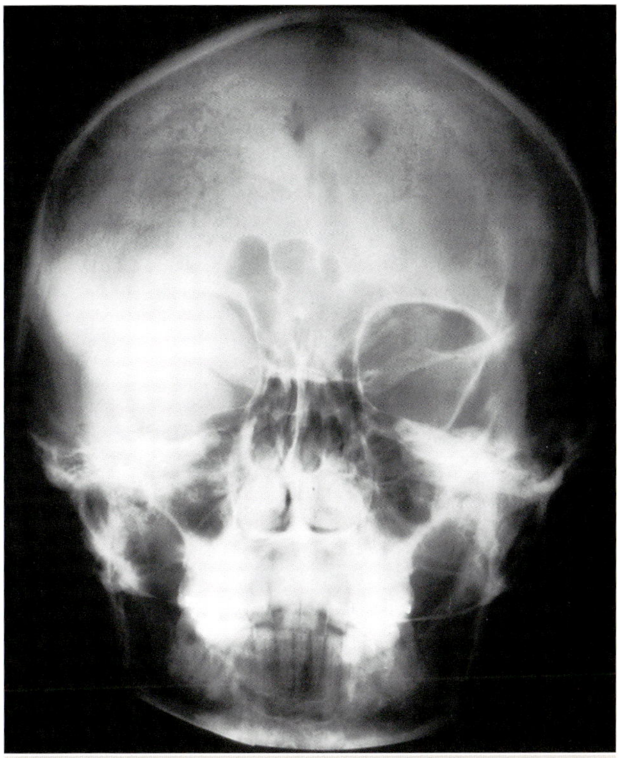

Figure 7.55 Hyperostosing meningioma of the sphenoid ridge involving the orbit, middle cranial fossa and roof of the infratemporal fossa. Although benign, this is a locally invasive neoplasm requiring aggressive resection.

to minimise the likelihood of postoperative limitation of mandibular opening. The zygomatic segment is fixed in place, usually with bone plates and screws. If these are not available, wire osteosynthesis is used.

A major disadvantage of this approach is that medial exposure becomes more restricted as the depth of dissection increases and the medial dissection is carried out 'blind'. If exposure of the carotid arteries and the internal jugular vein is necessary, these are identified in the neck in the usual manner (Fig. 7.7).

In view of the limited medial access, this approach should be limited to the treatment of benign pathology of the infratemporal fossa, particularly in the superior compartment. Malignant disease is better approached through either a transfacial or transmandibular approach as described above. As stated, the role of a lateral transzygomatic approach in the treatment of malignant disease involving the infratemporal fossa is principally that of a supplementary approach if required. In this context, the additional exposure it provides can prove valuable.[36]

Approach for pathology involving the infratemporal fossa and middle cranial fossa

The proximity of the superior compartment of the infratemporal fossa, middle cranial fossa and orbit frequently results in pathology involving these areas simultaneously (Fig. 7.55). Disarticulation of the zygoma and a (fronto)temporal craniotomy provides access to these adjacent areas with minimal brain retraction.

As described above, by displacing the zygoma inferiorly and moving the temporalis muscle in either a superior or inferior direction, direct access is provided to the superior compartment of the infratemporal fossa. The exposure extends inferiorly to the level of the maxillary teeth. An important additional benefit of displacing the zygoma inferiorly is that it allows a low (fronto)temporal craniotomy. This in turn provides access directly along the floor of the middle cranial fossa with minimal brain retraction, a fact recognised by Cushing in 1900.[37] If the lateral orbit is included (i.e. the entire zygoma mobilised), the greater wing of the sphenoid may be resected thereby exposing:

- the lateral margin of the superior orbital fissure, which transmits the third, fourth, ophthalmic division of the fifth and the sixth cranial nerves from the cavernous sinus to the orbital apex,
- the lateral margin of foramen ovale and foramen rotundum that transmit the mandibular and maxillary divisions of the trigeminal nerve from the inferior aspect of the cavernous sinus and which enter the superior pole of the infratemporal fossa and the pterygopalatine fossa, respectively.

The complete resection of the greater wing of the sphenoid therefore provides extradural exposure of the anterolateral and inferolateral boundaries of the cavernous sinus as well as direct access to the pterygomaxillary fissure (i.e. the medial limit of the infratemporal fossa), nasopharynx and complete exposure of the lateral, superior and inferior orbit. This wide exposure allows aggressive resection of neoplasms involving these sites.[30]

When accessing both the middle cranial fossa and the infratemporal fossa the lateral orbit rim is always mobilised with the zygoma. If the lateral orbital rim is left *in situ*, resection of the greater wing of the sphenoid back to the superior orbital fissure and beyond is not possible. This limits exposure of the middle fossa and restricts surgical flexibility.

Surgical technique

The zygoma and the capsule of the temporomandibular joint are exposed via a hemi- or bicoronal flap extended inferiorly to the level of the ear lobe and preserving the zygomatic branch of the facial nerve as described above (Fig. 7.56). The temporalis muscle is displaced inferiorly by detaching it from the temporal crest and the posterolateral aspect of the orbit with cutting diathermy. The orbital periosteum is elevated from the lateral orbital wall. The osteotomy in the lateral orbital wall is made at the junction of the zygoma and greater wing of the sphenoid. The osteotomy bone cuts are shown (Figs 7.57, 7.58). The inferior cut extends from the anterior end of the inferior orbital fissure to the lateral orbital rim. To avoid exposure of the

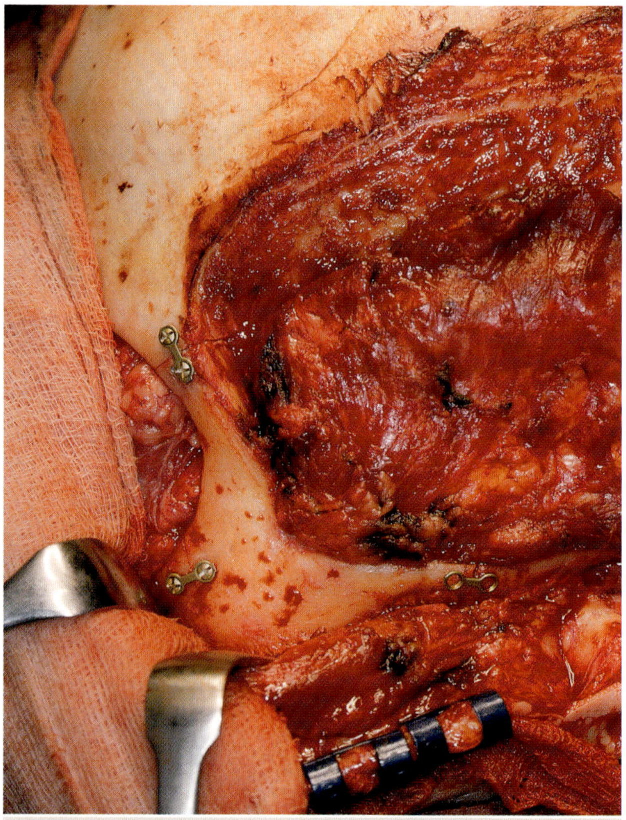

Figure 7.56 The zygoma and the capsule of the temporomandibular joint exposed. The bone plates placed to prelocalise the zygoma prior to completing the osteotomy bone cuts.

Figure 7.57 Osteotomy bone cuts outlined.

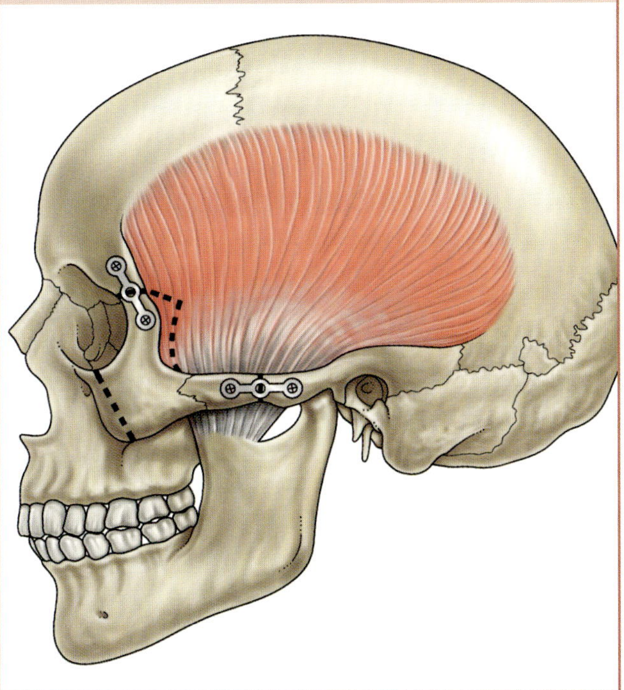

maxillary antrum, the inferior bone cut is kept as lateral as possible. The position of the zygoma is prelocalised with low profile bone plates and screws. This ensures accurate replacement of the zygoma. The osteotomy is completed with an osteotome and the zygoma displaced inferiorly, pedicled to the masseter muscle. This exposes the superior compartment of the infratemporal fossa (Fig. 7.59).

After displacing the zygoma and the temporalis muscle *inferiorly*, the vertical exposure may be increased by opening the mouth wide with a mouth prop. This manoeuvre may provide sufficient vertical

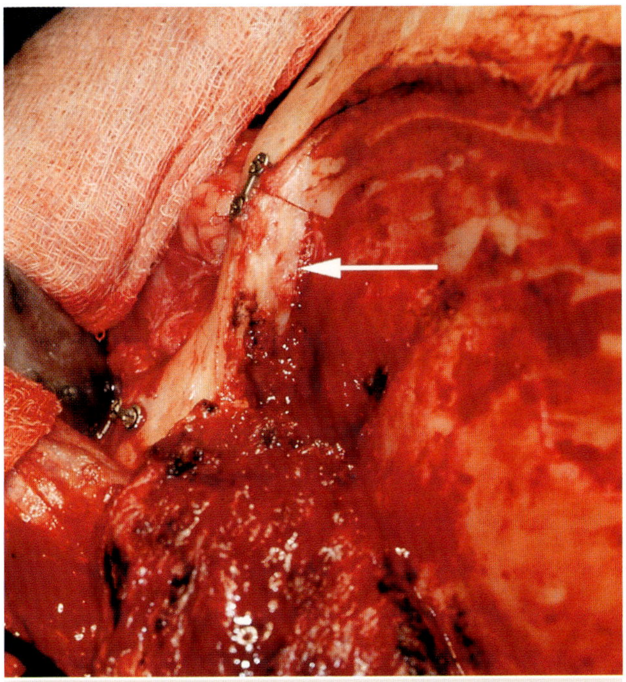

Figure 7.58 The osteotomy bone cut in the lateral orbit shown after inferior reflection of the temporalis. The arrow marks the osteotomy bone cut at the junction of the zygoma and the greater wing of the sphenoid. Left profile.

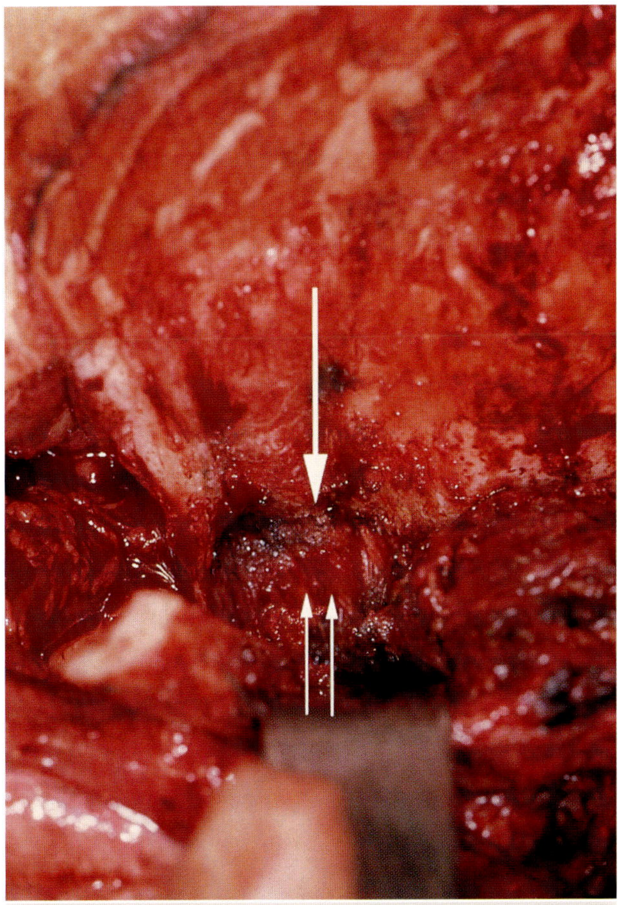

Figure 7.59 Lateral aspect of the superior compartment of the infratemporal fossa exposed following reflection of the zygoma inferiorly. The level of the infratemporal crest of the greater wing of the sphenoid is shown with the large arrow. The upper head of the lateral pterygoid muscle is shown with the small, double arrows. Left profile.

exposure without the need for a coronoidectomy although the latter is performed if required. As stated above, there is little point in carrying out a coronoidectomy with later replacement of the coronoid process as this merely increases the likelihood of postoperative limitation of mouth opening.

For complete exposure of the lateral orbit to its apex, a frontotemporal craniotomy is necessary (Fig. 7.60). This allows the safe removal of the greater wing of the sphenoid under direct vision and the exposure of the contents of the superior orbital fissure and the optic nerve if required (Figs 7.61, 7.62). The dissection is developed as necessary with exposure of the foramen ovale and foramen rotundum by resection of the greater wing of the sphenoid lateral to these foramina, i.e. floor of the middle fossa (Fig. 7.62).

Medial exposure may be increased by sectioning the mandibular nerve at the foramen ovale, and resecting the lateral and medial pterygoid plates at the skull base. Increased exposure medial to the cavernous sinus may be obtained by the removal of the anterior clinoid process and opening the optic canal (Fig. 7.63).

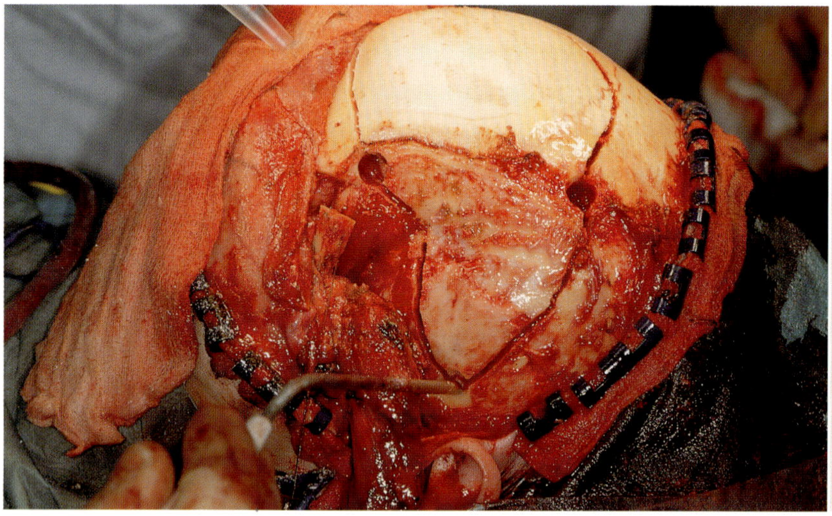

Figure 7.60 A frontotemporal craniotomy is necessary to allow the resection of the greater wing of the sphenoid for the exposure of the superior orbital fissure and optic nerve. The craniotomy also permits the safe resection of the floor of the middle fossa under direct vision. Left profile.

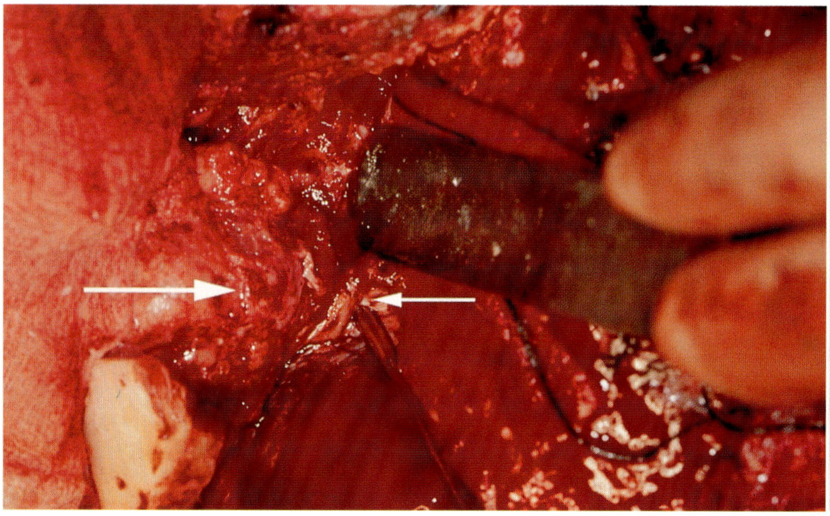

Figure 7.61 Retractor on the temporal lobe. The small arrow indicates the dura at the lateral aspect of the superior orbital fissure following the resection of the greater wing of the sphenoid in the lateral wall of the orbit. The large arrow indicates the orbital contents. Left profile.

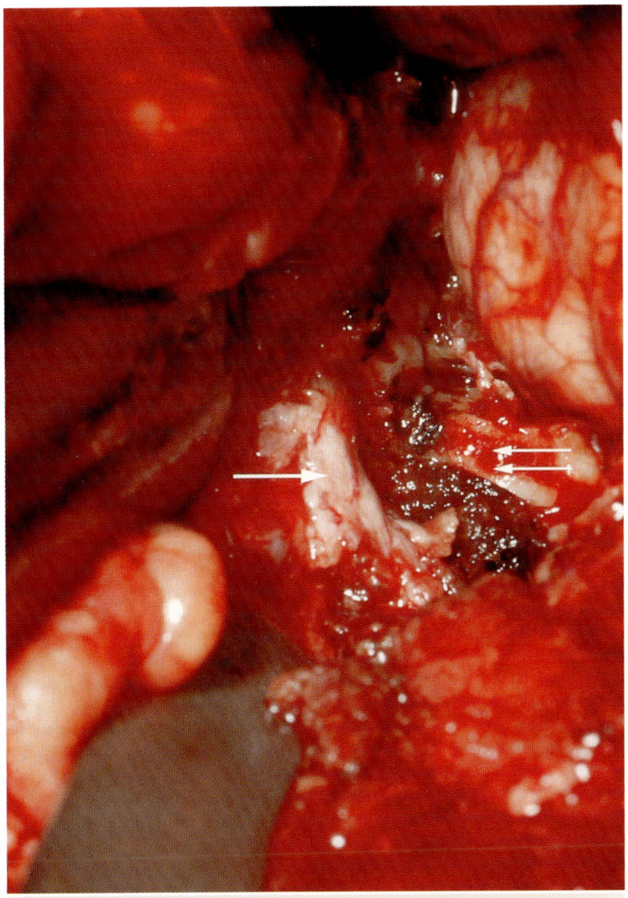

Figure 7.62 Mandibular nerve at the base of the skull following resection of the floor of the middle fossa. The retractor is depressing the lateral pterygoid muscle inferiorly. Single, large arrow indicates mandibular nerve. Double, small arrows indicate cut edge of middle fossa floor with the temporal lobe above. Left profile.

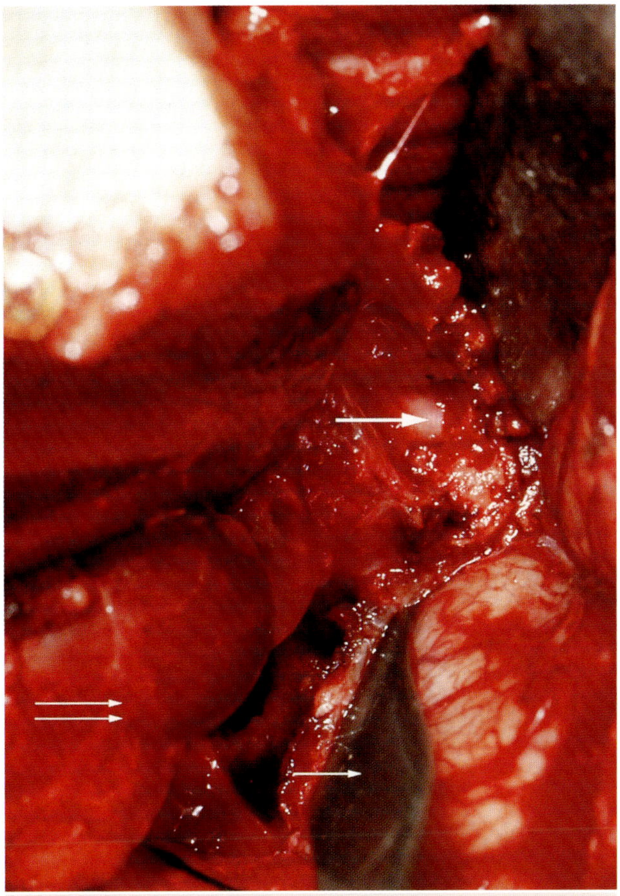

Figure 7.63 Optic nerve exposed by the resection of the lateral aspect of the optic canal. This provides increased exposure medial to the cavernous sinus. Removal of the anterior clinoid process will further increase exposure – great care is required in dissection in the region of the anterior clinoid process in view of the proximity of the internal carotid artery. Single, large arrow indicates the optic nerve. Single, small arrow indicates the retractor on the temporal lobe. Double, small arrows indicate the orbital contents. Left profile.

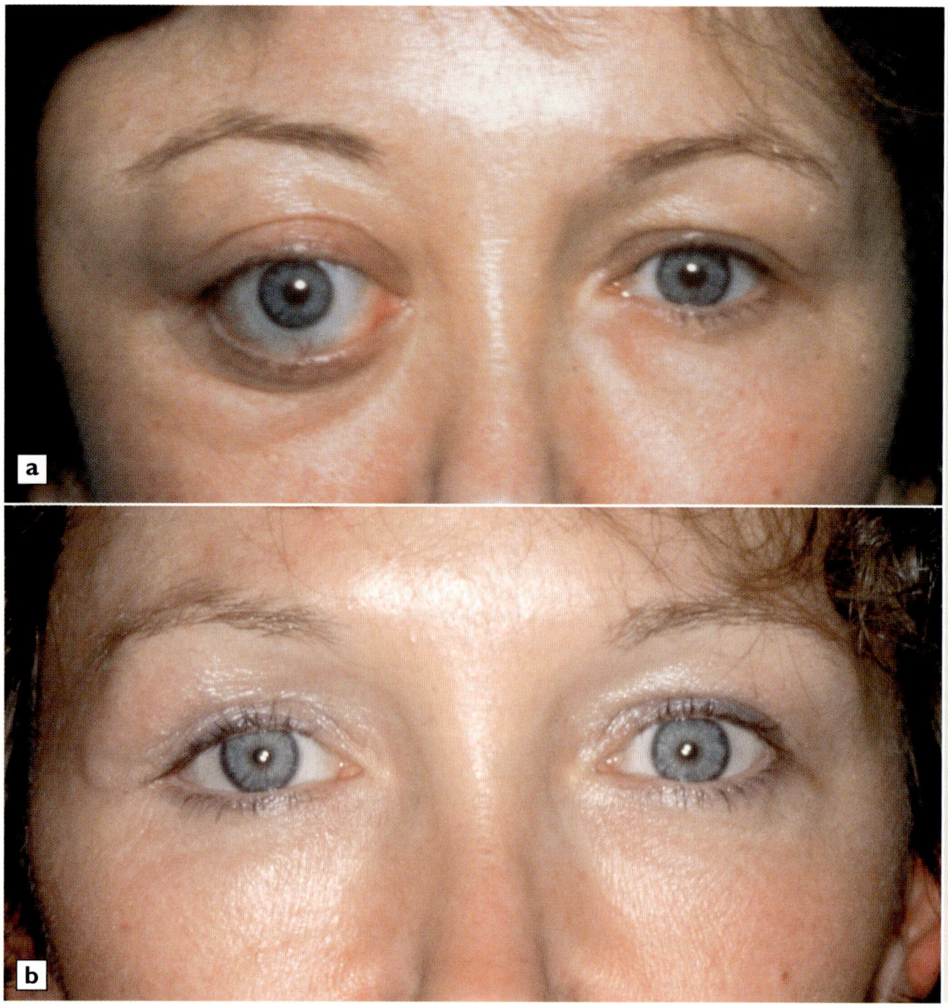

Figures 7.64 and 7.65 Pre- (a) and postoperative (b) appearance of a patient with hyperostosing meningioma of the sphenoid ridge.

Figures 7.58–7.65 are not all from the same case. All, however, relate to the resection of hyperostosing meningiomas of the sphenoid wing requiring varying degrees of orbital, middle and infratemporal fossa resection and reconstruction. These tumours are locally invasive. The access provided by the lateral transzygomatic approach in each case allowed aggressive resection of the tumour and immediate reconstruction. (Cases in Figs 7.58–7.65 operated jointly with either Mr G. Neil Dwyer or Miss D. Lang, Consultant Neurosurgeons, Southampton University Hospitals.) Figs 7.64 and 7.65 demonstrate the pre- and postoperative appearance of one such patient.

With both benign and malignant tumours involving the middle cranial fossa, the limit of the medial resection margin is the line running medial to the foramen rotundum, foramen ovale, foramen spinosum and lateral to the internal carotid artery and foramen lacerum (Fig. 7.66).

If intradural dissection is required, this can proceed with minimal brain retraction by dividing the Sylvian fissure, allowing the temporal lobe to fall posterolaterally (transsylvian approach) or subtemporally (along the floor of the middle fossa). The transsylvian approach provides a low anterolateral corridor of access directly along the floor of the middle cranial fossa, whereas the subtemporal approach provides a wide and shallow surgical field along the floor of the middle cranial fossa. Both approaches may be used simultaneously providing greater three-dimensional exposure. The distance between the surgeon and the edge of the tentorium may be halved – fulfilling the requirement of a short straight line between the surgeon and the pathology. Utilising either or both of these approaches provides exposure of the apex of the petrous part of the tempo-

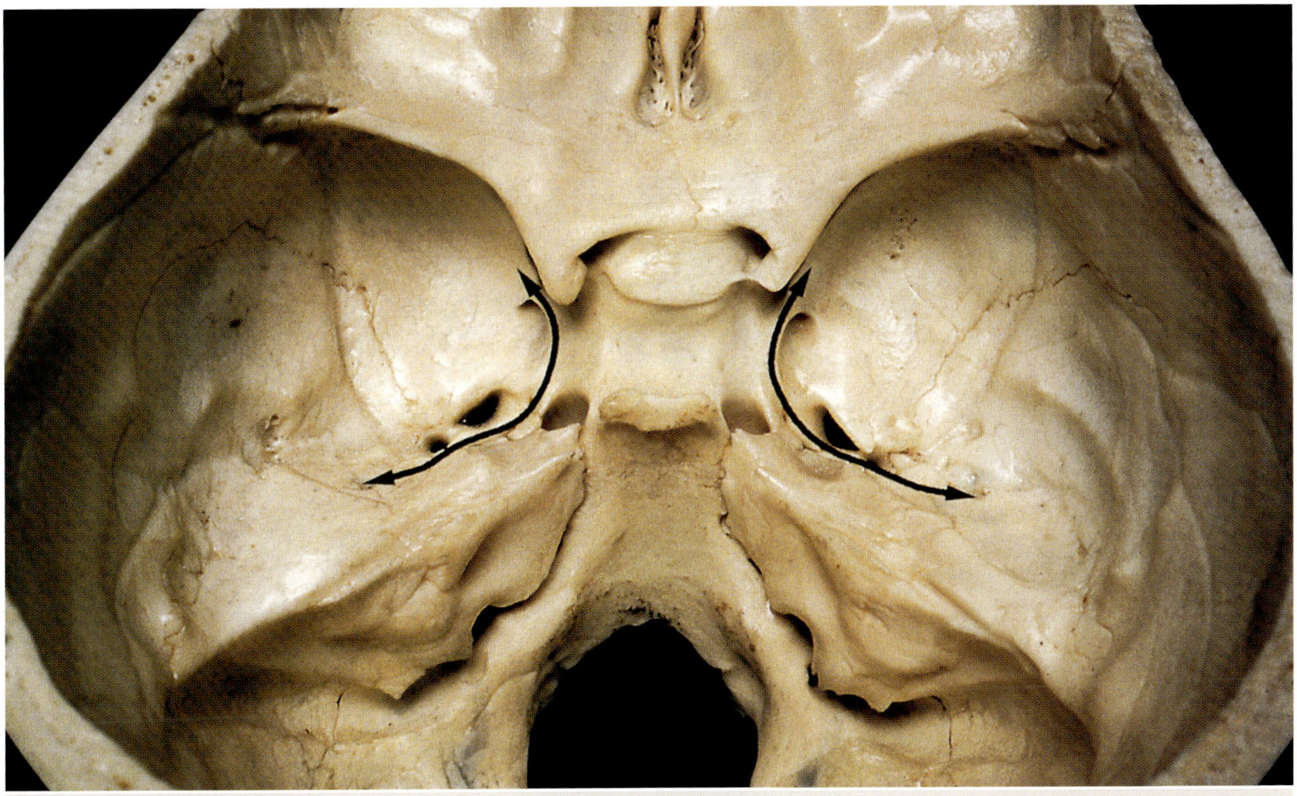

Figure 7.66 The limit of the medial resection margin for both benign and malignant tumours is the line running medial to the foramen rotundum, foramen ovale and foramen spinosum and lateral to the internal carotid artery (foramen lacerum).

ral bone, upper clivus, distal internal carotid and proximal middle cerebral arteries, optic chiasm and ipsilateral optic nerve. The oculomotor nerve is visualised early in the dissection. The isolation and protection of these structures is greatly facilitated by the increased exposure of the skull base.[28]

The advantages of the lateral transzygomatic approach combined with a frontotemporal craniotomy are well summarised by Mickey *et al.*[30] It is a technically straightforward technique providing wide exposure of the interface between normal and neoplastic tissues in the middle cranial fossa, infratemporal fossa and orbit. It defines the anterior and inferior boundaries of the cavernous sinus and may therefore aid the intraoperative assessment of the resectability of a tumour adjacent to the sinus. The combined exposure of the middle cranial fossa and the infratemporal fossa significantly facilitates the complete resection of benign neoplasms in this region and allows a maximally aggressive resection in carefully selected malignant neoplasms.

As with pathology at any site, the potential morbidity of the surgical approach and resection of tumours invading the middle cranial fossa must be weighed against the potential benefit. When dealing with aggressive malignant tumours such as squamous cell carcinomas, the aim must be *en-bloc* clearance. If the tumour involves the cavernous sinus for example, or is found to traverse the dura, i.e. involves the full thickness of the dura, it is unresectable. With benign or less aggressive malignant tumours, such as adenoid cystic carcinomas, microscopic disease could be left *in situ* in critical areas such as the cavernous sinus and lower cranial nerves to reduce morbidity to an acceptable level. This approach allows inspection of the dura along the floor of the middle cranial fossa and adjacent to the cavernous sinus early in the procedure. If the tumour is considered unresectable, surgery may therefore be discontinued with no morbidity to the patient.

The dissection of the temporalis, medial and lateral pterygoid muscles necessary for wide exposure of the infratemporal fossa inevitably impairs their vascular and nerve supply. This can result in restricted mandibular movement despite aggressive jaw exercises in the postoperative period. Resection of the greater wing of the sphenoid bone in the lateral orbit can result in mild enophthalmos but does not disturb ocular motility. The enophthalmos is usually not of concern to the patient.

The direction of displacement of the temporalis muscle is dictated to some extent by the site of the pathology. If the lesion is primarily intracranial with inferior extension into the infratemporal fossa (such as with neurofibromas and meningiomas), the temporalis muscle is usually displaced inferiorly. With subcranial lesions extending superiorly into the middle cranial fossa, such as with squamous cell carcinomas and nasopharyngeal angiofibromas, temporalis muscle displacement superiorly is more common.

As already noted when dealing with malignant lesions, the lateral transzygomatic approach is readily combined with other approaches such as the extended mandibular swing, a transfacial approach or a lateral mandibular approach.

The lateral transzygomatic approach provides access anterior to the petrous part of the temporal bone and therefore is not suitable for the lesions involving this bone. The resection of lesions involving the petrous temporal bone may require the isolation and control of the internal carotid artery from the carotid canal to the cavernous sinus. The petrous portion of the internal carotid artery must therefore be exposed along its length in these cases. The extent of the petrous resection is dependent upon the nature and extent of the tumour. The type C approach as described by Fisch[33] and the subtemporal preauricular approach described by Sekhar *et al.*[32] provide the exposure necessary in these cases. Fisch's type C approach results in mandatory conductive hearing loss, and a sensory and motor deficit as a result of sectioning the mandibular division of the trigeminal nerve. In addition, the mandibular fossa and disc of the temporomandibular joint are removed and, if required, the mandibular condyle is resected with a resultant malocclusion. The approach described by Sekhar *et al.*[32] avoids the conductive hearing loss unless the resection of the hearing conduction mechanism is required because of tumour involvement. These procedures have a mortality rate of between 2% and 10% and varying morbidity depending in part upon the preoperative status of the patient. These approaches will not be described here but relevant references can be found in the reference list[29,32–35].

Careful case selection is necessary when contemplating resections of high grade malignant tumours involving the floor of the middle fossa from below if unnecessary morbidity and mortality are to be avoided. When true *en-bloc* clearance of high grade malignant tumours with tumour-free margins is possible the results are encouraging. If the disease is extensive and invading multiple cranial base foramina making *en-bloc* clearance impossible, surgery in this critical area is not justified.

REFERENCES

1. Friedman W H, Katsantonis G P, Cooper M H *et al.* Stylohamular dissection: a new method for *en-bloc* resection of malignancies of the infratemporal fossa. *Laryngoscope* 1981; 91: 1869–1879
2. Iannetti G, Belli E, Ciconnetti A *et al.* Infratemporal fossa surgery for malignant diseases. *Acta Neurochirurgica* 1996; 138: 658–671
3. McGregor I A, MacDonald D G. Mandibular osteotomy in the surgical approach to the oral cavity. *Head Neck Surg* 1983; 5: 457–462
4. Spiro R H, Gerold F P, Strong E W. Mandibular 'swing' approach for oral and oropharyngeal tumours. *J Head Neck Surg* 1981; 3: 371–378
5. Biedlingnaier J F, Ord R A. Modified double mandibular osteotomy for tumours of the parapharyngeal space. *J Oral Maxillofac Surg* 1994; 52: 348–352
6. Attia E L, Bentley K C, Head T. Mulder D. A new external approach to the pterygomaxillary fossa and the parapharyngeal space. *Head Neck Surg* 1984; 6: 884–891
7. Blanchaert R H, Ord R A. Vertical ramus compartment resection of the mandible for deeply invasive tumours. *J Oral Maxillofac Surg* 1998; 56: 15–22
8. Bozetti A, Biglioli F, Gianni A B, Brusati R. Mandibulotomy for access to benign and deep lobe tumours with parapharyngeal extension: report of four cases. *J Oral Maxillofac Surg* 1998; 56: 272–276
9. Brusati R, Bozzeti A, Raffaini M. Access osteotomies to cranio-maxillo-facial oncological surgery. In: Caronni E (ed) *Craniofacial surgery: Proceedings of the Second International Congress of the International Society of Cranio-Maxillo-Facial Surgery.* Florence: Monduzzi Editore 1989
10. Flood T R, Hislop W S. A modified approach for parapharyngeal space tumours: use of the inverted 'L' osteotomy. *Br J Oral Maxillofac Surg* 1991; 29: 82–86
11. Seward G R. Nodular enlargements of the salivary glands. In Moore J R (ed) *Surgery of the mouth and jaws.* Oxford: Blackwell Scientific Publications, 1985; Chapter 38
12. Seward G R. Tumours of the parapharyngeal space. *J Roy Coll Surg Edinburgh* 1989; 82: 111–112
13. Shaheen O H. Swellings of the infratemporal fossa. *J Laryngol Otol* 1982; 96: 817–836
14. Altemir F H. Transfacial access to the retromaxillary area. *J Maxillofac Surg* 1986; 14: 165–170
15. Curioni C, Toscano P, Maraggia A. Il lembo Maxillo Genieno. Considerazioni su una nuova metodica di smotaggio dell infrastruttura, In: *Abstracts of the 2nd Congress of Maxillofacial Surgery,* Bologna 5–7th November 1981. Monduzzi: Bologna, 1981: 375–377
16. Curioni C, Padula E, Toscano P, Maraggia A. Nasopharyngeal angiofibromas: considerations about a new method of dismantling and reassembly of the inferomeso structure. *Chirurgia Della Testa e Del Collo* 1984; 1: 47–56
17. Curioni C, Clauser L. The facial split in tumour surgery: dismantling and reassembly of the face. In Caronni P (ed) Craniofacial Surgery 3. Monduzzi: Bologna, 1991: 223–224
18. Clauser L. 70 cases of facial splitting: planning, problems, results. An overview lecture presented at 'Dismantling and reassembly of the facial skeleton', course and workshop. Nantes, France, 16 September 1997
19. Grime P D, Haskell R, Robertson I, Gullan R. Transfacial access for neurosurgical procedures: an extended for maxillofacial surgeons. *Int J Oral Maxillofac Surg* 1991; 20: 285–295
20. Jackson I, Marsh W R, Bite U, Hide T A H. Craniofacial osteotomies to facilitate skull base tumour resection. *Br J Plast Surg* 1986; 39: 153–160
21. Janecka I P, Sen C, Sekhar L N, Arragia M A. Facial translocation: a new approach to the cranial base. *J Otolaryngol Head Neck Surg* 1990; 103: 413–419
22. Janecka I P, Sen C, Sekhar L N, Arragia M A. Facial translocation: approach to nasopharynx, clivus and infratemporal fossa. In: Sekhar L N, Janecka I P (eds) Surgery of cranial base tumours. Raven Press: Philadelphia, 1993: 245–259
23. Archer D, Young S, Utley D. Basilar aneurysms: a new transclival approach via a maxillotomy. *J Neurosurg* 1987; 67: 54–58
24. James D, Crockard A. Surgical access to the base of the skull and upper cervical spine by extended maxillotomy. *Neurosurgery* 1991; 29: 411–416
25. Wood G, Stell P M. Ostoteomy at the Le Fort I level. A versatile procedure. *Br J Oral Maxillofac Surg* 1989; 27: 33–38
26. Kornfehl J, Gstottner W, Kontrus M, Sedivy R. Transpalatine excision of a cavernous haemangioma of the infratemporal fossa. *Eur Arch Otolaryngol* 1996; 253: 172–175
27. Gibbons S D, Wiesenfeld D, Millar H, Busmanis I A. Removal of retromaxillary schwannoma via a temporal approach. *J Oral Maxillofac Surg* 1991; 49: 191–193
28. Honeybull S, Neil-Dwyer G, Lang D *et al.* The orbitozygomatic infratemporal fossa approach. A quantitative anatomical study. *Acta Neurochirurgia* 1996; 138: 255–264
29. Jenkins H A, Franklin D J. Infratemporal approaches to the skull base. In: Jackson C G (ed) *Surgery of skull base tumours.* Churchill Livinstone 1991; 121–139

30. Mickey B M, Close L, Schaefer S, Sampson D. A combined frontotemporal and lateral infratemporal fossa approach to the skull base. *J Neurosurg* 1988; 68: 678–683
31. Obwegesser H L. Temporal approach to the TMJ, the orbit and the retromaxillary-infracranial region. *J Head Neck Surg* 1985; 7, 185–199
32. Sekhar L N, Schramm V, Jones N. Subtemporal–preauricular infratemporal fossa approach to large lateral and posterior skull base neoplasms. *J Neurosurg* 1987; 67: 488–499
33. Fisch U. The infratemporal fossa approach for nasopharyngeal tumours. *Laryngoscope* 1983; 93: 36–44
34. Shahinian H, Dornier M D, Fisch U. Parapharyngeal space tumours; the infratemporal fossa approach. *Skull Base Surg* 1995; 5: 73–81
35. Suarez C, Garcia A, Fernandez de Leon R, Rodrigo J P, Ruiz B. Infratemporal approaches to nasopharyngeal tumours. *Eur Arch Otolaryngol* 1996; 254 (Suppl): 165–168
36. Evans B T. Access surgery. In: Langdon J, Patel M (eds) *Operative Maxillofacial Surgery*. Chapman and Hall Medical: London, 1998: 231–257
37. Cushing HA. A method if total extirpation of the Gasserian ganglion for trigeminal nerualgia. By a route through the temporal fossa and beneath the middle meningeal artery. *JAMA* 1900; 34: 1035–1041

Chapter **8** # The facial nerve and the parotid gland

B.K.B. BERKOVITZ, J.D. LANGDON AND B.J. MOXHAM

INTRODUCTION

Although not strictly part of the infratemporal fossa, an understanding of the surgical anatomy of the facial nerve is fundamental to all surgery in this region. The facial nerve trunk exits the posterior cranial fossa via the internal acoustic meatus and, after coursing through the middle ear, exits the skull at the stylomastoid foramen that lies just posterior to the styloid process. The styloid process itself forms part of the posterior boundary of the infratemporal fossa. More importantly, the trunk of the facial nerve then enters the posterior aspect of the parotid gland where it divides into its various branches. These branches define a plane between superficial and deep 'lobes' of the parotid gland. The deep lobe of the parotid gland extends into the infratemporal fossa behind the posterior border of the mandibular ramus. Therefore, it is essential for the surgeon to be familiar with the anatomy of the facial nerve and the parotid gland when considering the infratemporal fossa.

The facial nerve, the nerve of the second embryonic branchial arch, is the motor nerve to the muscles of facial expression, muscles of the scalp and external ear and to the buccinator, platysma, stapedius, stylohyoid and posterior belly of the digastric muscles. Its sensory fibres supply the anterior two-thirds of the tongue with taste and parts of the external acoustic meatus, soft palate and adjacent pharynx with general sensation. It also carries parasympathetic, secretomotor fibres to the submandibular, sublingual, lacrimal, nasal, pharyngeal and palatine glands.

As the facial nerve emerges from the stylomastoid foramen, the nerve runs anteriorly within the parotid gland, crossing the external carotid artery and, at the posterior border of the mandibular ramus, it divides into an upper temporofacial branch and a lower cervicofacial branch.

DEVELOPMENT OF THE PAROTID GLAND

The parotid gland develops as an outgrowth from the epithelium of the buccal cavity, which extends backwards towards the ear superficial to the facial nerve. From its deep surface, projections of the developing parotid gland develop between the branches of the facial nerve. These coalesce to form the deep lobe of the gland. The most extensive of these deep prolongations is that between the temporofacial and cervicofacial divisions of the facial nerve. In anatomical terms, therefore, the parotid gland comes to form a larger (80%) superficial lobe and a smaller (20%) deep lobe joined by an isthmus between the two major divisions of the facial nerve.[1] The branches of the facial nerve lie between these two lobes (Fig. 8.1), although embryologically the parotid gland is not truly a bilobed structure.

THE PAROTID GLAND

The parotid gland is the largest of the three paired major salivary glands, the others being the submandibular and sublingual glands. It may be classified as a compound, tubuloacinar, merocrine, exocrine gland whose duct opens into the oral cavity. The term compound is used because the gland has more than one tubule entering the main duct; tubuloacinar describes the morphology of the secreting cells; merocrine indicates that only the secretion of the cell is released; exocrine refers to a gland that secretes onto a free surface. In the adult, the

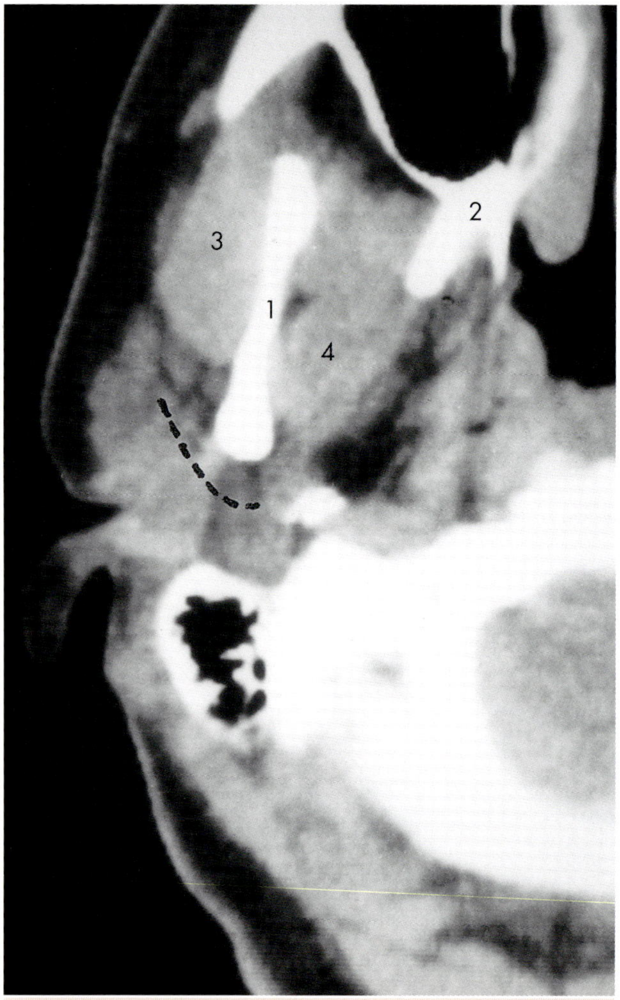

1 Ramus of mandible
2 Pterygoid process with pterygoid plates
3 Masseter muscle
4 Medial pterygoid muscle

Figure 8.1 A coronal CT scan showing the plane of the facial nerve (broken line).

parotid gland is composed entirely of serous acini.

The bony structures to which the parotid gland is related are best visualised in a lateral view of the skull (Fig. 8.2). The gland is wedged in the space between the posterior border of the ramus of the mandible in front and the mastoid process of the temporal bone behind. Above will lie the external acoustic meatus, the mandibular (glenoid) fossa housing the condyle of the mandible, and the zygomatic process of the temporal bone. More deeply (medially) the gland will be limited by the styloid process of the temporal bone. Below, the gland may overlap the angle of the mandible where its deep surface may be related to the transverse process of the first cervical vertebra or atlas.

The overall shape of the parotid gland is variable.[2] Viewed laterally, it appears roughly triangular in outline in approximately 50% of cases, with the base lying above and the apex being directed inferiorly (Fig. 8.3). In approximately 30% of cases, the gland is of more or less even width throughout, the upper and lower poles being rounded. Less common outlines encountered include: a triangular form with the apex superiorly and the base directed inferiorly (8%), and an inverted L-shape (8%). The average length of the gland is 6 cm (range 4.5–9.2) while the maximum width averages 3.3 cm (range 2.0–5.4).[1,2]

In about 20% of specimens, a small, detached part of the parotid gland tissue referred to as the accessory parotid gland may be seen in front of the main gland (Fig. 8.4).[1–3] The accessory parotid gland usually lies above the parotid duct, but may occasionally overlap the duct. It is separated from the main parotid gland by an average distance of 6 mm.

PAROTID CAPSULE

Standard textbooks describe the parotid gland as being surrounded by a fibrous capsule (the parotid capsule), this being a continuation of the investing layer of deep cervical fascia. This fascia is said to pass up from the neck and split to enclose the gland within a superficial and a deep layer. The superficial layer is attached above to the zygomatic process of the temporal bone, the cartilaginous part of the external acoustic meatus, and the mastoid process. The deep layer is attached to the mandible, and the tympanic plate and the styloid and mastoid processes of the temporal bone.[4–7]

There is general agreement that the deep layer of the parotid gland is derived from the investing layer of the deep cervical fascia, although there are some who believe that it is derived from the superficial cervical fascia and loose areolar tissue.[8,9] However, relatively recent investigations have produced data which require the origin of the superficial layer of the parotid capsule to be interpreted in association with the concept of the superficial musculo-aponeurotic system (SMAS).[8,10–15] From these studies, the superficial layer

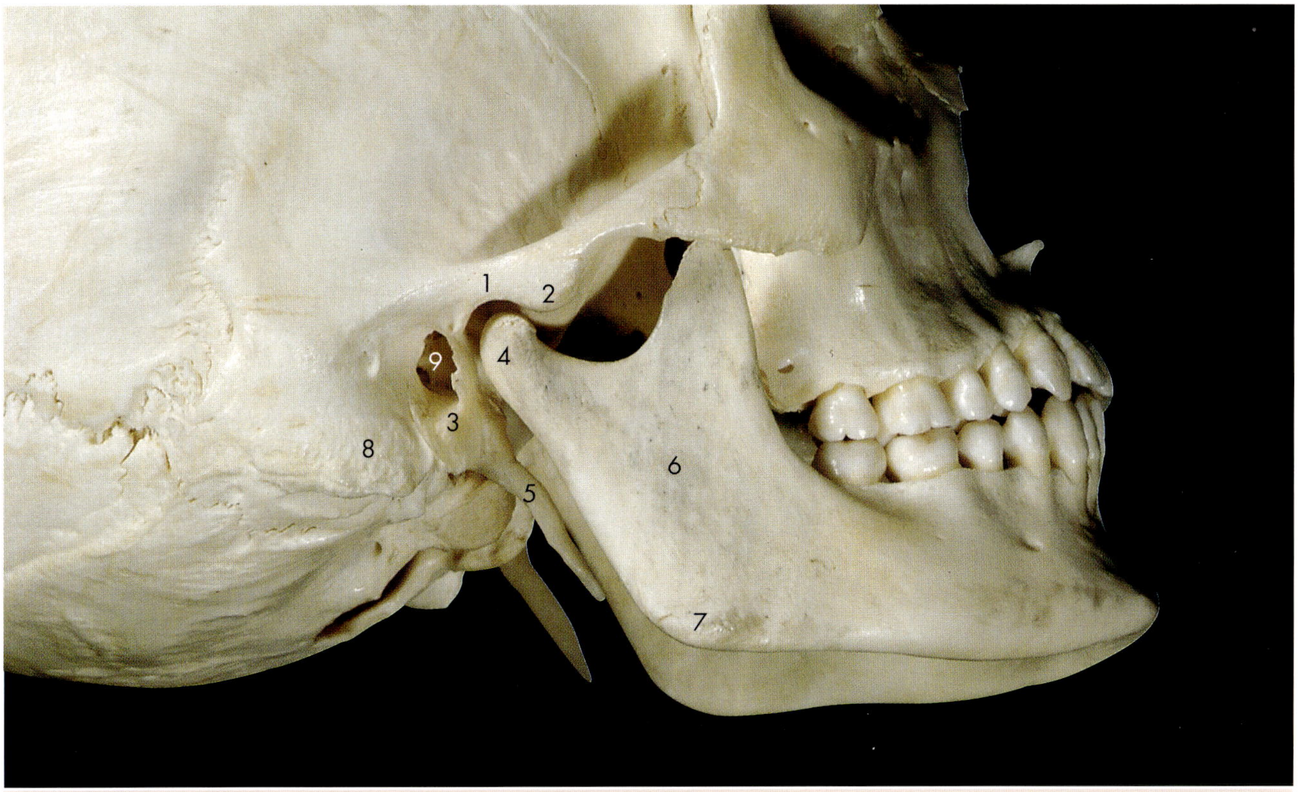

1 Mandibular fossa	6 Ramus of mandible
2 Articular eminence	7 Angle of mandible
3 Tympanic plate	8 Mastoid process
4 Mandibular condyle	9 External acoustic meatus
5 Styloid process	

Figure 8.2 A lateral view of the skull showing some of the bony features related to the bed of the parotid gland.

of the parotid capsule (also called parotid fascia) is seen to be of variable thickness, being a thick fibrous layer anteriorly and a thin transparent membrane posteriorly. Histological studies have shown that the parotid capsule is not a typical fascia as it contains muscle fibres that parallel those of platysma, especially in the lower part of the parotid capsule. The superficial layer of the parotid capsule appears to be continuous with the fascia associated with the platysma muscle. It may be traced forwards where it is seen as a separate layer passing over the masseteric fascia (itself derived from the deep cervical fascia) from which it is separated by a cellular layer containing branches of the facial nerve and the parotid duct. Between the skin and the superficial layer of the parotid capsule lies a subcutaneous layer containing variable amounts of adipose tissue. Although thin fibrous septa may be seen in the subcutaneous layer at the histological level, macroscopically there is little evidence of a distinct layer of superficial fascia. Functionally, therefore, the fascial layer associated with the platysma muscle and continuing as the superficial layer of the parotid capsule could be considered as the superficial fascia.

FORM AND RELATIONS OF THE PAROTID GLAND (Figs 8.3–8.5)

The parotid gland lies superficially on the side of the face in the space between the posterior border of the ramus in front and the mastoid process behind. Four

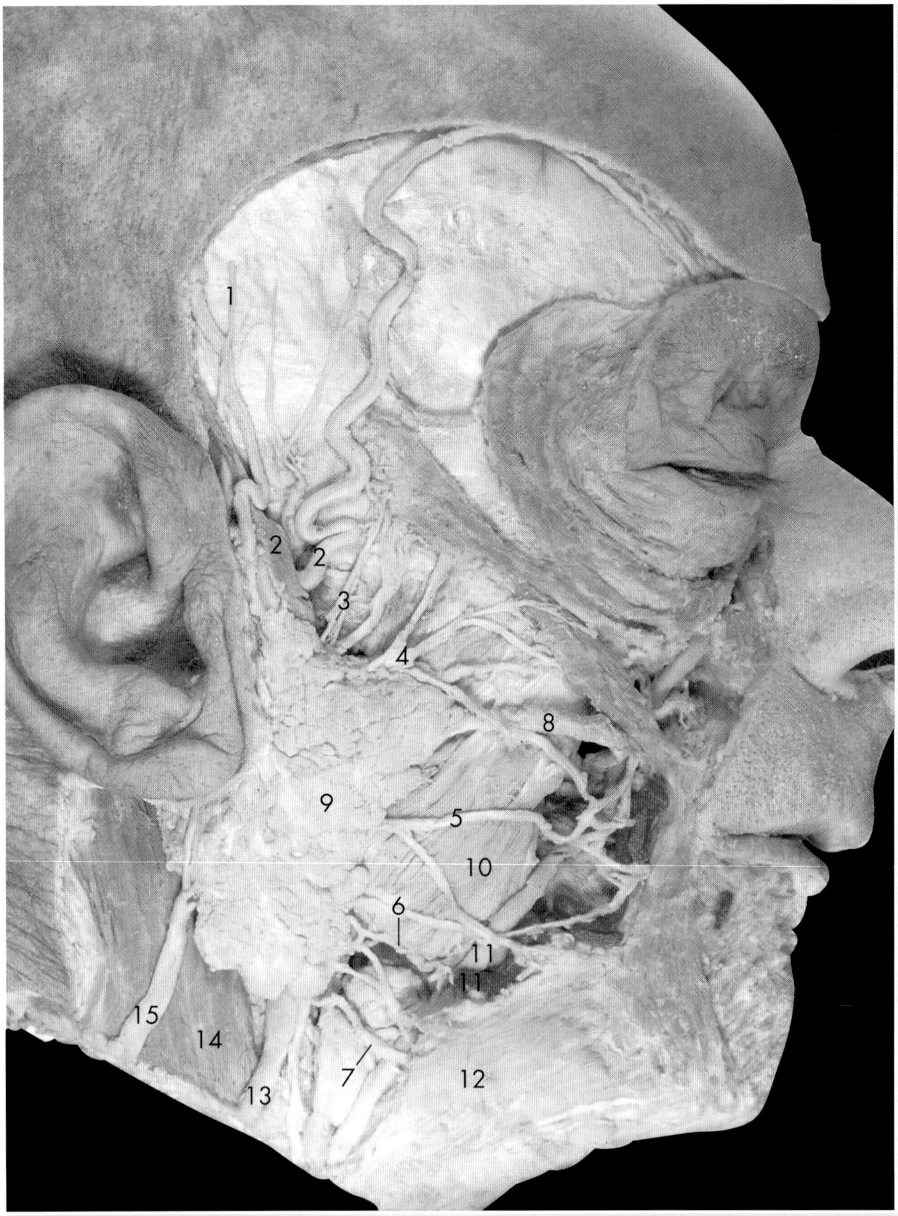

1 Auriculotemporal nerve
2 Superficial temporal vessels
3 Temporal branch of facial nerve
4 Zygomatic branch of facial nerve
5 Buccal branch of facial nerve
6 Mandibular branch of facial nerve
7 Cervical branch of facial nerve
8 Parotid duct
9 Parotid gland
10 Masseter muscle
11 Facial vessels
12 Platysma muscle
13 External jugular vein
14 Sternocleidomastoid muscle
15 Great auricular nerve

Figure 8.3 The parotid gland and associated structures. (Courtesy of Professor S. Standring, GKT School of Biomedical Sciences, London.)

borders (anterior, posterior, superior and inferior) and two surfaces (superficial and deep) can be defined.

The superior border (often corresponding to the base) is closely moulded around the external acoustic meatus and temporomandibular joint. The inferior border (often corresponding to the apex) is found in the region of the angle of the mandible, often extending beyond this to overlap into the digastric triangle (itself bounded by the anterior and posterior bellies of the digastric muscle). The anterior border extends onto the posterior part of the masseter muscle. The posterior border overlaps onto the anterior border of the sternocleidomastoid muscle.

As the parotid gland extends deep to the ramus, it has been divided for descriptive purposes into a large, superficial portion (lobe) and a small deep portion (lobe). Macroscopically, the region between the two portions is termed the isthmus. Consideration of

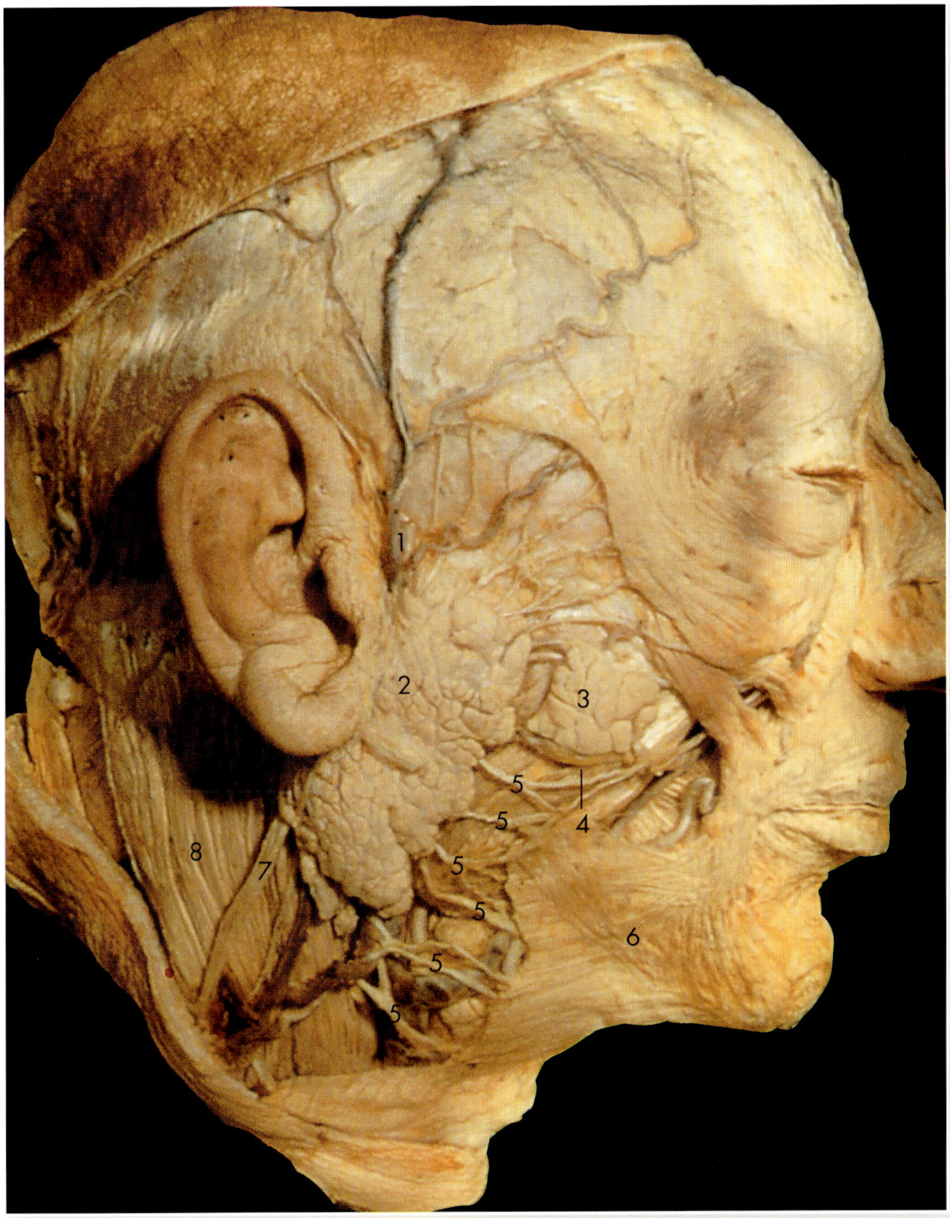

1 Superficial temporal vessels
2 Main parotid gland
3 Accessory parotid gland
4 Parotid duct
5 Branches of facial nerve
6 Platysma muscle
7 Great auricular nerve
8 Sternocleidomastoid muscle

Figure 8.4 The accessory parotid gland (Courtesy of the Royal College of Surgeons of England).

this subdivision of the parotid gland is also relevant to the clinical situation, as the approximate site of the junction between the two lobes corresponds to a plane in which lies the facial nerves and its branches, as well as some veins.[1,15–17] Thus, in many situations of parotid gland surgery, the aim is to carefully dissect away that part of the parotid gland overlying the facial nerve, roughly corresponding to the superficial portion. Some authors place considerable emphasis on the so-called bilobed nature of the parotid gland, purporting to show that it develops as a truly bilobed structure and that there is little glandular tissue between facial nerve branches, producing a specific cleavage plane. Others, however, are of the opinion that the gland is essentially unilobular and that the so-called superficial and deep lobes are in continuity between the branches of the facial nerve with no true cleavage plane existing.[1,16] The latter interpretation is probably more widely accepted.

Descriptive terms may be found in the anatomical literature for different parts of the parotid gland. For example, the superior part of the gland extends upwards behind the temporomandibular joint into the posterior part of the mandibular fossa and has been called the glenoid process or lobe. Where part of the anterior margin of the gland extends forward over the masseter muscle, it has been termed the facial process. The deep part of the gland extends forward between the medial pterygoid muscle and the ramus of the mandible and has been called the pterygoid process of the gland. The deep surface of the gland in close relationship to the internal carotid artery has been termed the carotid lobe.[15,19]

One classification, with a surgical emphasis, subdivides the gland into parts related to the plane created by the facial nerve and posterior facial vein.[9]

The superficial surface of the parotid gland is covered by the skin and the platysma muscle (Fig. 8.5).

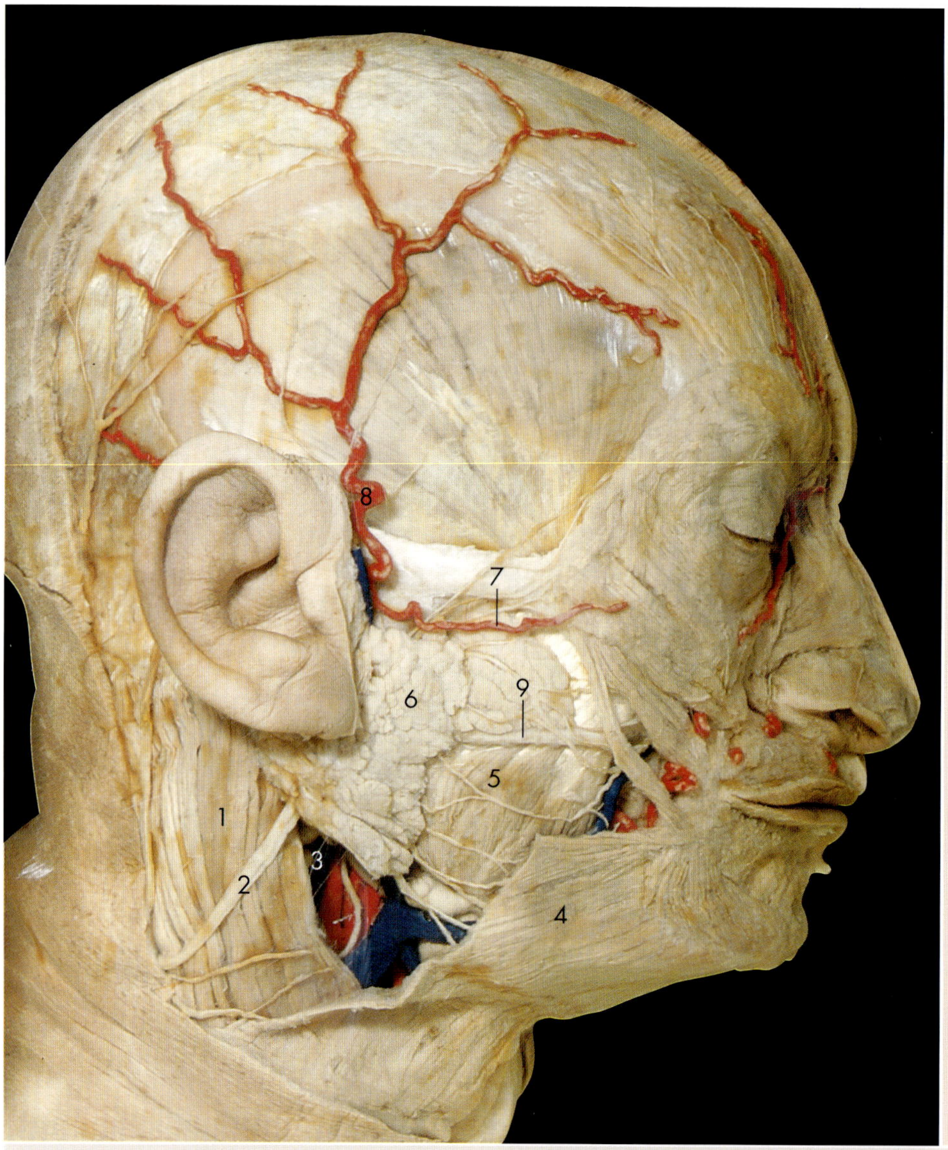

1 Sternocleidomastoid muscle
2 Great auricular nerve
3 Internal jugular vein in carotid sheath
4 Platysma muscle
5 Masseter muscle
6 Parotid gland
7 Transverse facial artery
8 Superficial temporal vessels and auriculo-temporal nerve
9 Parotid duct

Figure 8.5 Superficial and other relations of the parotid gland. (Courtesy of Professor L. Garey, Anatomy Department, Imperial College Medical School, London.)

Some additional fibres ascribed to an irregular facial muscle, the risorius muscle, may also be present, arising from the fascia in the area. Terminal branches of the great auricular nerve (a sensory component of the cervical plexus and derived from the anterior primary rami of the second and third cervical spinal nerves) also lie superficial to the gland and send branches through the substance of the parotid gland to communicate with the facial nerve.[19] Similar communications exist for the transverse cervical branch of the cervical plexus (also derived from the second and third cervical spinal nerves).

At the superior border of the parotid gland lie the superficial temporal vessels, the vein behind the artery. Posterior to the vessels, and at a slightly deeper level, is the auriculotemporal branch of the mandibular nerve.

The branches of the facial nerve are seen arising from the anterior border of the gland. Also arising from this surface is the parotid duct, above which may be seen the transverse facial artery, a branch of the superficial temporal artery (Fig. 8.5).

From the inferior border of the parotid gland may be seen the anterior and posterior branches of the posterior facial (retromandibular) vein (Fig. 8.18).

No vascular or neural elements are seen at the posterior border of the parotid gland.[20]

The medial or deep surface of the parotid gland lies adjacent to structures forming the so-called parotid bed (Figs 1.4, 8.19). In front, the gland is applied to the masseter muscle and the posterior border of the ramus, including its upper condylar process. As it passes around the ramus, the gland becomes related to the medial pterygoid muscle, where it inserts into the medial surface near the angle. Passing more posteriorly, the parotid gland becomes moulded around the styloid process and its group of muscles, namely the styloglossus (arising from the tip of the styloid process), the stylopharyngeus (arising from the base of the styloid process) and the stylohyoid (arising between the two previous muscles). More posteriorly, the parotid gland lies on the posterior belly of the digastric muscle (arising from the digastric notch at the base of the mastoid process) and finally on the sternocleidomastoid muscle. The digastric and styloid group of muscles separate the gland from the internal jugular vein, the external and internal carotid arteries, the glossopharyngeal, vagus, accessory and hypoglossal nerves and the sympathetic trunk.

The fascia covering the muscles forming the parotid bed shows two specialisations. The stylomandibular ligament passes from the styloid process to the angle of the mandible. The more extensive mandibulostylohyoid ligament (angular tract) passes between the angle of the mandible and the stylohyoid ligament for varying distances, generally reaching the hyoid bone (Fig. 8.6). It is thick posteriorly but thins anteriorly in the region of the angle of the mandible.[21–24] Whereas some authors regard the mandibulostylohyoid ligament as part of the deep cervical fascia,[21] others regard it as lying deep to it.[23] As the stylomandibular and mandibu-

Figure 8.6 The mandibulostylohyoid ligament.

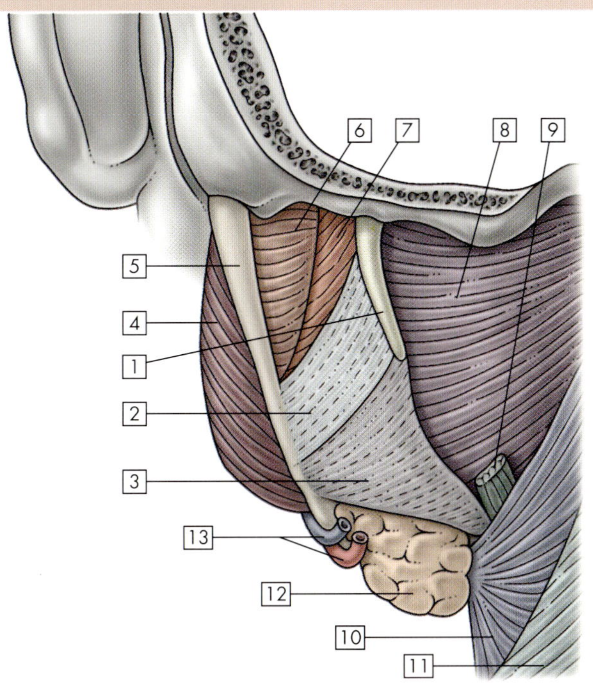

1 Styloid process
2 Stylomandibular ligament
3 Mandibulostylohyoid ligament
4 Masseter muscle
5 Posterior border of ramus
6 Lateral pterygoid muscle
7 Medial pterygoid muscle
8 Superior constrictor muscle
9 Cut end of stylopharyngeus muscle
10 Middle constrictor muscle
11 Inferior constrictor muscle
12 Submandibular gland
13 Facial vessels

Adapted from ref. 23.

lostylohyoid ligaments separate the parotid gland region from the superficial part of the submandibular gland, they are landmarks of surgical interest.

CONTENTS OF THE PAROTID GLAND

Four major structures lie within the substance of the parotid gland. From superficial to deep, these structures are: the facial nerve, the auriculotemporal nerve, the retromandibular vein and the external carotid artery. Each of these structures has branches or tributaries within the gland.

FACIAL NERVE[25,26] (Figs 8.3, 8.4, 8.7–8.10, 8.20)

The facial nerve exits the skull through the stylomastoid foramen. To expose this site surgically, one may dissect down between the parotid gland and the external acoustic canal until the junction of the cartilaginous and bony canals can be palpated (Fig. 8.7). A small triangular extension of the cartilaginous canal points towards the facial nerve as it exits the stylomastoid foramen.[27] At the stylomastoid foramen, the facial nerve lies about 9 mm from the posterior belly of the digastric muscle and 11 mm from the bony external acoustic meatus.[28] The facial nerve then passes downwards and forwards over the styloid process and the attached muscles for an average distance of 1.3 cm before entering the substance of the parotid gland.[29] The first portion of the facial nerve gives off the posterior auricular nerve to supply the auricular muscles and branches to supply the posterior belly of the digastric and stylohyoid muscles. There are also communicating branches with the transverse cervical nerve (Fig. 8.20).

On entering the substance of the parotid gland, the facial nerve soon separates into its two primary components, the temporofacial and cervicofacial divisions,

Figure 8.7 Surgically uncovering the facial. The nerve is found 5 cm below the skin incision. It crosses the lateral surface of the styloid process. The sternocleidomastoid muscle is retracted posteriorly just below its attachment to the mastoid process. The meatal cartilage narrows and points to the facial nerve. The tympanomastoid suture also points to the nerve. The facial nerve rapidly ascends on the posterior surface of the parotid gland, which is retracted forwards so that the nerve quickly approaches the skin of the face.

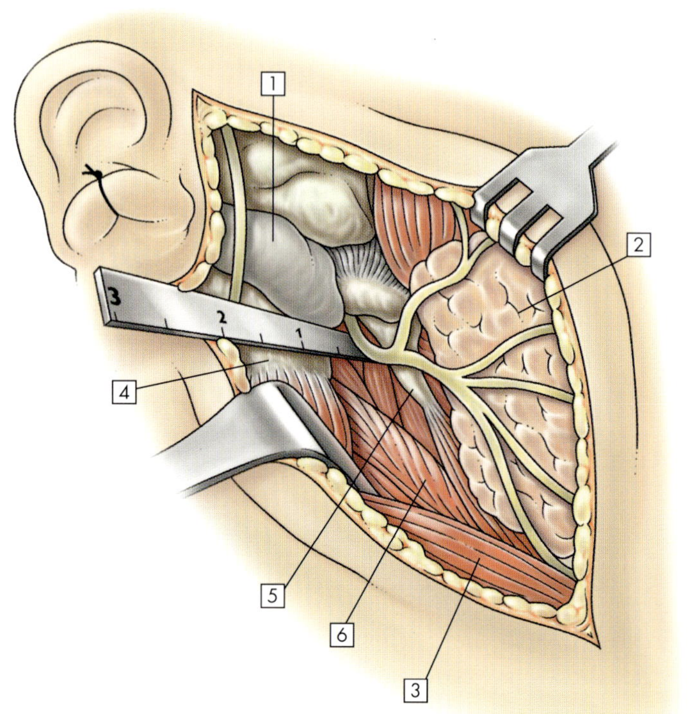

1 Cartilaginous external acoustic meatus
2 Parotid gland
3 Sternocleidomastoid muscle
4 Tip of the mastoid process
5 Styloid process
6 Posterior belly of digastric muscle

the former being the larger (Fig. 8.8). The point of this division has various landmarks. It lies an average of 2.3 cm vertically below the lowest point of the external acoustic meatus,[29] and an average of 3.2 cm above the angle of the mandible (two-thirds of the distance from the angle to the temporomandibular joint).[1] Where the nerve separates into its two main divisions, it has been termed the pes anserinus due to its resemblance to a

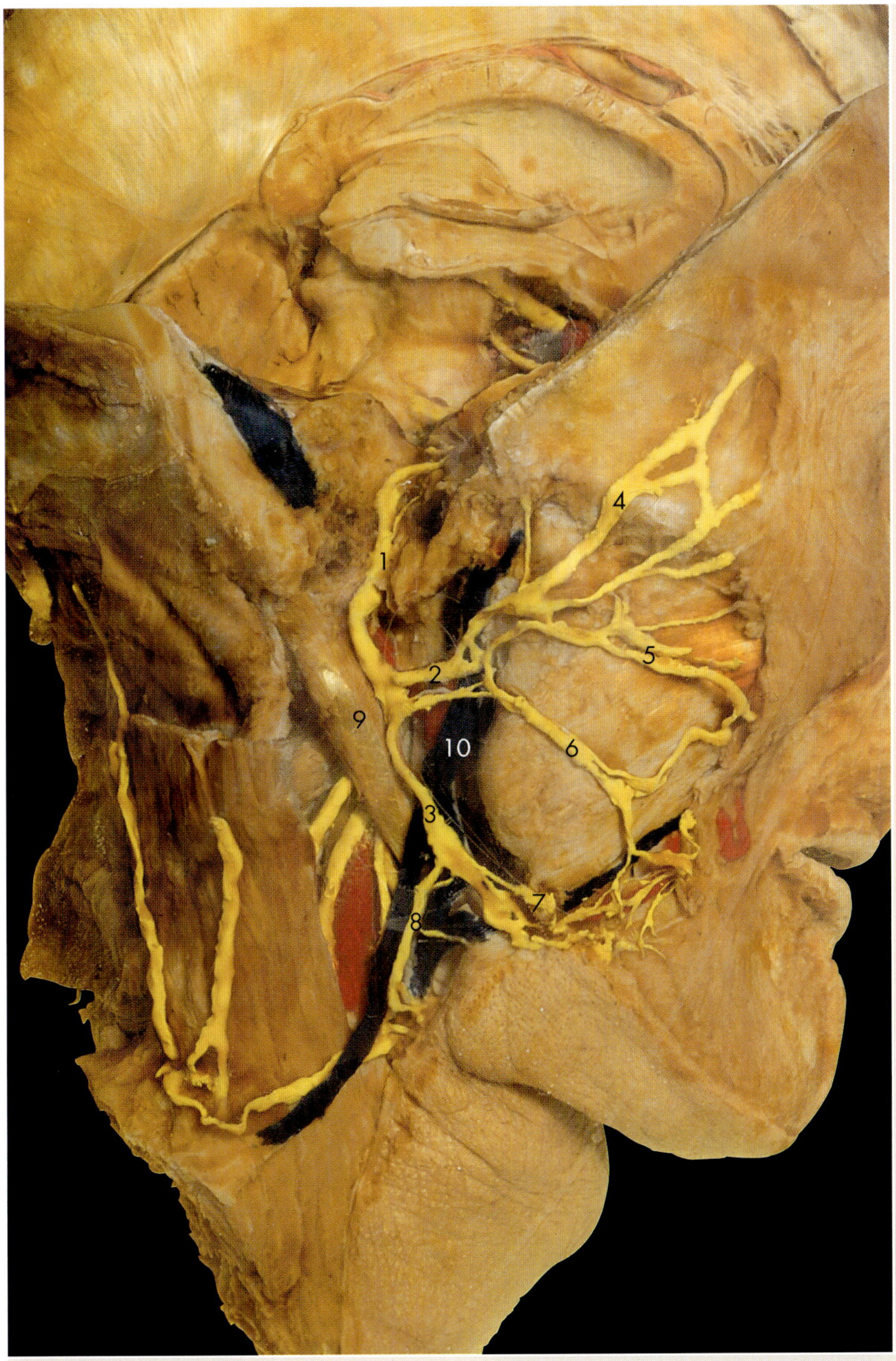

1 Facial nerve at stylomastoid foramen
2 Temporofacial branch of facial nerve
3 Cervicofacial branch of facial nerve
4 Temporal branch of facial nerve
5 Zygomatic branch of facial nerve
6 Buccal branch of facial nerve
7 Mandibular branch of facial nerve
8 Cervical branch of facial nerve
9 Posterior belly of digastric muscle
10 Retromandibular vein and external carotid artery

Figure 8.8 The facial nerve within the parotid gland.

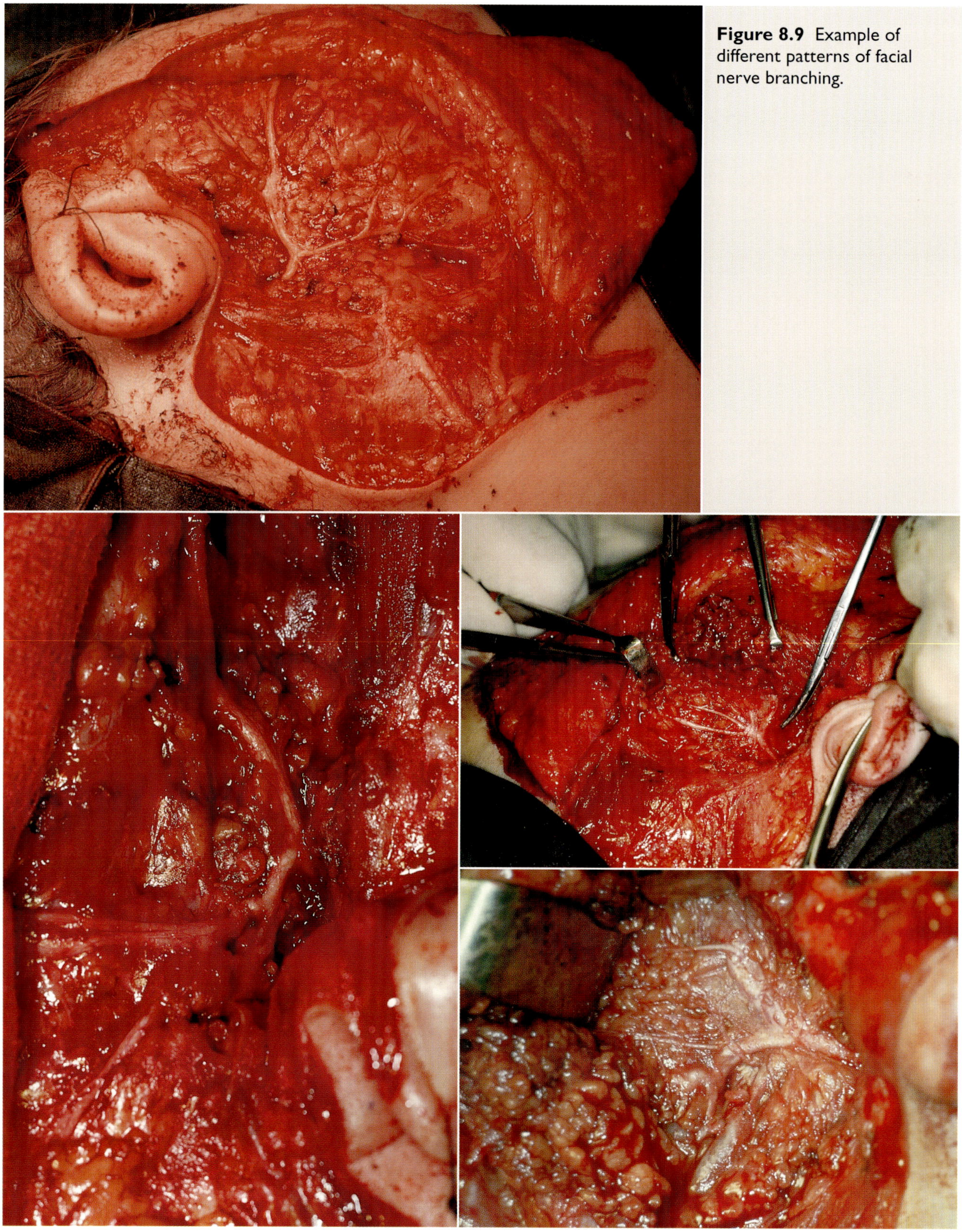

Figure 8.9 Example of different patterns of facial nerve branching.

goose's foot. Variations of these two main divisions as seen during surgery are illustrated in Fig. 8.9.

From its two main divisions (temporofacial and cervicofacial), the facial nerve gives rise to five sets of branches, named according to their region of distribution. These branches are temporal, zygomatic, buccal, mandibular and cervical. The temporal and zygomatic branches are generally multiple and arise from the temporofacial division. The mandibular (generally one or two branches) and cervical (generally one branch) nerves arise from the cervicofacial division. The buccal branch(es) has a variable origin.

The peripheral branches of the facial nerve described above form anastomotic arcades between adjacent branches to give the parotid plexus. In surgical terms, these anastomoses are important as accidental or essential division of a small branch often fails to result in the expected facial nerve weakness due to the dual innervation afforded by these anastomoses.

Davis *et al.* studied these patterns following the dissection of 350 facial nerves.[1] They found that the anastomotic relationships between the branches of the temporofacial and cervicofacial divisions resulted in the formation of six distinctive patterns (Fig. 8.10). Of facial nerves, 13% formed the *Type I* pattern. In this group, no anastomoses occur between adjacent branches. The five major branches (pes anserinus) spread out radially like the spokes of a wheel. From the surgical aspect, this is the most unforgiving pattern. Damage to any branch will result in paralysis of the muscles supplied by that branch. *Type II* nerves comprise 20% of the 350 facial nerve dissections. It is characterised by an arcade of anastomoses between the various branches of the temporofacial division. The anastomoses usually occur beyond the anterior border of the parotid where the nerve is very superficial. *Type III* specimens (28%) are characterised by a single relatively large anastomosis between the temporofacial and cervicofacial divisions again beyond the anterior margin of the gland crossing the parotid duct. *Type IV* nerves account for

Figure 8.10 Chief types of branching encountered in 350 cervicofacial halves. Schematically shown, as of the right side of the head. Note that types III, IV, V and VI together represent almost 70 per cent of specimens.

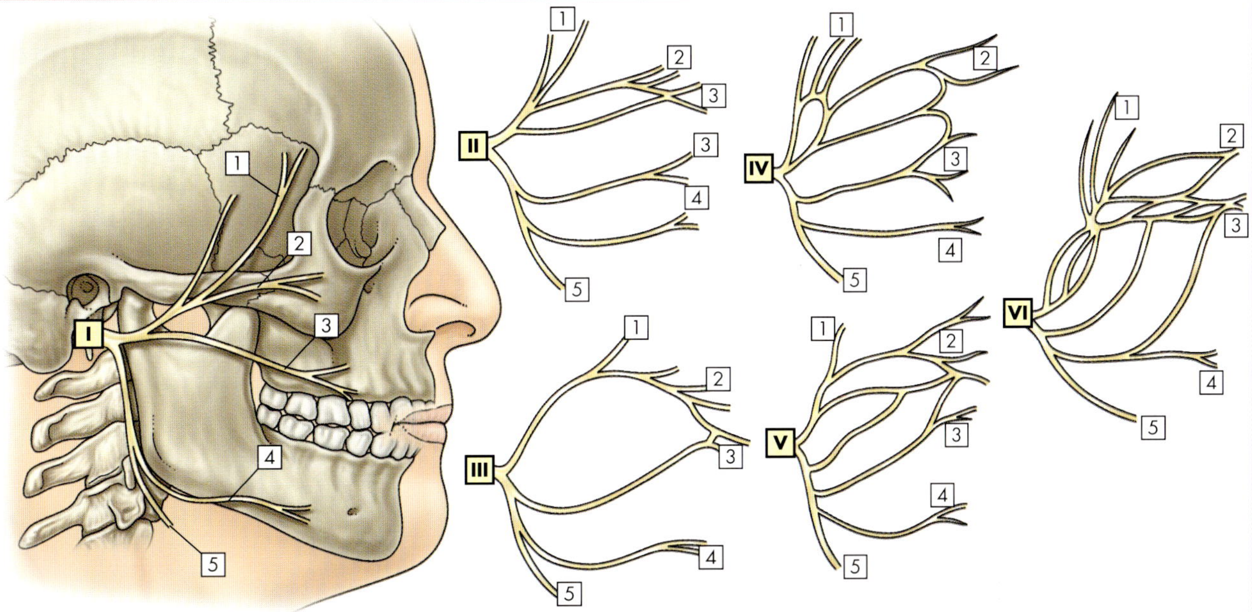

I Type I, 13%
II Type II, 20%
III Type III, 28%
IV Type IV, 24%
V Type V, 9%
VI Type VI, 6%
1 Temporal branch
2 Zygomatic branch
3 Buccal branch
4 Mandibular branch
5 Cervical branch
Adapted from ref. 1

24% of dissections. In this pattern, there are anastomotic loops between the temporal and zygomatic branches, as well as connections between the cervicofacial division and the buccal and zygomatic branches. These anastomotic loops may occur within the substance of the parotid gland in the plane between the two 'lobes'. Nerves classified as *Type V* account for 9% of the total. In this group, there were two anastomotic rami passing from the cervicofacial division to branches of the temporofacial division. The anastomosis arising from the cervicofacial division may arise from the buccal branch or directly from the point where the facial nerve trunk divides. Only 6% of nerves conform to the *Type VI* pattern, in which there was a rich plexiform arrangement of anastomoses. In this type, all the branches and filaments are small in diameter but because of the complicated pattern of anastomotic arcades, denervation of any muscle group is unlikely. Uniquely, this is the only pattern of the facial nerve in which the (marginal) mandibular branch is reinforced by an anastomosis from an adjacent branch. This finding certainly accounts for the observation that, when transient facial nerve weakness occurs following facial nerve surgery, it is usually the (marginal) mandibular branch that is affected. In Langdon's experience (unpublished data) of more than 250 facial nerve dissections as part of parotidectomy procedures, the patterns of the facial nerve branching do conform to these six main groups.

A similar pattern of branching of the facial nerve has been described by others, although some variation in frequency has been reported.[30,31]

The temporofacial division of the facial nerve does not usually pass lower than the level of the tip of the ear lobe and crosses the superficial temporal vessels and auriculotemporal nerve at the level of the intertragic notch of the ear.[32]

Temporal branch(es) of facial nerve

The temporal branches usually number three or four. They cross the zygomatic arch anteriorly to the superficial temporal vessels at a region intersected by a perpendicular from the anterior temporal hairline.[32,33] The main areas of anastomoses are between the middle and posterior branches.[34] The nerve contributes to the supply of the auricular muscles and supplies muscles of the forehead and orbicularis oculi.

Zygomatic branch(es) of the facial nerve

The zygomatic branches may be up to three in number and run below the lower border of the zygomatic arch.[27,35] They contribute to the innervation of the orbicularis oculi, to muscles of the nose and to muscles associated with the upper lip.

Buccal branch of the facial nerve

This nerve may originate from either the cervicofacial division (35%), the temporofacial division (21%), both the cervicofacial and temporofacial divisions (34%), or from both the cervicofacial and temporofacial divisions with additional anastomoses between these branches (10%).[36] Generally there is only one branch, but there may be two.[27,37]

A straight line drawn from the oral commissure to the tragus approximately identifies the level of the buccal branch (and the parotid duct). Where the nerve is single, it is usually (75%) found below the origin of the parotid duct, but it may cross the duct in 25% of cases in its anterior course. Where there are two branches (15%), one branch passes above and one below the parotid duct.[37] Passing beneath the zygomaticus muscle, the buccal branch of the facial nerve supplies the buccinator muscle and muscles of the upper lip.

(Marginal) mandibular branch(es) of the facial nerve

This nerve has two branches in 55–65% of cases, the lower branch being the larger. A single branch occurs in 25% of cases and three branches in about 10% of cases. Rarely (3%), four branches may be present.[31,38,39] Anastomoses with the buccal and with the cervical branches occur in between 6–15% of cases.[1, 30, 31]

The mandibular branch at the antero-inferior pole of the parotid gland is often closely accompanied by the cervical branch below, but the nerves have diverged by the level of the angle of the mandible. An important surgical relationship exists between the mandibular branch and the lower border of the mandible. In one study, this nerve ran below the mandible in 20% of cases,[38] while another study reported a higher incidence of 53%,[39] the furthest distance being 1.2 cm. Whereas the former study reported that the mandibular branch of the facial nerve was always above the lower border of the mandible distal to the point where the facial artery crossed the mandible at the anterior border of the masseter muscle, the latter study noted that the nerve was still below the mandible in 6% of cases. The mandibular branch supplies the muscles of the lower lip and, like the cervical branch below, lies in a plane deep to the platysma muscle but superficial to the investing layer of deep cervical fascia.

Cervical branch of facial nerve

In 80% of cases, this appears as a single branch at the lower pole of the parotid gland and always passes downwards at a variable distance behind the angle of the mandible, the furthest being 1.4 cm. The nerve passes behind the superficial portion of the submandibular gland at the level of the hyoid bone where it divides into a number of branches which supply the platysma muscle. In the remaining 20% of cases, the nerve is represented by two branches.[40]

As mentioned earlier, the studies described in this chapter have been undertaken on preserved human material. That this can differ from the situation seen in surgery is evident in the case of the facial nerve. For example, it has been reported that, at operation, the mandibular branch of the facial nerve is 1–2 cm below the lower border of the mandible in every case and, as the neck is extended, the nerve may be drawn even lower.

Surgical exposure of the facial nerve

The key to safe dissection of the facial nerve is the exposure of the trunk of the nerve as it leaves the stylomastoid foramen. Essentially, there are four techniques in regular use. The surgeon needs to be familiar with each as often the approach used in a particular case is dictated by the location of, for example, a large parotid tumour obstructing the surgeon's usual approach.

Approach 1

The method preferred by Langdon of exposing the trunk of the facial nerve is to dissect the parotid capsule from the cartilaginous part of the external acoustic meatus. Following a preauricular incision which curves gently posteriorly below the mastoid process and then down into a neck crease just behind the anterior border of the sternocleidomastoid (Fig. 8.11), the skin flap is raised in the plane of the superficial parotid fascia. The parotid fascia is then separated from the cartilage of the external acoustic meatus. This plane which is bloodless is opened up by sharp dissection with scissors (Fig. 8.12). Then with blunt dissection, the cartilage of the external acoustic meatus is separated from the parotid capsule medially until the base of the styloid process can be palpated in the depth of the dissection 4–5 cm from the skin surface (Fig. 8.13). In the infant or child, this distance is very much reduced as the mastoid process has not developed. Having defined the depth of the dissection, the area is then opened up inferiorly to expose the anterior border of the mastoid process and the anterior border of the sternocleidomastoid at its insertion (Fig. 8.14). Some small vessels are encountered at this stage and should be carefully controlled using bipolar diathermy. At this stage, the anterior branch of the great auricular nerve must be divided but, where possible, the posterior branch should be preserved as this supplies sensation to the ear lobe. Gentle retraction anteriorly on the posterior capsule of the parotid gland and posterior traction on the sternocleidomastoid exposes the trunk of the facial nerve as it emerges from the posterolateral aspect of the styloid process (Fig. 8.14). At this point, the facial nerve is covered by a dense cellular and fibrous tissue that impedes the dissection. If the surgeon is uncertain, advantage must be taken of the cartilaginous 'pointer' of the external acoustic meatus. The deep extension of the cartilage tapers to a point as it approaches the bony meatus and itself points to the facial nerve. This technique is particularly useful when a large parotid tumour is located in the lower pole of the gland.

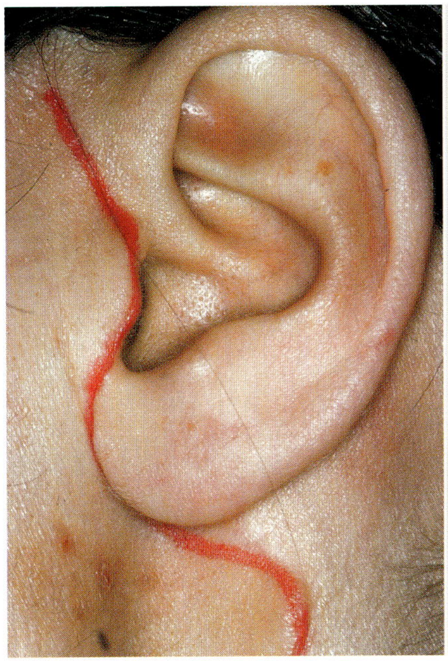

Figure 8.11 The skin incision for facial nerve exposure.

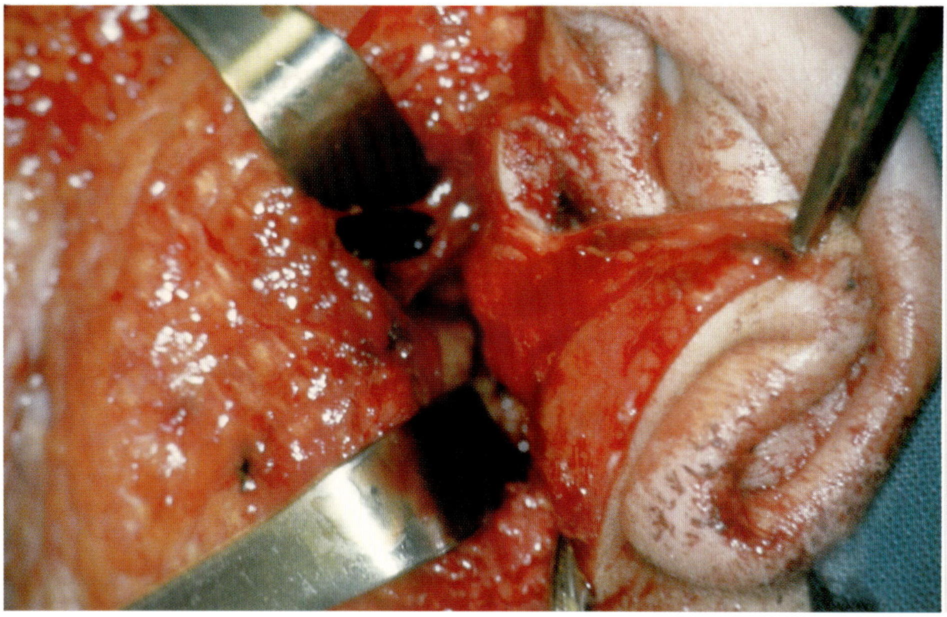

Figure 8.12 Surgery for the facial nerve showing the avascular plane in front of the external acoustic meatus.

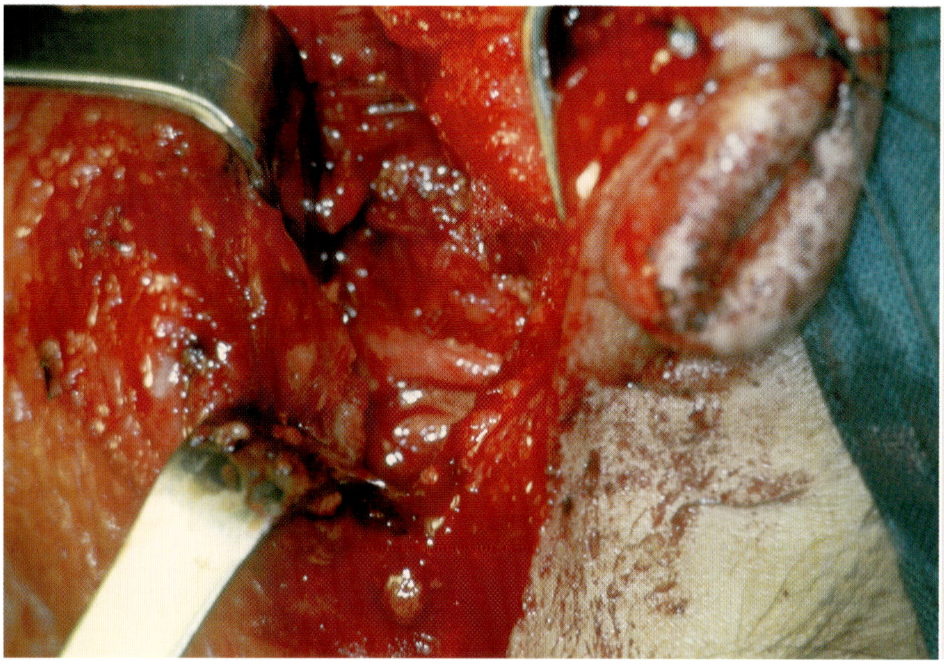

Figure 8.13 Surgery for the facial nerve. The trunk of the facial nerve is emerging from the posterolateral aspect of the styloid process.

Approach 2

An alternative approach to the facial nerve trunk is to rely on the fact that the digastric notch on the medial side of the mastoid process points to the stylomastoid foramen, which lies just anterior to the notch. In this technique, the anterior border of the sternocleidomastoid is approached from below and the muscle is retracted posteriorly. The anterior border of the posterior belly of the digastric above the transverse process of the atlas is defined (Figs 8.15, 8.16). Using blunt dissection the facial nerve trunk is exposed between the anterior border of the mastoid process and the styloid process. This technique is indicated when a bulky parotid tumour lies posteriorly in the preauricular region of the gland.

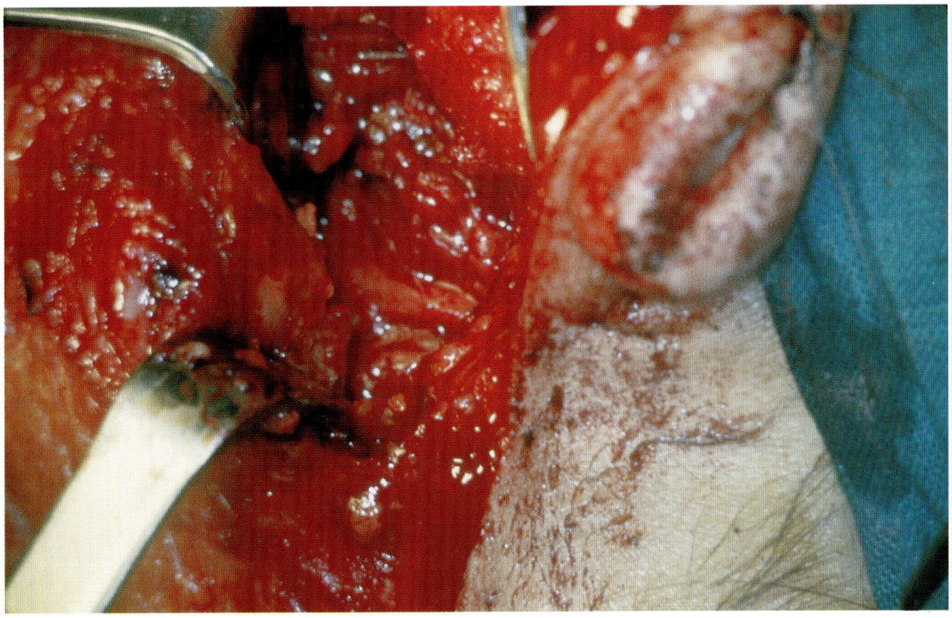

Figure 8.14 Surgery for the facial nerve. The surgical area is extended to expose the anterior borders of the mastoid process and the sternocleidomastoid muscle.

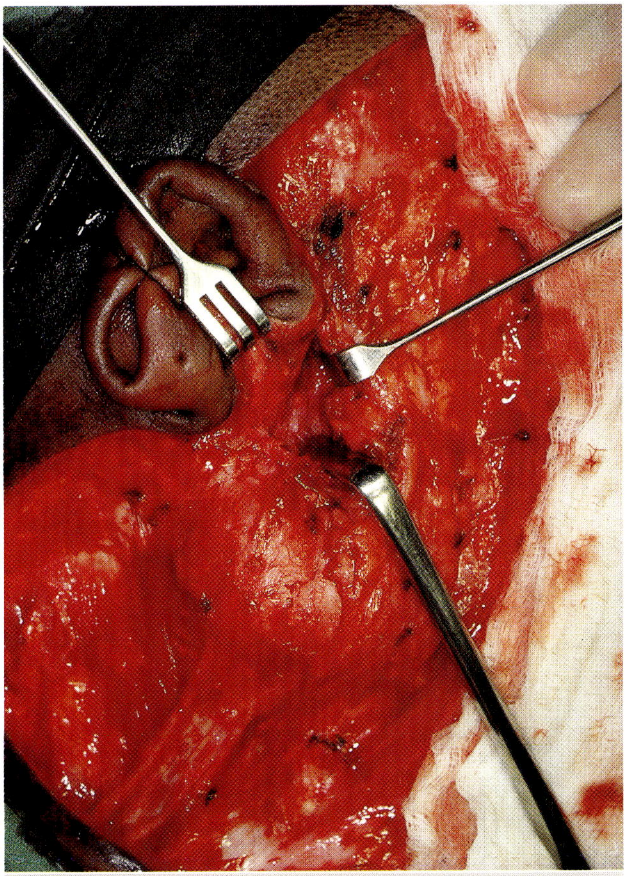

Figure 8.15 Surgery for the facial nerve. The anterior border of the sternocleidomastoid muscle is retracted posteriorly.

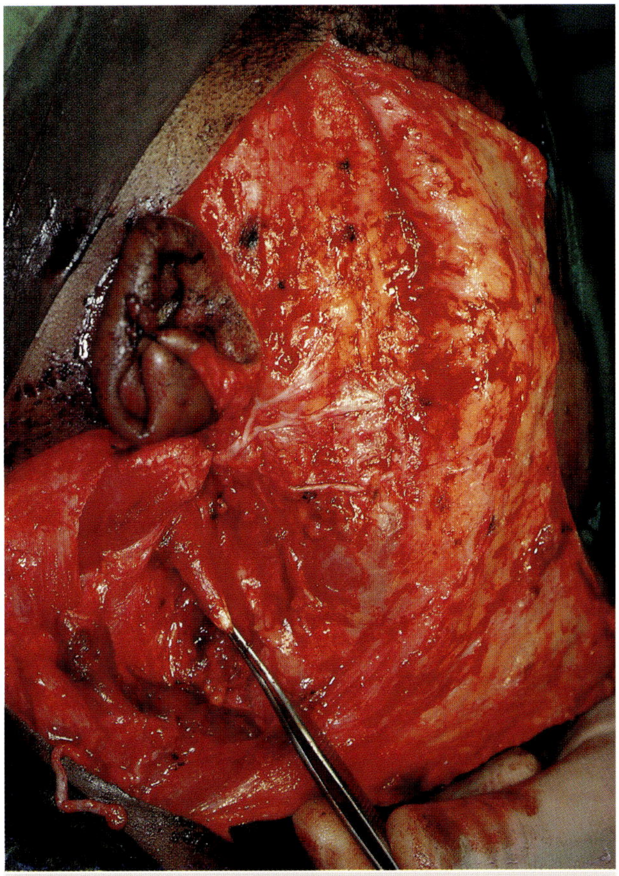

Figure 8.16 Surgery for the facial nerve. The insertion of the posterior belly of the digastric muscle is just inferior to the trunk of the facial nerve.

Approach 3

This method utilises the tympanomastoid suture. The posterior part of this fissure is almost subcutaneous and is rapidly palpated after the skin flap has been raised. When followed medially, the suture points the way to the facial nerve as it lies just inferior to the stylomastoid foramen.

Approach 4

In this technique, one or more peripheral branches of the facial nerve is traced centripetally. Usually the marginal mandibular branch is identified and followed centrally to the cervicofacial division and then back to the facial nerve trunk. The marginal mandibular nerve is located at the lower border of the mandible as it crosses the facial artery at the anterior border of the masseter muscle. Having identified the nerve, it is then traced backwards in the neck where it approaches the deep aspect of the retromandibular (posterior facial) vein.

THE AURICULOTEMPORAL NERVE

This nerve arises from the posterior division of the mandibular division of the trigeminal nerve in the infratemporal fossa. Running backwards beneath the lateral pterygoid muscle, the nerve passes between the medial surface of the condylar process of the mandible and the sphenomandibular ligament (Fig. 8.17). It enters the anteromedial surface of the parotid gland, passing upwards and outwards to emerge at the superior border of the gland between the temporomandibular joint and the external acoustic meatus (Fig. 8.19). The auriculotemporal nerve communicates with the facial nerve. The communications with the temporofacial division anchor the facial nerve close to the lateral surface of the condylar process of the mandible, limiting its mobility during surgery.[15] Communications with the temporal and zygomatic branches loop around the transverse facial and superficial temporal vessels.[34]

THE RETROMANDIBULAR VEIN

The retromandibular vein is formed within the parotid gland by the union of the superficial temporal vein, that enters at the superior border, and the maxillary vein, that enters the posteromedial surface. The retromandibular vein passes downwards and, near to the inferior pole of the parotid gland, often divides into two branches which pass out of the gland (Fig. 8.8). The posterior branch passes backwards to unite with the posterior auricular vein on the surface of the sternocleidomastoid muscle to form the external jugular vein. The anterior branch passes forwards to join the facial vein (Fig. 8.18).

Being of a reasonable size, the retromandibular vein is an important landmark for the facial nerve. The separation of the nerve into its two main divisions (temporofacial and cervicofacial) occurs just posterior to (within about 5 mm) the retromandibular vein. In nearly 90% of cases, the two divisions of the facial nerve are seen to lie superficial to the vein and in intimate contact with it (Fig. 8.8). Occasionally (9%), the temporofacial division is seen to pass deep to the retromandibular vein, while only very rarely (2%) does the cervicofacial division of the facial nerve pass deep to the retromandibular vein.[41]

THE EXTERNAL CAROTID ARTERY

The deepest structure within the parotid gland is the external carotid artery (Fig. 8.8). This appears from behind the posterior belly of the digastric muscle and grooves the posteromedial surface of the parotid gland before entering it. The first branch given off by the external carotid artery within the parotid compartment is the posterior auricular artery.[42] Ascending to a position medial to the neck of the condylar process of the mandible, the external carotid artery then divides into its two terminal branches, the superficial temporal and maxillary arteries (Figs 8.17, 8.19). The superficial temporal artery continues upwards to appear at the superior border of the gland and crosses the zygomatic arch to the temporal region. Within the substance of the gland, it gives off the transverse facial artery (Figs 8.5, 8.20), which emerges at the anterior border of the gland and runs across the face above the parotid duct. The maxillary artery emerges from the anteromedial surface of the gland to enter the infratemporal fossa (see pages 14–15). The maxillary artery gives off two branches within the substance of the gland:

- the deep auricular artery, which passes through the cartilaginous portion of the external acoustic meatus to supply the lateral surface of the tympanic membrane; and

- the anterior tympanic artery which passes through the petrotympanic fissure to supply the medial surface of the tympanic membrane.

The major arteries within the substance of the parotid gland also give off small branches to supply the gland tissue itself.

Sections of the head displaying the parotid gland are illustrated in Figs 8.21–8.23.

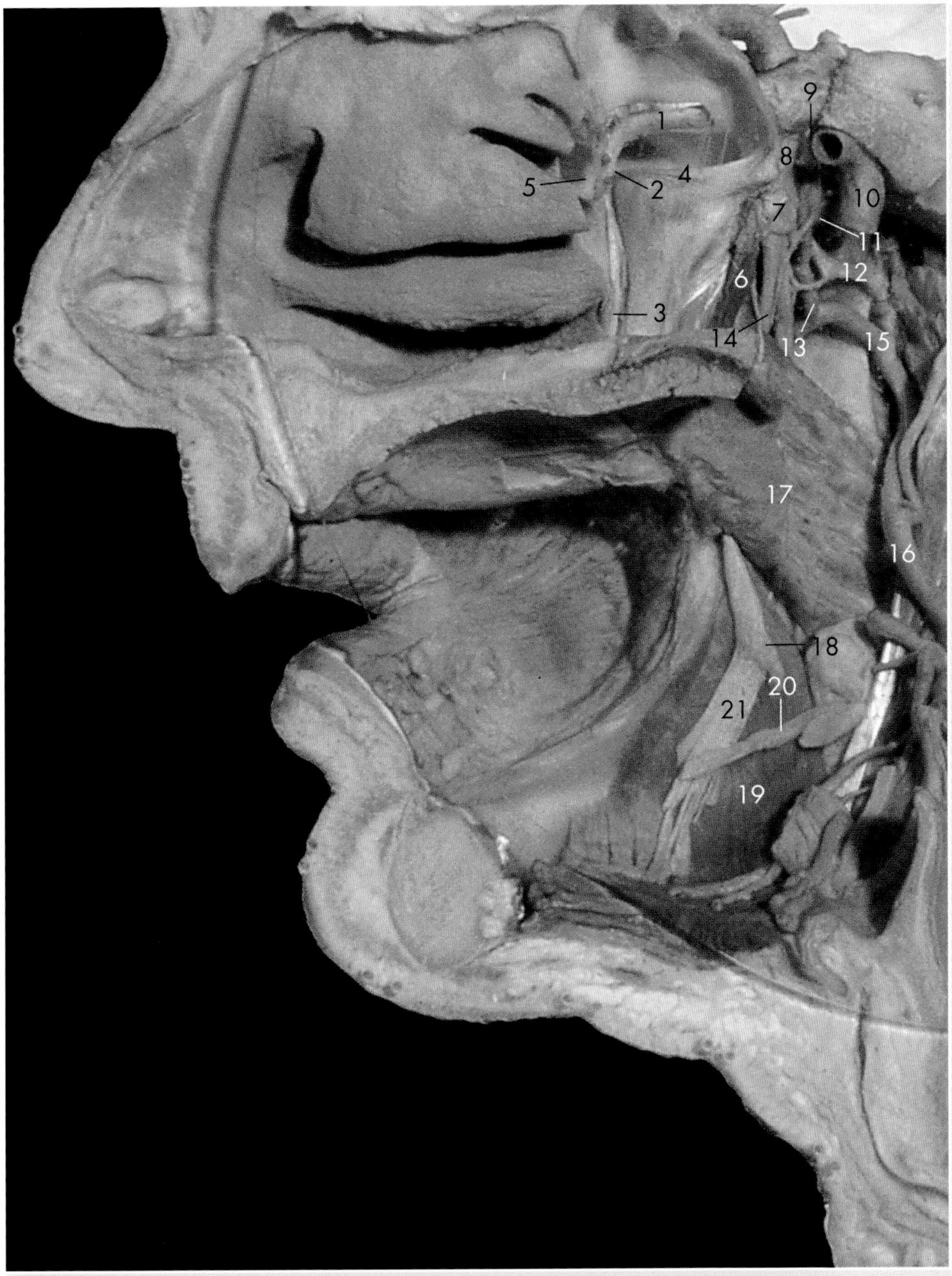

1 Maxillary nerve
2 Pterygopalatine ganglion
3 Palatine nerves
4 Nerve of pterygoid canal
5 Sphenopalatine artery at sphenopalatine foramen
6 Nerve to medial pterygoid muscle
7 Otic ganglion
8 Mandibular nerve
9 Greater petrosal nerve
10 Internal carotid artery
11 Chorda tympani nerve
12 Auriculotemporal nerve
13 Middle meningeal artery
14 Inferior alveolar nerve
15 Maxillary artery
16 External carotid artery
17 Medial pterygoid muscle
18 Submandibular ganglion
19 Mylohyoid muscle
20 Submandibular duct
21 Lingual nerve

Figure 8.17 The auriculotemporal nerve, viewed medially. (Courtesy of the Royal College of Surgeons of England.)

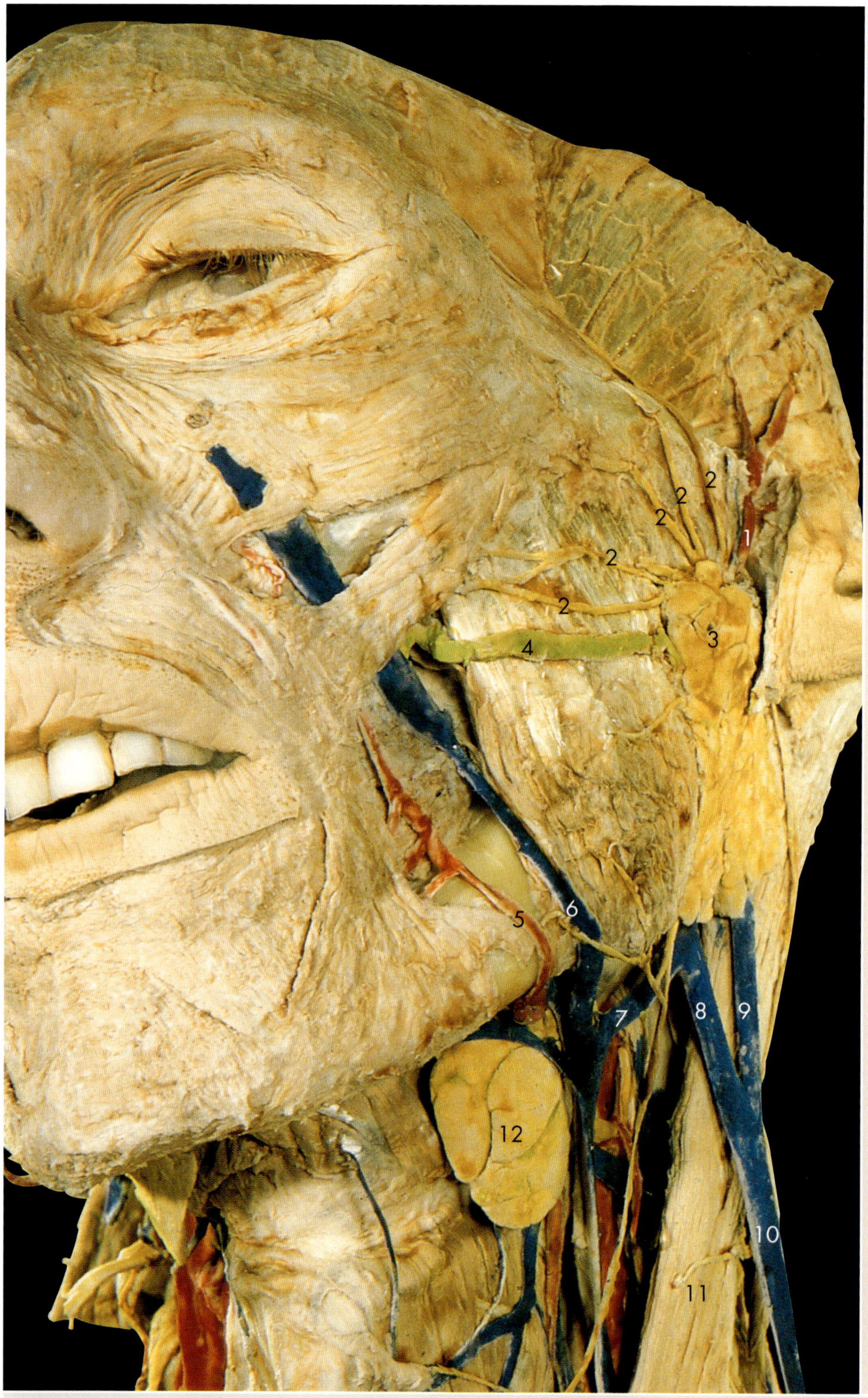

1 Superficial temporal vessels
2 Branches of facial nerve
3 Parotid gland
4 Parotid duct crossing masseter muscle
5 Facial artery
6 Facial vein
7 Anterior branch of retromandibular vein
8 Posterior branch of retromandibular vein
9 Posterior auricular vein
10 External jugular vein
11 Sternocleidomastoid muscle
12 Submandibular gland

Figure 8.18 The retromandibular vein dividing at the apex of the gland. (Courtesy of Professor C. Dean, Department of Anatomy and Developmental Biology, University College London.)

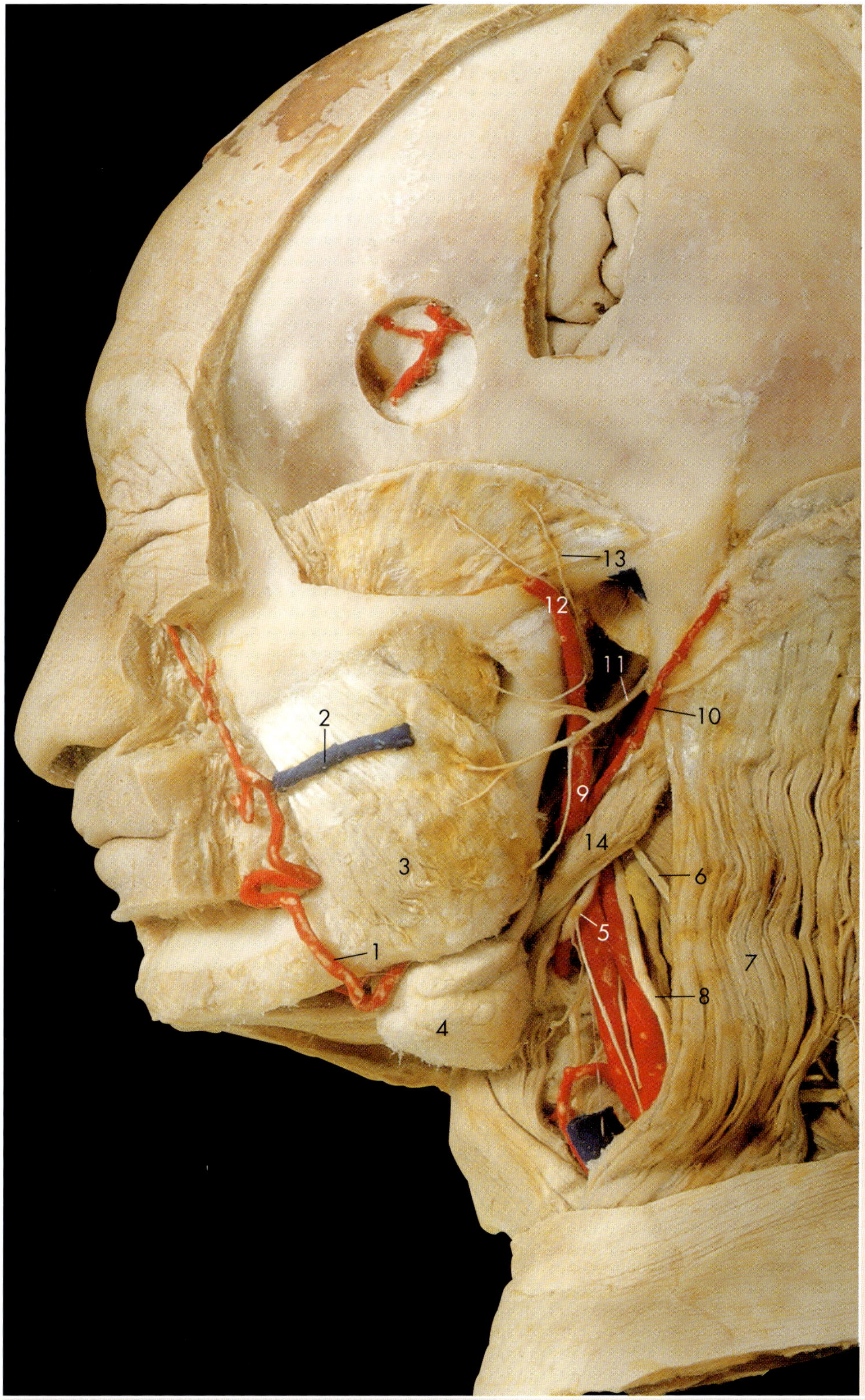

1 Facial artery
2 Parotid duct
3 Masseter muscle
4 Submandibular gland
5 Hypoglossal nerve
6 Spinal accessory nerve
7 Sternocleidomastoid muscle
8 Vagus nerve
9 External carotid artery
10 Posterior auricular artery
11 Facial nerve
12 Superficial temporal artery
13 Auriculotemporal nerve
14 Posterior belly of digastric muscle

Figure 8.19 The external carotid artery. (Courtesy of Professor L. Garey, Anatomy Department, Imperial College Medical School, London.)

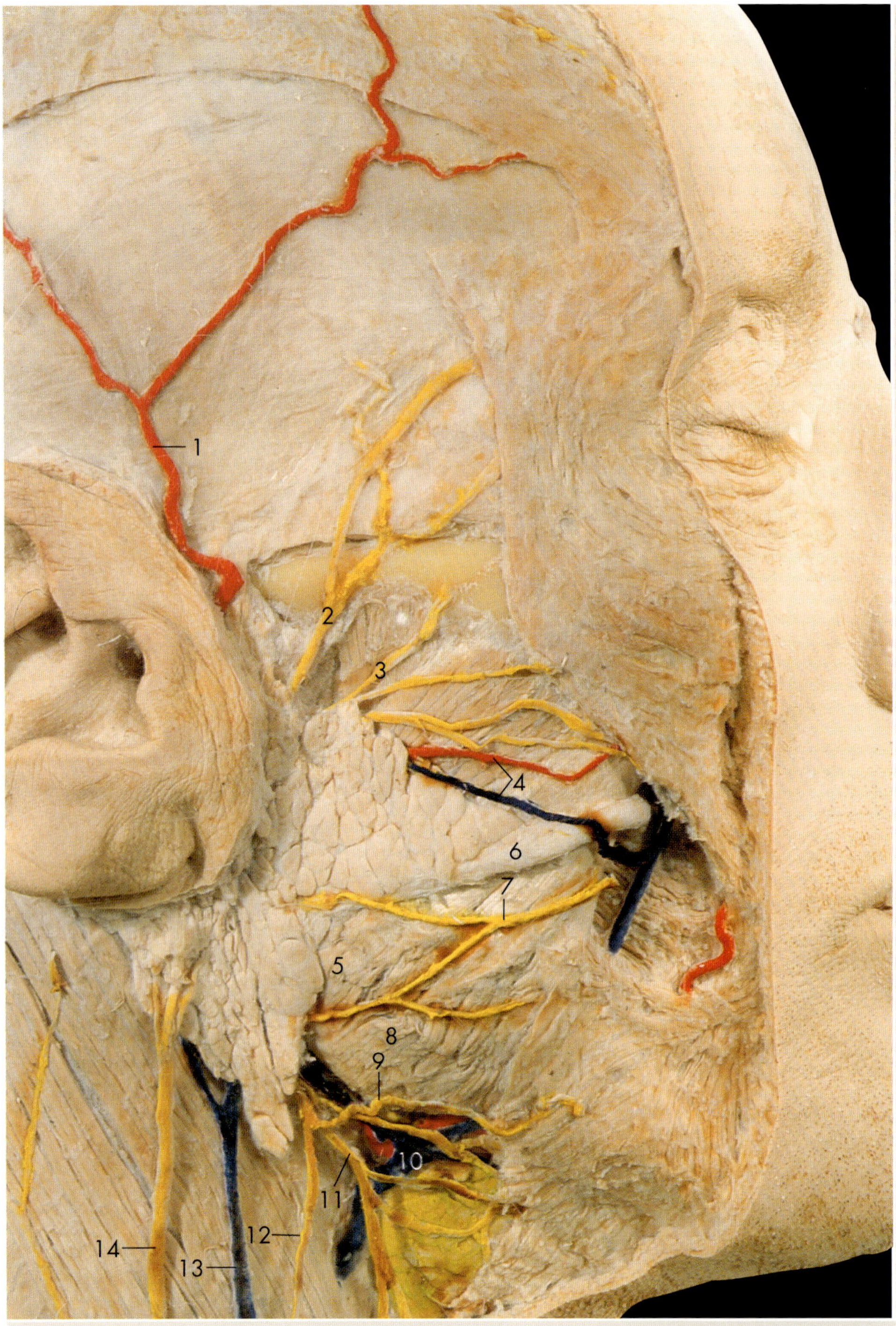

1 Superficial temporal artery
2 Temporal branch of facial nerve
3 Zygomatic branch of facial nerve
4 Tranverse facial vessels
5 Parotid gland
6 Parotid duct
7 Buccal branch of facial nerve
8 Masseter muscle
9 Mandibular branch of facial nerve
10 Facial vein
11 Cervical branch of facial nerve
12 Branch connecting facial and transverse cervical nerves
13 External jugular vein
14 Great auricular nerve on sternocleidomastoid muscle

Figure 8.20 The transverse facial artery. The specimen also shows the connection between the facial and transverse cervical nerves. (Courtesy of Professor C. Dean, Department of Anatomy and Developmental Biology, University College London.)

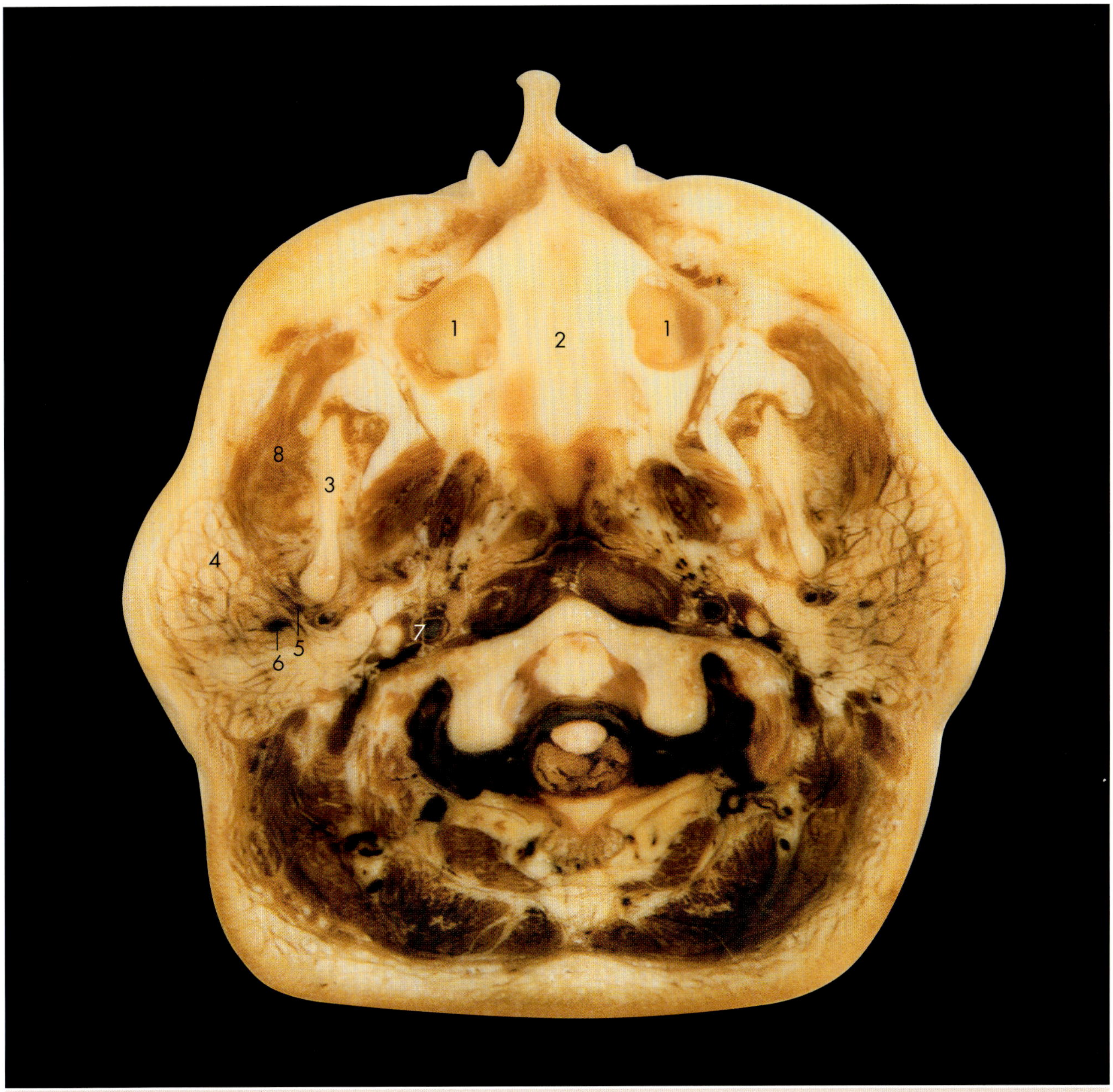

1	Maxillary sinus	5	External carotid artery
2	Hard palate	6	Retromandibular vein
3	Ramus of mandible	7	Carotid sheath
4	Parotid gland	8	Masseter muscle

Figure 8.21 Horizontal section of the head at the level of the floor of the maxillary sinus showing the parotid gland and its relations.

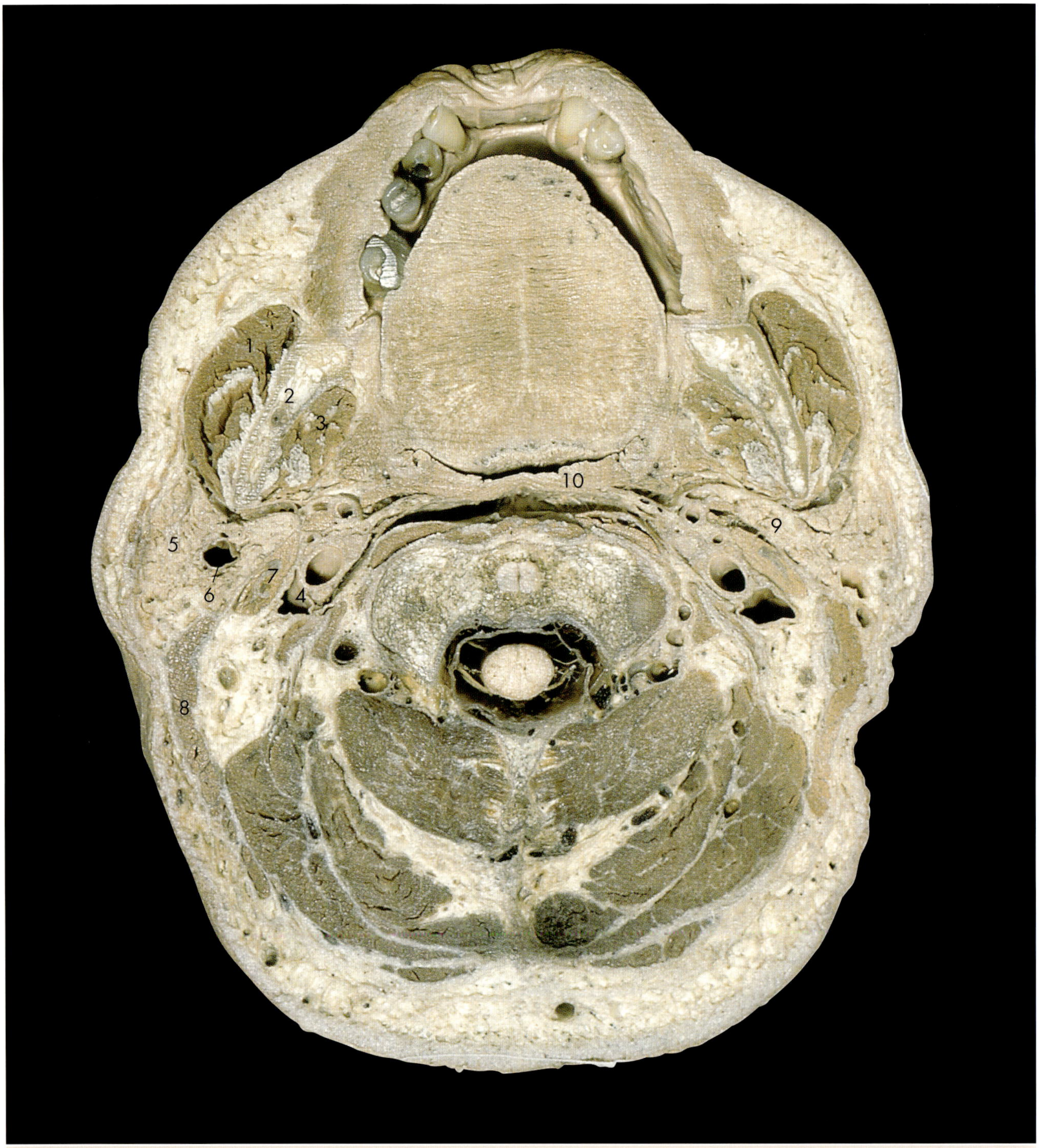

1 Masseter muscle
2 Ramus of mandible
3 Medial pterygoid muscle
4 Carotid sheath
5 Parotid gland
6 Retromandibular vein
7 Posterior belly of digastric muscle
8 Sternocleidomastoid muscle
9 'Deep' lobe of parotid gland directed towards superior constrictor muscle (10)
10 Superior constrictor muscle

Figure 8.22 A horizontal section of the head at the level of the tongue showing the parotid gland and its relations.

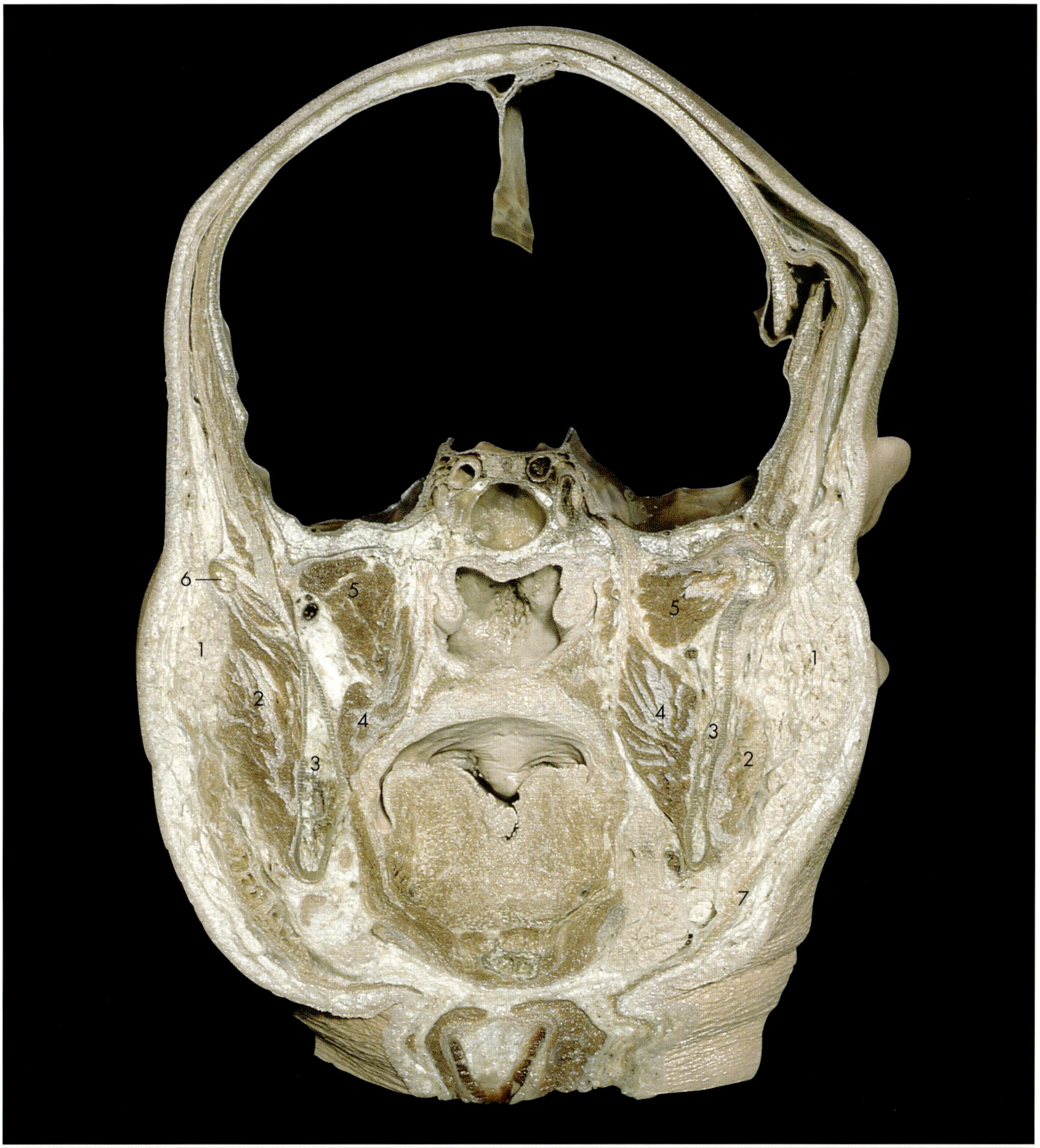

1 Parotid gland
2 Masseter muscle
3 Ramus of mandible
4 Medial pterygoid muscle
5 Lateral pterygoid muscle
6 Zygomatic arch
7 Platysma muscle

Figure 8.23 A coronal section of the parotid gland towards the back of the ramus showing the parotid gland and its relations.

LYMPH NODES ASSOCIATED WITH THE PAROTID GLAND

Lymph nodes are found both in the subcutaneous tissue overlying the parotid gland (preauricular nodes) and in the substance of the parotid gland itself. The lymph nodes present in the parotid gland average about 10, the majority of which are found in the superficial part of the gland, lying above the plane related to the facial nerve. The deeper part of the parotid gland beneath the branches of the facial nerve contains one or two lymph nodes.[43–45] Lymph from the parotid gland drains to the upper deep cervical lymph nodes.

PAROTID DUCT (Figs 8.32, 8.4, 8.20)

This duct appears at the anterior border of the upper part of the parotid gland and passes horizontally across the masseter muscle, approximately at the level midway between the angle of the mouth and the zygomatic arch. If the duct arises lower down, it may run obliquely upwards.[2] The duct lies below the transverse facial vessels and receives one or more ducts from the accessory parotid gland when present. The parotid duct may also be crossed by anastomosing branches between the zygomatic and buccal branches of the facial nerve. The duct bends sharply around the anterior border of the masseter to pierce the buccal pad of fat and the buccinator muscle at the level of the upper third molar tooth. A further bend in the duct is found as it passes forwards beneath the oral mucosa before opening into the vestibule opposite the crown of the upper second molar tooth. Adjacent to the opening is usually a small elevation of the mucosa termed the parotid papilla.

THE INNERVATION OF THE PAROTID GLAND

The innervation is related to the otic parasympathetic ganglion (Fig. 8.24). The parasympathetic secretomotor supply is from the inferior salivatory nucleus of the brainstem. Passing with the glossopharyngeal nerve, the fibres run in the tympanic branch which contributes to the tympanic plexus on the promontory of the middle ear. The lesser petrosal nerve arises from this plexus and exits the middle ear. It runs in a groove on the petrous portion of the temporal bone in the middle cranial fossa (Fig. 8.25) and passes through the foramen ovale (or more rarely the canaliculus innominatus medial to the foramen spinosum) to the otic ganglion lying on the medial surface of the main trunk of the mandibular nerve (Fig. 8.17). After synapsing within

Figure 8.24 The otic parasympathetic ganglion and innervation of the parotid gland.

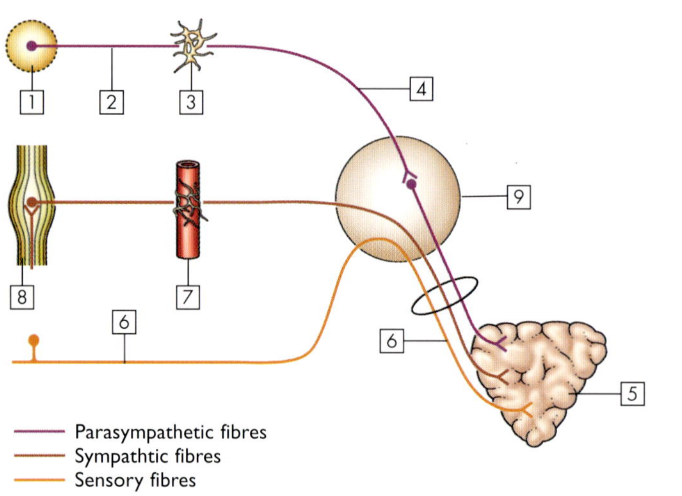

1 Inferior salivatory nucleus in brainstem
2 Tympanic branch of glossopharyngeal nerve
3 Tympanic plexus in middle ear
4 Lesser petrosal nerve
5 Parotid gland
6 Auriculotemporal branch of mandibular nerve
7 Middle meningeal artery
8 Superior cervical ganglion
9 Otic ganglion

the ganglion, postganglionic fibres leave to join the nearby auriculotemporal nerve which distributes the fibres to the parotid gland. There is some evidence that secretomotor fibres from the chorda tympani branch of the facial nerve may also reach and supply the parotid gland.

The sympathetic supply to the parotid gland is derived initially from the superior cervical sympathetic ganglion. From this ganglion, the innervation reaches the gland via the plexus around the middle meningeal artery, the otic ganglion (without synapsing) and eventually the auriculotemporal nerve. It seems likely that an alternative source of sympathetic fibres is derived directly from the sympathetic plexuses accompanying the many blood vessels supplying the parotid gland.

Sensory fibres to the connective tissue within the parotid gland are derived directly from the auriculotemporal nerve. In passing back to the parent mandibular nerve, they run through the otic ganglion (without synapsing) via a connecting branch.

The sensory innervation to the parotid capsule is via the great auricular nerve (Fig. 8.3).

Knowledge of the secretomotor innervation to the parotid gland has clinical significance in explaining the redness and sweating that may develop in the temple region following parotidectomy (Frey's syndrome) when eating. The explanation forwarded to account for these symptoms is that there is abnormal regeneration of nerves such that secretomotor fibres, which normally pass directly to the parotid gland, regenerate beyond to the auriculotemporal nerve. Thus, the sweat glands and blood vessels in the skin in the region of distribution of this nerve respond to impulses that should reflexly stimulate parotid secretion (gustatory sweating).

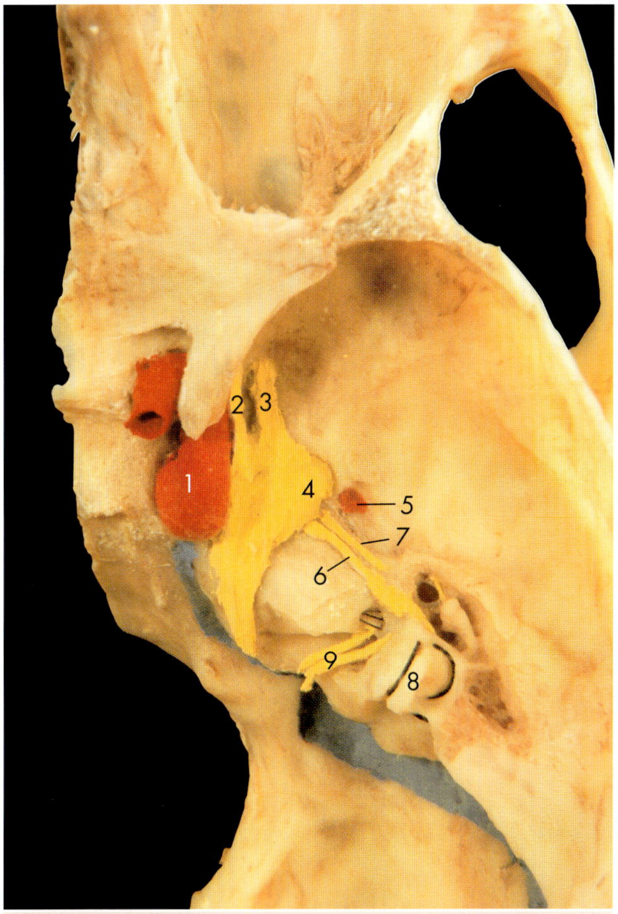

1 Internal carotid artery
2 Ophthalmic nerve
3 Maxillary nerve
4 Mandibular nerve
5 Cut end of middle meningeal artery
6 Greater petrosal nerve
7 Lesser petrosal nerve
8 Semicircular canal of inner ear
9 Facial and vestibulocochlear nerves

Figure 8.25 The lesser petrosal nerve in the middle cranial fossa. (Courtesy of Professor S. Standring, GKT School of Biomedical Sciences, London.)

REFERENCES

1. Davis R A, Anson B J, Budinger J M, Kurth L E. Surgical anatomy of the facial nerve and parotid gland based upon a study of 350 cervicofacial halves. *Surg Gynecol Obstet* 1956; 102: 385–412
2. McCormack L J, Cauldwell E W, Anson B J. The surgical anatomy of the facial nerve. *Surg Gynecol Obstet* 1945; 80: 620–630
3. Fromer J. The human accessory parotid gland: its incidence, nature and significance. *Oral Surg Oral Med Oral Pathol* 1977; 43: 671–676
4. McMinn R M H, Hutchings R T, Logan B M. *A colour atlas of applied anatomy.* London: Wolfe 1984
5. Berkovitz B K D, Moxham B J. *A textbook of head and neck anatomy.* London: Wolfe Medical 1988

6. Williams P L. (ed). *Gray's anatomy* (38th edn). Oxford: Blackwell 1995
7. Ellis H. *Clinical anatomy* (9th edn). Oxford: Blackwell 1997
8. Zigiotti G L, Liverani M B, Ghibellini D. The relationship between parotid and superficial fasciae. *Surg Radiol Anat* 1991; 13: 293–300
9. Patey D A, Ranger I. Some points in the surgical anatomy of the parotid gland. *Br J Surg* 1957; 45: 250–259
10. Mitz V, Peyronie M. The superficial musculo-aponeurotic system (SMAS) in the parotid and cheek area. *Plast Reconstr Surg* 1976; 58: 80–88
11. Jost G, Levet Y. Parotid fascia and face lifting: a critical evaluation of the SMAS concept. *Plast Reconstr Surg* 1983; 74: 42–51
12. Wassef M. Superficial fascial and muscular layers in the face and neck: a histological study. *Aesthetic Plast Surg* 1987; 11: 171–176
13. Thaller S R, Kim S, Patterson H *et al.* The submuscular aponeurotic system (SMAS): a histologic and comparative anatomy evaluation. *Plast Reconstr Surg* 1989; 86: 691–696
14. Gosain A K, Yousif N J, Madiedo G *et al.* Surgical anatomy of the SMAS: a reinvestigation. *Plast Reconstr Surg* 1993; 92: 1254–1263
15. Flatau A T, Mills P R. Regional anatomy. In: Norman J E de B, McGurk M. (ed) *Color atlas and text of the salivary glands.* London: Mosby Wolfe 1995; 13–39
16. Winsten J, Ward G E. Recent advances in surgery. The parotid gland: an anatomic study. Surgery 1956; 10: 585–606
17. McWhorter G L. The relations of the superficial and deep lobes of the parotid gland to the ducts and to the facial nerve. *Anat Rec* 1917; 12: 149–154
18. Snell R S. *Clinical anatomy for medical students.* Boston: Little, Brown and Co. 1995
19. Anderson R, Byars L T. *Surgery of the Parotid Gland.* St Louis: Mosby 1965
20. Saracco C G, Crabill E V. Anatomy of the human salivary glands. In: Dobrosielski-Vergona K (ed) *Biology of the salivary glands.* Boca Raton: CRC Press 1993; 1–14
21. Ziarah H A, Atkinson M E. The angular tract: an anatomical structure of surgical significance. *Br J Oral Surg* 1981; 19: 116–120
22. Shimada K, Davis W J, DiDio L J A. The mandibulo-stylohyoid ligament. *Surg Radiol Anat* 1986; 8 67–70
23. Shimada K, Gasser R F. Morphology of the mandibulo-stylohyoid ligament in human adults. *Anat Rec* 1988; 222: 207–210
24. Jovanovic M S. The mandibulo-stylohyoid ligament (tractus angularis). *Surg Radiol Anat* 1990; 12: 91–95
25. Conley J. *Salivary glands and the facial nerve.* Stuttgart: Thieme 1975
26. May M. *The facial nerve.* Stuttgart: Thieme 1986
27. Keyes G R, Tenta L T. Surgery of the parotid gland and the viscerovertebral angle. *Clin Plast Surg* 1985; 12: 323–330
28. Holt J J. The stylomastoid area: anatomic–histologic study and surgical approach. *Laryngoscope* 1996; 106: 396–399
29. Hawthorn R, Flatau A. Temporomandibular joint anatomy. In: Norman J E deB, Bramley P (eds) *A textbook and colour atlas of the temporomandibular joint.* London: Mosby Wolfe 1990; 1–51
30. Park I Y, Lee M E. A morphological study of the parotid gland and the peripheral branches of the facial nerve in Koreans. *Yonsai Med J* 1977; 18: 45–51
31. Myint K, Azian A L, Khairul F A. The clinical significance of the branching pattern of the facial nerve in Malaysian subjects. *Med J Malaysia* 1992; 47: 114–121
32. Furnas D W. Landmarks for the trunk and the temporofacial division of the facial nerve. *Br J Surg* 1965; 52: 694–696
33. Ammirati M, Spallone A, Ma J *et al.* An anatomicosurgical study of the temporal branch of the facial nerve. *Neurosurgery* 1993; 33: 1038–1044
34. Bernstein L, Nelson R H. Surgical anatomy of the extraparotid distribution of the facial nerve. *Arch Otolaryngol* 1984; 110: 177–183
35. Baker D C, Conley J. Avoiding facial nerve injuries in rhytidectomy. *Plast Reconstr Surg* 1979; 64: 781–795
36. Miehlke A, Stennert E, Chilla R. New aspects in facial nerve surgery. *Clin Plast Surg* 1979; 6: 451–459
37. Pogrel M A, Schmidt B, Ammar A. The relationship of the buccal branch of the facial nerve to the parotid duct. *J Oral Maxillofac Surg* 1996; 54: 71–73
38. Dingman R O, Grabb W C. Surgical anatomy of the mandibular ramus of the facial nerve based on the dissection of 100 facial halves. *Plast Reconstr Surg* 1962; 29: 266–272
39. Ziarah H A, Atkinson M E. The surgical anatomy of the mandibular distribution of the facial nerve. *Br J Oral Surg* 1981; 19: 159–170
40. Ziarah H A, Atkinson M E. The surgical anatomy of the cervical branch of the facial nerve. *Br J Oral Surg* 1981; 19: 171–179
41. Laing M R, McKerrow W S. Intraparotid anatomy of the facial nerve and retromandibular vein. *Br J Surg* 1988; 75: 310–312
42. Gaughran G R L. The parotid compartment. *Ann Otol Rhinol Laryngol* 1961; 70: 31–51
43. Garatea-Crelgo J, Gay-Escoda C, Bermejo B, Buenechea-Imaz R. Morphological study of the parotid lymph nodes. *J Cranio-Maxillo-Facial Surg* 1993; 21: 207–209
44. McKean M E, Lee K, McGregor I A. The distribution of lymph nodes in and around the parotid gland: an anatomical study. *Br J Plast Surg* 1985; 38: 1–5
45. Marks N J. The anatomy of the lymph nodes of the parotid gland. *Clin Otolaryngol* 1984; 9: 271–275

Chapter **9** # Trigeminal pain

C. SHIEFF AND J. ALLIBONE

INTRODUCTION

The face has a dense sensory innervation which is conveyed by the trigeminal nerve. Combined with the emotional overlay which often accompanies facial pain this may confuse interpretation of symptoms and make it difficult to localise the pain or to determine its source. In turn this can lead to a delay in diagnosis and, possibly, to inappropriate treatment.

Although many patients can be treated medically, facial pain often presents a surgical challenge. Ideally any intervention will be simple and effective. It should also anticipate possible further change in the primary disease or in its symptoms. In devising treatment, remedies directed at the cause are logical. These will be based on assumptions that:

1. Pain triggered by stimulation of the peripheral nerve can be relieved by reducing its sensitivity
2. The nociceptive symptoms of pathological compression will resolve following decompression
3. Neural augmentation will abolish pain resulting from deafferented and disorganised central neural processing.

Clinical experience, however, has demonstrated that these do not always succeed, probably because the actual experience of pain is a complex summation of multiple influences involving anatomy, physiology, pathology and psychology.[1] Some procedures result only in temporary improvement while others worsen the original condition by inducing deafferentation. In consequence even more intricate approaches have been developed for some facial pain syndromes, especially those associated with orofacial denervation. Identification of the spinal nucleus of the trigeminal nerve as the primary relay for nociception in this region has led to surgical techniques in the brain stem and upper cervical cord for the treatment of facial pain. Unfortunately even these may fail.

To facilitate what might otherwise be an overwhelming task, we have chosen the syndrome of primary trigeminal neuralgia as a template upon which to base additional comments relating to other painful afflictions of the face and the trigeminal pathways.

CLASSIFICATION

It is simplest to consider the trigeminal pain syndromes in the groupings primary, secondary, and atypical (Table 9.1). It is logical first to consider the normal anatomy of the trigeminal nerve and its pathways, then the patho-

Table 9.1 Classification of trigeminal pain syndromes (trigeminal and facial pain syndromes which may present as, or mimic, trigeminal pain)

Primary	Secondary	Atypical
Idiopathic	Intracranial vascular lesion:	Acute herpes zoster
Neurovascular conflict	aneurysm	Postherpetic neuralgia
Root entry zone pathology:	arteriovenous malformation	Post-traumatic neuralgia
multiple sclerosis	Cerebellopontine angle tumours:	Deafferentation
	acoustic neuroma	Central (thalamic) pain
	dermoid	Raeder's syndrome
	meningioma	SUNCT syndrome
	Skull base pathology:	Musculoskeletal pain
	chordoma	Oral and ENT conditions
	cavernous sinus disease	'Headache' syndromes

physiology of trigeminal mediated pain, followed by an overview of the various pathologies from which these conditions arise. In turn this should lead us to identify simple protocols for diagnosis and management.

- *Primary neuralgia.* This comprises neuralgia without an identifiable structural lesion ('idiopathic') as well as that now accepted by many as due to neurovascular contact. The trigeminal pain experienced by multiple sclerosis patients is included here as their pathophysiology and treatment are almost identical.
- *Secondary neuralgia.* Resulting from nociceptive phenomena due to adjacent structural disease either within the cranial cavity or at the skull base or below.
- *Atypical and non-neuralgic.* These include those syndromes that are not neuralgic although occasionally confused with such. They may arise from primary disease of the oral cavity or from musculoskeletal disorders of adjacent structures. The group includes those conditions mediated by autonomic dysfunction.

For those who seek a comprehensive listing of pain syndromes of the head and neck, the classification proposed by the International Association for the Study of Pain (IASP) is useful.[2] Unfortunately, this will offer little help in arriving at a diagnosis in an individual case nor does it direct one to causation and hence to therapy.

ANATOMY OF THE TRIGEMINAL NERVE

The trigeminal nerve is the largest cranial nerve. It subserves sensation of the face and the greater part of the scalp, the cornea and conjunctiva, the teeth, the mucosa of the oral and nasal cavities, the dura mater and the cerebral blood vessels. Motor fibres supply the masticatory muscles, tensor tympani, tensor veli palatini, and the anterior belly of digastric and mylohyoid. It contains proprioceptive fibres from these muscles and from the facial and extraocular muscles.

TRIGEMINAL NERVE ROOTS AND DIVISIONS

The trigeminal nerve contains approximately 125,000 fibres when it emerges as two or three roots from the ventrolateral surface of the mid-pons. The smaller motor root, in which 10–20% of the fibres appear to have a sensory function, is slightly more rostral and medial to the larger sensory root. An additional intermediate root may branch off the main sensory root and enter the brainstem between the two other roots.

The roots pass forwards and upwards in the subarachnoid space, under the superior petrosal sinus and tentorium cerebelli and over the petrous ridge, into the middle cranial fossa where a dural envelope forms Meckel's cave in a depression in the petrous temporal bone. It is here that the trigeminal ganglion of the sensory root contains the cell bodies of the primary afferent fibres of the trigeminal nerve. Its medial relation is the internal carotid artery within the posterior part of the cavernous sinus. Inferiorly are situated the motor root, the greater petrosal nerve, the apex of the petrous temporal bone, the foramen lacerum and the internal carotid artery in its bony canal. From the ganglion the sensory root separates into its three divisions.

- The *ophthalmic nerve* arises from the anteromedial end of the ganglion passing forward in the lateral wall of the cavernous sinus below the oculomotor and trochlear nerves. Just before its entry into the orbit through the superior orbital fissure it divides into lacrimal, frontal and nasociliary branches. It supplies the forehead, upper eyelid and the cornea.
- The *maxillary nerve*, also wholly sensory, passes forward and medially in the lateral wall of the cavernous sinus to enter the pterygopalatine fossa via the foramen rotundum. It supplies sensation over the middle third of the face and nose, including the palate and upper jaw and buccal mucosa.
- The *mandibular nerve* leaves the lateral part of the ganglion and passes downwards through the foramen ovale into the infratemporal fossa. It supplies the lower jaw and tongue with the overlying skin to a variable distance from the lower border of the mandible to the tragus. The only predictable area of skin receiving innervation from the third division is the vermilion border of the lower lip. The motor root passes diagonally under the ganglion and through the foramen ovale to unite with the mandibular nerve just outside the skull.

Brainstem nuclei

The trigeminal root penetrates 4–5 mm at the level of the middle cerebellar peduncle. Approximately 50% of fibres divide, the rest ascending or descending without division in the lateral portion of the tegmentum.

The ascending fibres terminate primarily in the principle nucleus of the trigeminal nerve making them unique as the only primary sensory neurones whose somata lie within the central nervous system. They carry sensory modalities such as fine touch and two-point discrimination. Proprioceptive information passes to the mesencephalic nucleus in large myelinated fibres from trigeminal innervated muscle spindles, facial and extraocular muscles and from tooth pressure receptors in the periodontal membrane. After synapsing these send collaterals to the motor nucleus and to the spinal trigeminal tract.

The descending fibres synapse within the spinal trigeminal nucleus and sometimes within the solitary tract. The spinal nucleus is subdivided cytoarchitectonically into the subnuclei oralis, interpolaris and caudalis, which are interconnecting. The subnucleus caudalis extends from the level of the obex to the level of the C4 root. It is analogous both structurally and electrophysiologically to the grey matter of the dorsal horn of the spinal cord and resembles the substantia gelatinosa containing synapses of A-δ and C fibres and axo-axonal contacts. Somaesthetic inputs from the seventh, ninth and tenth cranial nerves and inputs from adjacent reticular formation, medial and lateral cuneate nuclei, contralateral subnucleus caudalis, cerebral cortex, and first-order neurones of the upper levels of the cervical cord also reach the subnucleus caudalis. It is here that most integration and processing of nociception, including presynaptic inhibition, from facial structures takes place.

The motor nucleus lies in the upper pons. Its descending input is from both corticonuclear tracts. Local inputs are received from the sensory nuclei, the locus coeruleus, the medial longitudinal fasciculus and other brainstem nuclei. Collectively these represent pathways by which salivary secretion and mastication may be co-ordinated.

Central connections

The rostral projection is predominantly to the contralateral ventro-posteromedial nucleus of the thalamus (VPM) via the trigeminal lemniscus and onwards to the sensory cortex. The trigeminal nuclei also project to the nucleus centralis lateralis (medial to VPM), a projection that probably contains unmyelinated nociceptive fibres. A significant projection to the contralateral superior colliculus is present and nucleus subcaudalis also projects to the periaqueductal grey matter of the midbrain. Projections to other brainstem nuclei, such as the nucleus of the tractus solitarius, coordinate control of phonation, salivation and swallowing. Less well-defined central connections to elements of the limbic circuit are also recognised.

Somatotopy

There is a somatotopic arrangement of the sensory fibres throughout their path. In the region of the ganglion the maxillary fibres lie between the ophthalmic fibres medially and the mandibular fibres laterally. Considerable rotation occurs as the root travels towards the root entry zone where the ophthalmic fibres lie caudally and the mandibular fibres rostrally. Within the brainstem the mandibular fibres are dorsomedial and the ophthalmic fibres ventrolateral. A circumferentially orientated 'onion skin' pattern of cutaneous representation exists centred on the mouth and nose whereby the central area of the snout is represented in the upper spinal nucleus and the peripheral portions in the lower nucleus.

PRIMARY (IDIOPATHIC) TRIGEMINAL NEURALGIA

We must assume that trigeminal neuralgia has existed as long as mankind although the earliest clear description of trigeminal neuralgia was not provided until 1671 when the unfortunate sufferer was Johannes Laurentius Bausch, a physician and municipal councillor in Franconia. In 1677 Locke reported the trigeminal neuralgia suffered by the countess of Northumberland who was the wife of the English ambassador to France. Fothergill, in a detailed account published in 1773, described the sudden onset and paroxysmal nature of the attacks and also noted the tendency for the condition to affect the elderly and its predilection for women. The term *tic douloureux* was originated by André when he described five cases of facial pain in 1776. The disorder remained poorly understood until the description of the anatomy and the distinction between the separate functions of the trigeminal and facial nerves by Bell permitted localisation of *tic douloureux* to the trigeminal nerve. This led to the name trigeminal neuralgia and later to our understanding of the condition and to the various approaches that we can offer for its management.

CLINICAL FEATURES

Approximately 4 per 100,000 population are affected with a male:female ratio between 1:1.5 and 1:2. It is primarily a disease of the middle to later years as 70% of those with the condition are over fifty. Its prevalence has probably increased with the rise in life expectancy.[3]

The diagnosis of idiopathic or primary trigeminal neuralgia is made by obtaining a full history and by examination of the patient. The pain typically occurs in paroxysms confined to one or more divisions of the trigeminal nerve. It is usually pattern-specific for an individual patient, triggered by light mechanical contact in a constant and well circumscribed location peculiar to the attack, and felt superficially in the skin or buccal mucosa. It is unusual for more than one site to act as trigger during a single paroxysm although multiple triggers are encountered. The duration is characteristically brief, lasting seconds only, with repetition occurring in bursts over several minutes. Episodes may occur infrequently or many times in a day and possibly so frequently that they appear to merge. They may prevent facial and oral hygiene, ingestion of food or drink, or the wearing of dentures. There may be refractory periods which can last many years.

Isolated first-division pain is seen in only 4% of patients. Between 2 and 4% of those suffering from trigeminal neuralgia will have multiple sclerosis, which is in excess of the incidence of this condition in the general population. Synchronous bilateral pain is diagnostic of multiple sclerosis although asynchronous attacks are compatible with 'normal' trigeminal neuralgia.[4,5]

The patient may demonstrate the lancinating character of the pain by flinging out the fingers from a clenched position in a characteristic manner even when unable to speak by virtue of the pain. Examination is normal other than in confirming a trigger on the skin or the mucous membrane. If sensory loss or abnormality of any other cranial nerve is demonstrated an alternative diagnosis should be sought.

By definition, simple investigations, including imaging by standard diagnostic CT scan and MRI scan, are negative in idiopathic trigeminal neuralgia. It is generally acknowledged that up to 3% of patients diagnosed to have the condition will be found to have a structural lesion in the posterior cranial fossa which makes imaging essential (Fig. 9.1). Sophisticated MR sequences are now proving helpful in revealing neurovascular compression at the root entry zone (see below). Magnetic resonance tomographic angiography (MRTA) uses a standard angiography sequence to provide high-definition MR

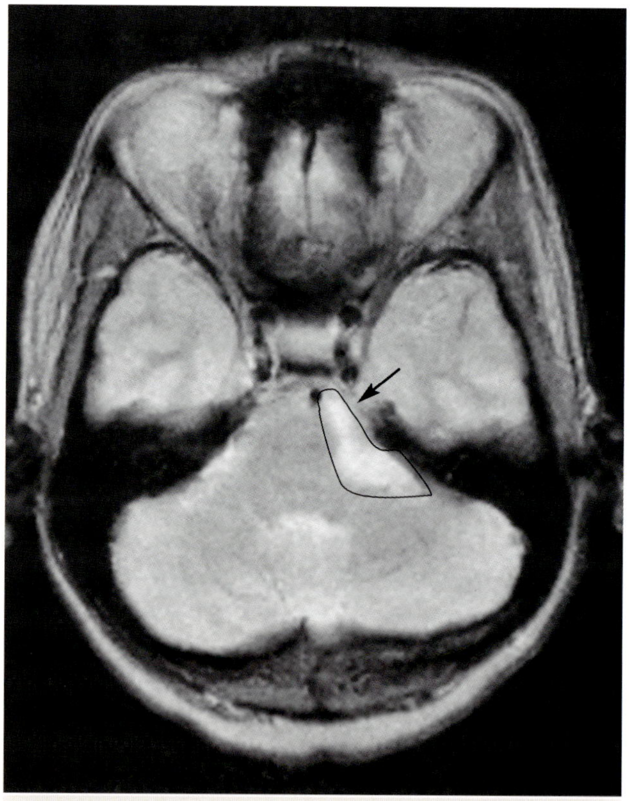

Figure 9.1 Axial MRI scan of a 35-year-old woman with an 8-year history of typical third-division left trigeminal neuralgia that initially occurred during a pregnancy. Examination was normal. Three previous percutaneous procedures by glycerol, thermocoagulation and glycerol provided periods of remission. The pain was incompletely controlled on carbamazepine (maximum 2400 mg). CT scans at onset and after 4 years were normal. MRI had revealed this small cerebellopontine angle epidermoid tumour (arrowed). Now pain free 6 years after middle fossa root section.

images of both brain parenchyma and the posterior fossa vasculature. This selectively visualises fast-flowing (arterial) blood; contrast enhancement with gadolinium is required to demonstrate venous compression. The technique has 96% specificity and 100% sensitivity for the identification of vascular compression.[6] We have found MR cisternography using very heavily T2-weighted images to examine the cerebello-pontine angle helpful (Figs 9.2, 9.3). Others use a combination of 3D reconstructions of imaging using spoiled gradient recalled acquisition in the steady state (SPGR) and MR angiography.[7] As some neurosurgeons believe that demonstration of a compressive lesion on MRI increases the chance of a good outcome from microvascular decompression and reduces the likelihood of response to non-surgical management, imaging in this manner may improve surgical planning.

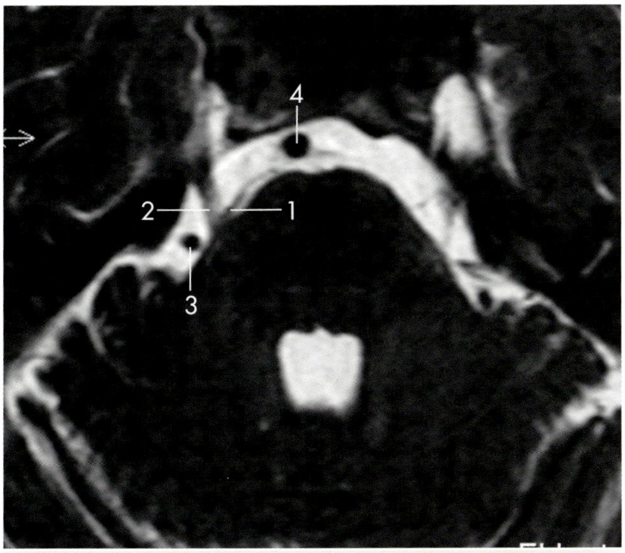

Figure 9.2 MR cisternogram of a 47-year-old woman with a 12-year history of continuous atypical symptoms characterised by 'fizzing' sensations in right third division. Diagnosed as functional, the pain was controlled intermittently by carbamazepine. A small unnamed vessel (1) is shown in the expected position at the axilla of the trigeminal nerve (2) adjacent to the root entry zone. The superior cerebellar artery (3) and basilar artery (4) are also shown. Absence of classic features led to treatment by glycerol injection with immediate cessation of symptoms.

PATHOPHYSIOLOGY OF TRIGEMINAL NEURALGIA

For many years trigeminal neuralgia was labelled idiopathic when no anatomical or pathophysiological cause could be identified. More recently some understanding of the underlying mechanism has been possible following experimental and clinical observation. The phenomena of spontaneous generation of pain and of

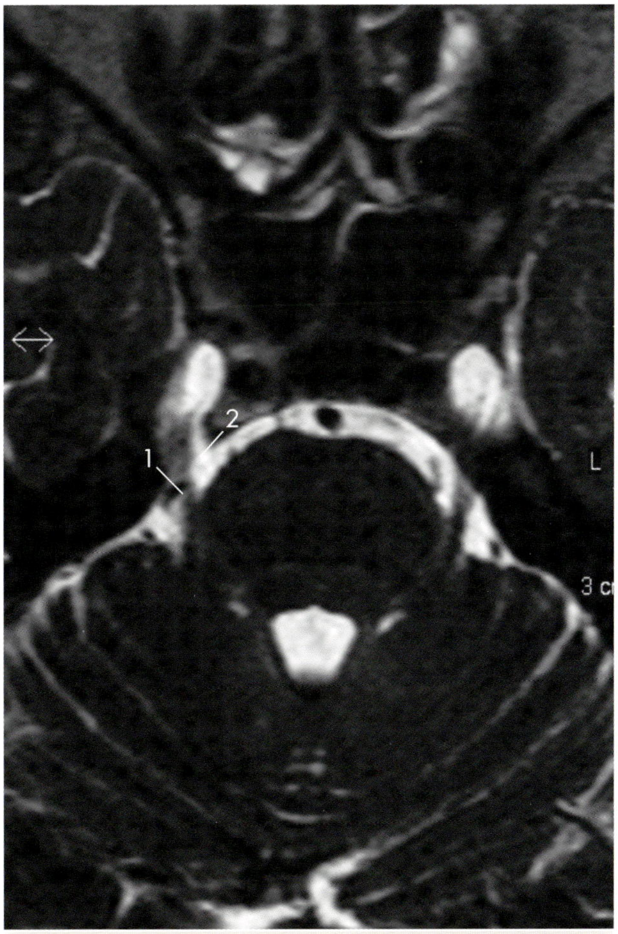

Figure 9.3 MR cisternogram of a 59-year-old woman with a 7-year history of right second- and third-division trigeminal neuralgia treated initially by glycerol injection with 5 years relief before recurrence. She was unable to tolerate carbamazepine. A looping branch of the superior cerebellar artery (1) was in 'conflict' with the trigeminal nerve (2), which was atrophic at this point. Pain resolved after repeat glycerol injection.

triggering, whereby an innocuous stimulus elicits severe pain, still require complete elucidation. Theories of partial denervation, peripheral sensitisation and intrinsic pathological change within the nerve or its roots have all been considered and seem to provide feasible hypotheses. As it passes outward from the pons the root is myelinated to a variable extent by central (oligodendroglial) myelin that renders it more susceptible to damage. Anatomical and histological changes have been identified in this trigeminal root entry zone. Segmental demyelination has been seen in the roots of patients with trigeminal neuralgia, even in those who do not have multiple sclerosis.[8,9] Although the exact mechanism by which pain is generated is unclear, ephaptic transmission (short-circuiting) from tactile to small nociceptive fibres within the areas of demyelination has been proposed to explain triggering.[10] Another explanation is that a trigger stimulus can initiate bursts of activity in a small cluster of trigeminal ganglion neurones previously rendered excitable by root damage. It is also possible that spontaneous impulse generation occurs in these areas.[11,12] The increased susceptibility of the lower two divisions is explained by their larger representation within the nucleus and by an experimentally demonstrated lower threshold for discharge.[13] Much that has been proposed for central causation is speculative although the concept of 'peripheral cause and central pathogenesis' remains popular.[14]

The aetiology of any demyelination is less certain but chronic pressure on the root may be sufficient. Tumours and vascular malformations in contact with the root are known to cause trigeminal neuralgia. Dandy reported finding vascular compression of the root in 30% of the cases that he explored for root section and suggested that this might be a cause of neuralgia.[15] Despite his observation the petrous ridge remained popular as the suspected site of compression of the trigeminal root although early attempts at decompression at this location were not very successful until Malis reported only a 9% rate of recurrence during 18 months after decompression of a band crossing from the anterior clinoid process to the petrous ridge.[16] Kerr suggested that the proximity of the pulsating carotid to the ganglion might be responsible.[17] Dandy's theory was later supported by Gardner[18] and confirmed by the work of Janetta (Table 9.2).[19] The reported incidence of arterial or venous compression of the nerve root or the adjacent root entry zone of the pons varies between 11 and 97% as does the apparent local effect of (visible) atrophy of the root.[20] Although the concept of microvascular compression is widely accepted there are apparent inconsistencies as arterial contact may also be seen in patients without trigeminal neuralgia – 60% in one study[21] and in 35% of cadavers.[22]

Table 9.2 Identification of compressive vessel in microvascular decompression of trigeminal nerve

Compressive vessel	Frequency (%)
Superior cerebellar artery	75
Anterior inferior cerebellar artery	10
Posterior inferior cerebellar artery	1
Vertebral artery	2
Basilar artery	1
Labyrinthine artery	<1
Other unspecified small artery	15
Vein and artery	56
Vein alone	13

Theories of non-compressive central mechanisms have also been considered. These depend mainly upon the concept of 'short-circuiting' within the brainstem, with or without demyelinating lesions.[23] Recent work in chronic pain has shown rewiring of large A-β fibres subserving tactile sensation into lamina II of the dorsal horn, which is the usual destination of A-δ and C fibres conveying nociceptive information (fluidity of dendritic connections). This might explain why simple tactile sensation is perceived as painful and why pain may be provoked by non-noxious stimuli in deafferentation states. Similar changes may also be happening in subnucleus caudalis in patients with trigeminal neuralgia when meticulous sensory examination may reveal small areas of hypoaesthesia.

TREATMENT

The ideal treatment provides relief of pain with minimum side-effects. The major choices lie between medical and interventional but it must be remembered that spontaneous remissions can occur in patients with idiopathic trigeminal neuralgia. Of patients in one review, 50% had spontaneous remission lasting 6 months or more while 25% experienced a remission greater than 1 year. Conversely, because the disease can be protracted, patients who remain symptomatic may require medication for a long time, which will increase the possibility of side-effects.

Symptomatic compression of the trigeminal nerve (from which sensory loss is more likely than pain) inside the cranial cavity, at the skull base or externally should be dealt with in a manner appropriate to the primary pathology. Definitive approaches for the treatment of trigeminal neuralgia can be divided broadly into peripheral, percutaneous and operative. The latter category includes open operations on the middle or posterior cranial fossae, complex procedures on the brainstem, and chronic electrical stimulation of cerebral structures. That no individual procedure has become universally accepted is indication that all procedures should be considered in treating an individual patient. Treatment should only be offered after considering all relevant medical factors, the clinician's experience and preference, and the views of the patient (Tables 9.3 and 9.4a–d).

MEDICAL TREATMENT

It is now obvious that the earliest medical treatments with various poisons, analgesics and counter-irritants were ineffective in treating neuralgic pain. Hutchinson in the 1820s claimed some success with ferrous carbonate whilst Plesner, a century later, reported good results with the use of the industrial solvent trichlorethylene after observing trigeminal sensory loss in workers who had inhaled it. Potassium bromide and stilbamidine, an antihelminthic with the known side-effect of selective trigeminal sensory neuropathy, were used for a time but discontinued as the benefits were outweighed by the side-effects.[24]

Effective drug therapies date from the early 1940s when hydantoin preparations were tried. In 1942 the use of phenytoin was first reported by Bergouignan. Since then a variety of anticonvulsants and muscle relaxants have been tried. In 1969 White and Sweet reported their use of phenytoin in 70 patients of whom 33 had temporary relief of their pain. Intravenous phenytoin, acting as a first pass agent, can be used effectively to abort an atypical and prolonged acute paroxysm (Hitchcock E R, unpublished observation).

Table 9.3 Current medical treatment of trigeminal neuralgia

Carbamazepine
Phenytoin
(Valproate)
Baclofen
Amitryptyline

Table 9.4a Peripheral procedures in treatment of trigeminal neuralgia

Local anaesthetic block
Cryotherapy
Neurolytic injection
Avulsion

Table 9.4b Percutaneous procedures in treatment of trigeminal neuralgia

Balloon compression
Glycerol injection
Differential radiofrequency thermocoagulation
Cryosurgery

Table 9.4c Open procedures on the trigeminal nerve and roots in treatment of trigeminal neuralgia

Middle cranial fossa	Posterior cranial fossa
Root decompression	(Partial) root section
(Partial) root section	Microvascular decompression
Root/nerve stimulation	

Table 9.4d Other procedures in treatment of trigeminal neuralgia

Nucleus caudalis DREZ lesioning
Trigeminal tractotomy
Stereotactic mesencephalotomy
(Stereotactic thalamotomy)
(Stereotactic cingulotomy–psychosurgery)
Chronic electrical stimulation of thalamus
Chronic stimulation of pre-motor cortex

Carbamazepine had been reported to be more effective than phenytoin by Blom in 1962 and is now the first-choice medication as 80% of patients will respond initially. The dose is increased as tolerated until the pain is relieved. It should also be reduced after a pain-free period of 1–2 months. The anticonvulsant therapeutic range is a guide only and can be exceeded if the patient tolerates the drug. Over 50% of patients will be controlled outside this range without problems. Monitoring is important especially early in therapy because of potential idiosyncratic and toxic reactions, and the side-effects of drowsiness, poor job performance and motor impairment even at low doses.

Those patients who become less responsive to medication at the maximum tolerated dose can try a second drug, either alone or in conjunction with carbamazepine. The commonest second-line drug is phenytoin. Baclofen, clonazepam and the newer anti-epileptics such as lamotrigine have also been reported to be useful although polypharmacy should be avoided if possible on empirical grounds.

Peripheral procedures

These may appear trivial and are no longer widely performed but this should not devalue such procedures. The supraorbital, infraorbital and mental nerves can be approached directly for simple local anaesthetic nerve block. The effect on a patient with severe debilitating facial pain of any cause is dramatic. It affords immediate relief and demonstrates the suitability of local denervation by neurolytic injection, cryotherapy or surgical avulsion. Neurolysis by alcohol injection will not be permanent but it can be repeated if necessary although local anaesthesia will result. Avulsion of the supraorbital nerve requires only a small incision at the eyebrow through which the several branches must be identified for complete success. It obtains pain relief without corneal anaesthesia which cannot be guaranteed by any more invasive technique. For the frail patient with a first division trigger this approach has much to commend it.[25] Avulsion of the infraorbital nerve and division of the mental nerve are both as effective and readily performed under local anaesthetic. The disadvantages are the anaesthesia produced, the recurrence rate due to regeneration or incomplete section (up to 40% at 2 years) and deafferentation, which might amount to frank anaesthesia dolorosa.

Percutaneous procedures

These all necessitate puncture of the cavum trigeminale (Meckel's cave) by a needle passed into the foramen ovale and can be undertaken freehand or stereotactically. The anterior freehand approach of Härtl[26] is probably the most widely used and is easily accomplished using a 90 mm 20 gauge spinal needle. It can be performed under local anaesthesia, neuroleptanalgesia or full general anaesthetic. Three anatomical landmarks must be identified: the entry point is approximately 2.5 cm lateral to the oral commissure, while a point 3 cm anterior to the external acoustic meatus and a point beneath the medial aspect of the pupil together indicate the position of the foramen ovale. The needle is passed backwards in the cheek, medial to the angle of the mandible, until it encounters the lateral pterygoid plate and the posterior edge of the hard palate. A gloved finger within the mouth will guide the needle and should detect any breach of the oral mucosa. The initial trajectory is slightly anterior to that intended until the needle contacts the base of the skull when it can be 'walked' backwards into the foramen ovale, which it enters with a characteristic give. At this time the patient may wince or bite on the finger. Hypertension or a bradyarrhythmia may be observed and should be anticipated by the attending anaesthetist. Cerebrospinal fluid flowing from the needle will confirm penetration of the cave although radiological confirmation, usually by image intensifier, is also required before proceeding further. Lateral views alone may suffice if the line of the needle appears otherwise satisfactory. Longitudinal adjustment to ensure correct position of the needle tip follows. The ideal trajectory places the needle tip 5–10 mm below the intersection of a line drawn from the floor of the sella turcica to the clival line (Fig. 9.4a). Supplementary transorbital or submento-vertical views may be helpful if placement is difficult (Fig. 9.4b).[27] If edentulous patients are allowed to wear their dentures during the procedure this will reduce possible penetration of the oral cavity.

The approach was originally used for neurolytic injection with alcohol. This is no longer used because of the potential risks of subarachnoid injection and the probability of frank deafferentation. It remains appropriate for all of the currently utilised percutaneous procedures: glycerol rhizolysis, balloon compression, cryosurgery and thermocoagulation.

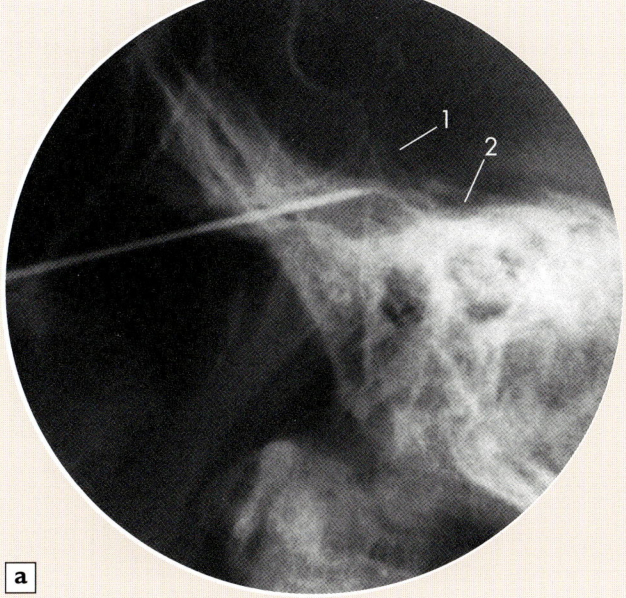

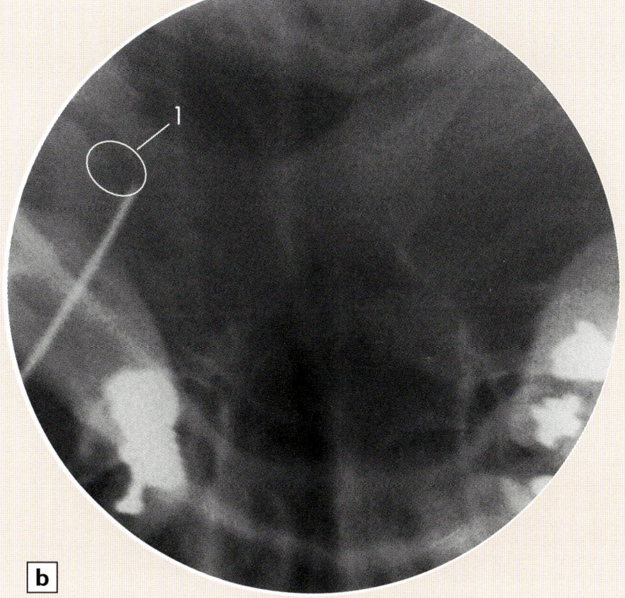

Figure 9.4 (a) Lateral skull film (coned view). This shows placement of an electrode tip within the region of Meckel's cave at junction of clivus (1) and petrous bone (2). (b) Submento-vertical skull film (coned view). The electrode is shown passing through the medial part of the foramen ovale (1).

Balloon compression

Percutaneous balloon compression of the trigeminal ganglion was introduced in 1983 by Mullan and Lichtor.[28] Compression selectively injures medium and large myelinated fibres in preference to small myelinated and unmyelinated fibres. A 14 gauge introducer cannula is passed to the foramen ovale through which a size 4 Fogarty type balloon catheter can be placed at the triangular plexus level. The balloon is inflated with contrast medium, using between 0.4 and 0.7 ml for between 1 and 9 minutes. The best results are obtained when the balloon becomes pear shaped during inflation, presumably because this conforms to the outline of Meckel's cave and results in equal pressure distribution. Immediate 100% relief of trigeminal neuralgia has been reported. Long-term results are also satisfactory; two-thirds of a series of 150 patients were pain free at a mean follow-up of 4 years (20% recurring within the first year). The technique is not division sensitive but major sensory disturbance is uncommon providing the procedure is not prolonged. Because the corneal reflex is subserved by small fibres which are less likely to be damaged by the method it is suitable for the ophthalmic territory. Otalgia is a specific complication; this is presumed to be due to damage to the innervation of the tensor tympani or to perforation of the membranous portion of the Eustachian tube.[29]

Retrogasserian glycerol injection

Jefferson used glycerol as an agent to carry phenol for chemical ablation of the trigeminal ganglion but was unaware that glycerol alone would be sufficient to relieve tic without causing deafferentation. The use of glycerol alone in the treatment of trigeminal neuralgia results from the serendipitous observation made by Håkanson who was planning stereotactic irradiation of the ganglion to treat trigeminal neuralgia. In order to provide a permanent radiographic target, tantalum dust suspended in glycerol was injected into the trigeminal cistern. Relief of tic occurred prior to irradiation.[30] The mode of action is uncertain although it appears that glycerol causes demyelination and axonal fragmentation.

With the patient in the sitting position water-soluble contrast is injected in 0.05 ml aliquots until on a lateral radiograph contrast is seen running into the posterior fossa. This estimates the volume of the trigeminal cistern (usually <0.4 ml)(Fig. 9.5a, b). The contrast is allowed to clear by gravity before anhydrous glycerol (99.9%) is injected. Selective glycerol rhizolysis for the ophthalmic division is accomplished by allowing glycerol to float on the surface of the contrast after part drainage of the cistern. Following the procedure the patient should remain sitting for 2 hours to allow fixation. Some authors have reported omitting cisternography suggesting that this does not reduce the success rate or increase the complication rate. Patients often experience diminishing ipsilateral forehead pain for several days after the injection during which time the neuralgia will gradually abate.

Patients respond well whatever the underlying cause although the results are variable. Between 72 and 99% can expect complete relief of tic after glycerol rhizolysis, 50% within 24 hours. In others the onset of pain relief usually occurs within 10 days but it may take even longer. Early recurrence occurs within the first 6 months in 10% and within 2 years in 26% while the long-term recurrence rate is 40–45%. The recurrence rate with multiple sclerosis is higher; in one series 75% of patients had recurrence of their trigeminal neuralgia at 1 year.[31] Technical failure may be expected in up to 15%; this is usually because the anatomy is such that intracisternal injection is not possible. Second injections lead eventually to lasting remission in approximately 70% but many can be rendered pain-free with supplemental drug therapy. Surgery will be required for the remainder.

The incidence of facial sensory loss is quite high if patients are carefully questioned and examined. Reported rates of facial hypoaesthesia range from 17 to 70%. Corneal sensory loss may occur in up to 9% but corneal keratitis is rare. Håkanson found 30% had sensory loss after one injection, 50% after two and 70% after three injections. In most cases this will be mild, not disturbing to the patient and resolves over days to weeks although there is a risk of complete facial sensory loss. We have observed one patient who had symptoms on the side of an exceptionally small (and presumably relevant) foramen ovale who subsequently developed complete second division anaesthesia after injection through a 24 gauge needle. Anaesthesia dolorosa has not been reported although up to 60% may experience (mild) dysaesthesia. In Gybels and Sweet's series the incidence of dysaesthesia with glycerol was 20% compared with only 6% with thermal rhizotomy. Other complications are uncommon but

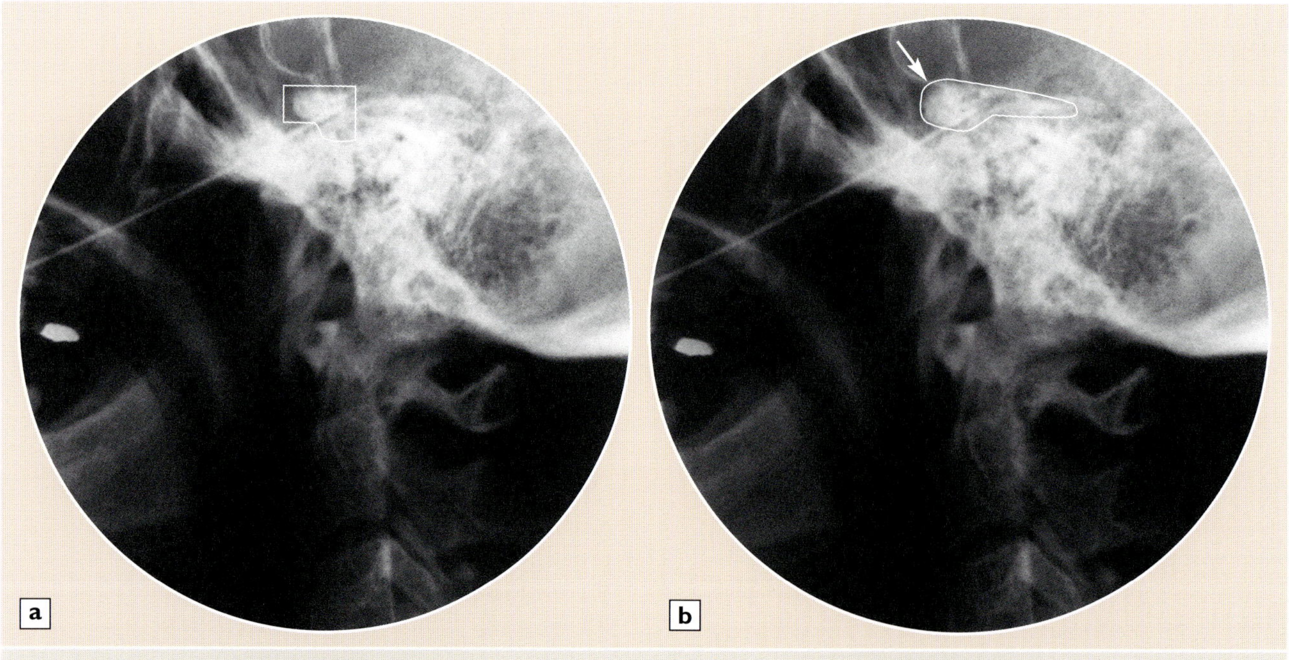

Figure 9.5 (a, b) Lateral view, contrast cisternogram (Iopamadol). Shown is progressive filling of trigeminal cistern (0.1 ml increment [a], final volume 0.5 ml) until contrast overflows into the subarachnoid space (b) (arrow).

include meningitis (bacterial and aseptic), post-procedure headache (10%), herpes simplex eruption (10% between days 3 and 5), and damage caused by the needle (carotid bleeding, carotido-cavernous fistula, visual loss and other cranial neuropathies).

Thermocoagulation

Thermal rhizotomy has reliably been used as an analgesic procedure in neoplastic facial pain although its primary indication is in the treatment of trigeminal neuralgia, for which it remains the most commonly performed percutaneous procedure. It was introduced in 1932 by Kirschner who used the Bovie apparatus. By 1967 over 2000 patients had been treated in this way; there were no deaths but the recurrence rate was 80%. Sweet introduced controlled differential thermocoagulation in 1965.[32] With precise physiological localisation of the needle tip this facilitated selection of the division treated and improved the results. The mechanism of action is still not clear although in theory the A-δ and C fibres are more susceptible to thermal damage.

This requires collaboration between the surgeon, the anaesthetist and the patient to obtain the best results. After needle placement the patient must be able to communicate. Accurate responses are necessary to localise the sensation produced by high frequency electrical stimulation of the trigeminal fibres at the needle tip as this is then adjusted until these correspond to, or include, the trigger zone. Once this has been confirmed serial thermal lesions are created by application of a constant current. The precise protocol will be determined by the clinician and by the configuration of the thermal (radiofrequency) lesion generator used and by the characteristics of the electrode which may be a thermocouple or thermistor tipped. Some advocate the use of a flexible string electrode to minimise the production of a trigeminal motor palsy. Simple abolition of the trigger will suffice to relieve pain but this will not produce long-term benefit. Ideally the production of hypoalgesia to pinprick in the relevant area provides the best result with long-term benefit and minimal prospect of deafferentation.

Numerous series have been published reviewing individual, institutional and collated results of this procedure. Perhaps the most authoritative is that of Sweet who reviewed 14,000 cases from 33 series which included only two operative deaths. In evaluating the long-term results of radiofrequency rhizotomy he reported 99% immediate response rate and up to 25% recurrence at 14 years (15% at 5 years, 7% between 5 and 10 years, 3% between 10 and 15 years); 30% did not require further surgery. Tew was able to report 96% pain-free at 1 year and 77% at 10 years.[33] Hypoalgesia and hypoaesthesia will be present after effective thermocoagulation but these are the best predictors of long-term relief. Of those with hypoalgesia in the affected division, 60% will remain pain-free, compared with 20% of those without. Unfortunately corneal anaesthesia can occur in up to 8% and may lead to keratitis in 2% although both will be minimised by careful technique. Painful dysaethesiae occur in as many as 13%, of whom a third might require medication. Clinically detectable motor palsy is apparent in 12%, of which most resolve fully over 12 months.

Cryosurgery

This requires the use of a dedicated small diameter probe which is inserted in routine fashion to permit freezing of the ganglion until the sensory trigger has been abolished. Generally, however, the long-term results are poor and the technique has been superseded by other procedures. Permanent damage to the nerve and adjacent structures occurs if the cryoprobe is removed before reattaining body temperature.

OPEN PROCEDURES – MIDDLE CRANIAL FOSSA APPROACH

Retrogasserian ganglionectomy and neurectomy

In 1890 Rose developed a procedure whereby the trigeminal ganglion was avulsed through an enlarged foramen ovale. For technical reasons this procedure was unsatisfactory and was abandoned. In 1892 an extradural approach to the ganglion was described by Hartley in performing trigeminal neurotomy.[34] In ignorance of this Krause described the same approach for ganglionectomy later that year.[35] Cushing modified the Hartley–Krause approach and by 1910 he had reduced the operative mortality of ganglionectomy from 10% to 5%.

In 1891 Horsley had described the exposure and division of the trigeminal root via an intradural approach through the middle cranial fossa to the trigeminal ganglion. Unfortunately his first patient died and in the light of early successes with ganglionectomy further attempts at neurotomy were not made until 1901 when the operation was performed again by Frazier using the

Hartley–Krause approach. His subsequent successes established the procedure and gradually ganglionectomy was abandoned in favour of retrogasserian neurotomy. Although Tiffany had suggested motor root preservation in 1896[36] he did not achieve this and it was not until 1918 that Peet reported sparing of the motor root for the first time. It was Frazier, however, with a large series of cases performed over 14 years and described in 1927, who truly developed the technique for differential root section. This approach has been superseded by the posterior fossa route (see below) and is now almost obsolete although still finding favour for its ability to spare the hazards of surgery in the posterior fossa. Frazier's approach required that the patient, often elderly and infirm, be in the sitting position during the procedure – a phenomenon almost guaranteed to induce cardiac irregularity in the anaesthetist if not in the patient!

Middle fossa decompression

Prior to the advent of theories of neurovascular conflict the petrous ridge was considered a likely area for compression of the root. In 1937 Lee suggested that excision of the petrous ridge would relieve neuralgia. Following this, procedures to decompress the root via the middle cranial fossa were described with good immediate results but late recurrence rates of up to 40% in some series. This led the procedure into disrepute until Malis was able to report a recurrence rate at 18 months of only 9%.[16] It appears, therefore, that manipulation at the petrous apex can give results that compare favourably with posterior fossa microvascular decompression in respect of recurrence. The procedure was recently described again by Vengsarkar who used an extradural approach to decompress the nerve from dural attachments over the petrous ridge and interposed a thin layer of fat between the root and the petrous ridge; 20 of 23 patients were rendered pain free.[37]

OPEN PROCEDURES – POSTERIOR CRANIAL FOSSA APPROACH

Trigeminal root section

Dandy is credited with the development of the posterior fossa approach to the trigeminal nerve. Whilst others were dividing the nerve via a subtemporal approach he developed the posterior cranial fossa approach to the trigeminal nerve.[38] The pain relief that he achieved by dorsal root section at least equalled that of others using middle fossa approaches. In addition it was easier to preserve the motor root. Although many other options are now available this remains a very effective way to deal with trigeminal neuralgia, and with severe facial pain of other causes, although it does result in anaesthesia. We continue to offer this operation for patients with multiple sclerosis whose neuralgia recurs frequently after lesser procedures. It guarantees future freedom from pain at the cost of anaesthesia and with the potential risk of bilateral motor palsy.

Microvascular decompression (MVD)

In 1934 Dandy had observed vascular compression in 45% of 215 cases undergoing trigeminal neurectomy in the posterior cranial fossa and postulated this as a major cause of trigeminal neuralgia.[15] In 1959 Gardner and Miklos reported moving a compressive artery away from the nerve, maintaining the separation using gelatin sponge.[18] Since 1967 this operation, developed and popularised by Janetta in many publications, has become the main operative approach for trigeminal neuralgia.[19]

A small retromastoid craniectomy is performed with the patient in the lateral decubitus position. The cerebellum falls away under gravity as cerebrospinal fluid is released. The trigeminal nerve is exposed by following the junction of the petrous ridge with the tentorium cerebelli and superior petrosal sinus anteromedially. The superior petrosal vein may be compressing the nerve or it may impede access; it can be mobilised and, if necessary, coagulated and transected. The entire nerve from the entry zone at the pons to Meckel's cave should be inspected for any vessel in contact as it may be compressed by an artery, a vein or by both (Table 9.2). Compressing veins may be mobilised or coagulated. Compressive arteries are mobilised and separated from the nerve using an implant, which is most commonly shredded Teflon felt.

Janetta has reviewed the long-term outcome of over 1000 patients after microvascular decompression of the trigeminal nerve.[39] There was complete relief of tic in 82% and partial relief in 16%. After 1 year 75% still had complete relief and 9% partial relief. At 10 years 64% had complete relief of pain, a further 4% had partial relief. Of the recurrences, 60% occurred within the first 6 months and 90% within the first 2 years.[40] Factors reported to predispose to recurrence include: sex (female), youth (<35 years), symptom duration greater than 8 years, venous compression of the root entry zone, and no immediate response to decompression. Previous destructive procedures made no difference in Janetta's series in comparison although others regard it as significant.[41] Re-operation is not as successful; after 10 years only 42% are pain-free.

In Janetta's series there were two deaths (0.2%). Other complications included 1% ipsilateral deafness, 1% severe facial numbness, 0.3% burning or aching facial pain (more likely if there has been a previous radiofrequency lesion, but no anaesthesia dolorosa), 0.1% cerebellar infarction and 5% transient cranial nerve disturbances. Aseptic meningitis characterised by headache, malaise, meningism and sterile cerebrospinal fluid may occur during the first week. In a recent review 98% of patients had immediate relief of symptoms, 80% were pain-free at 1 year and 77% at 10 years. Persistent sensory loss was suffered by 5% of patients and the mortality rate was 0.15%.[41]

OPEN PROCEDURES – THE NUCLEUS CAUDALIS DORSAL ROOT ENTRY ZONE (DREZ) PROCEDURE

The nucleus caudalis harbours second-order neurones from the fifth, seventh, ninth and tenth cranial nerves subserving pain, temperature and crude touch from the ipsilateral face. It has been suggested that these second-order neurones have a role in the generation of central pain especially when there is deafferentation. DREZ lesioning affects these second-order neurones in the sensory pathway. Open trigeminal tractotomy was first performed by Sjöquist in 1938[42] and has subsequently been refined by a stereotactic approach.[43] DREZ lesion of the trigeminal nerve was first reported by Nashold who has subsequently modified the operation.[44,45] The procedure coagulates the nucleus caudalis: some advocate a localised coagulation, some near complete destruction of the whole nucleus and some combining nucleotomy with tractotomy. Whilst trigeminal neuralgia responds well to the interventions previously mentioned, most other facial pain syndromes do not. This technique is also of value in treating postherpetic neuralgia, atypical facial pain, post-traumatic pain, pain following cerebrovascular accident, deafferentation pain and trigeminal pain in multiple sclerosis. The results of the first caudalis DREZ lesions showed that 17 of 18 patients initially had a good result although this later fell to 58%.[46] After revising the procedure a later series reported good or excellent pain relief in 74%.[47] Both series included patients with a variety of aetiologies for their facial pain and acknowledge the risk of sensory loss in the distribution of the ninth and tenth nerves as well as sensory loss and ataxia in the limbs.

OTHER PROCEDURES

Stereotactic ablation

These procedures are seldom indicated for trigeminal mediated pain as they serve only to increase deafferentation, with the sole exception of rostral mesencephalic tractotomy, which is a non-specific ablative technique but has the additional although delayed benefit of alleviating the emotional component of chronic pain.[48] It appears inappropriate to suggest that any psychosurgical procedure might be a valid option in the treatment of chronic pain but cingulotomy can abolish the secondary obsessive features and morbid depressive state that infrequently result from such conditions.

Stereotactic radiosurgery

This is undertaken in the Leksell gamma knife (cobalt source) irradiator or using a linear accelerator. The target is the proximal root entry zone. The typical dose is 70 Gy. One study has shown 86% initial response rate and in another study 75% of 12 patients showed complete relief or improvement of their trigeminal neuralgia after 3 years follow-up. At this time there is limited experience with the technique, particularly in respect of long-term sequelae.[49] Most centres will only consider it following the failure of more conventional treatment.

Electrical stimulation

The rationale behind chronic deep brain stimulation is well established.[50,51] Large series of patients with various pain syndromes so treated have been accumulated but few, if any, patients with trigeminal neuralgia are reported. Stimulation in the somatosensory thalamus is now reserved for the treatment of deafferentation pain whereas, because of its proven effect on the descending pain modulating pathways, stimulation in the periventricular and periaqueductal grey matter is preferable for nociceptive cancer pain. Overall less than 50% with neuropathic pain will benefit compared with 70% of those with pain due to other causes. Clearly, considering the considerable effort and expense involved, these are not appropriate treatments for trigeminal neuralgia.

Recent work, however, suggests that chronic electrical stimulation of the pre-motor cerebral cortex might be of benefit in intractable trigeminal deafferentation pain. Clinical results are awaited.

SUMMARY

There are advocates for all of these techniques. Unfortunately, there are only a handful of comparative studies and no prospective randomised comparisons of the available treatments.

The ideal treatment considers patients' preferences as well as their age, physical condition and symptoms. It should be both as simple and as reliable as possible. Assuming that medical treatment has been ineffective, peripheral measures and definitive treatment of any structural lesion should be considered first, which requires that appropriate imaging is obtained. For the elderly patient with first-division pain, glycerol injection or balloon compression have little competition although supraorbital nerve avulsion may be an option. In this age group, second- or third-division symptoms probably do better in the long term after thermocoagulation. Younger patients have a greater choice and must balance the low risk of sensory loss associated with any of the percutaneous procedures against the greater hazards of posterior cranial fossa microvascular decompression. Patients with multiple sclerosis are best treated by the percutaneous techniques unless the rapidity of symptom recurrence becomes unacceptable, when they should be offered trigeminal root section via the approach with which the surgeon is most familiar. Recurrent symptoms can be treated in the same way as the first occurrence. This will be subject to any revised preference of the patient, acknowledging that the risks of sensory loss and of incidental complications are greater when either thermocoagulation or microvascular decompression are repeated. Patients experiencing unequivocal deafferentation pain *should not* be treated by these techniques; in these cases one must decide between the nucleus caudalis DREZ procedure and chronic brain stimulation, the choice being determined as much by patient comprehension and acceptance of risk in the former, and their ability to use a stimulator effectively and the cost implications of the latter.

Pain is a subjective experience and the interpretation of response to treatment is inherently flawed when it is submitted to selective reporting. Perhaps, until comparative studies which address both pathophysiology and different modalities of treatment are available, we should restrict our interventions to those that have a well substantiated reason to work with the caveat 'first do no harm'.

REFERENCES

1. Bonica J J. Definitions and taxonomy of pain. In: Bonica J J (ed) *The management of pain*. Baltimore: Lea & Febiger, 1990; 18–27
2. Merksey H, Bogduk N (eds) *Classification of chronic pain*. (2nd edn). Seattle: IASP Press 1994, 8–10
3. Katusic S, Beard M, Bergstralh E, Kurland C T. Incidence and clinical features of trigeminal neuralgia, Rochester, Minnesota, 1945–1984. *Ann Neurol* 1990; 27: 89–95
4. Rushton J G, Olafson R A. Trigeminal neuralgia associated with multiple sclerosis. *Arch Neurol* 1965;13: 383–387
5. Hooge J P, Redekop W K. Trigeminal neuralgia in multiple sclerosis. *Neurology* 1995; 45: 1294–1296
6. Meaney J F, Eldridge P R, Dunn L T *et al.* Demonstration of neurovascular compression in trigeminal neuralgia with magnetic resonance imaging. *J Neurosurg* 1995; 83: 799–805
7. Kumon Y, Sakaki S, Kohno K *et al.* Three dimensional imaging for presentation of the causative vessels in patients with hemifacial spasm and trigeminal neuralgia. *Surg Neurol* 1997; 47: 178–184
8. Beaver D L. Electron microscopy of the Gasserian ganglion in trigeminal neuralgia. *J Neurosurg* 1967; 26: 138–150
9. Kerr F W L. Pathology of trigeminal neuralgia: light and electron microscopic observations. *J Neurosurg* 1967; 26: 151–156
10. Seltzer Z, Devor M. Ephaptic transmission in chronically damaged peripheral nerves. *Neurology* 1979; 29: 1061–1064
11. Burchiel K. Abnormal impulse generation in focally demyelinated trigeminal roots. *J Neurosurg* 1980; 53: 674–683
12. Calvin W H, Devor M, Howe J F. Can neuralgias arise from minor demyelination? Spontaneous firing, mechanosensitivity and after discharge from conducting axons. *Expl Neurol* 1982; 75: 755–763
13. King RB. Evidence for a central etiology of tic douloureux. *J Neurosurg* 1967; 26: 175–180
14. Fromm G H, Terrence C F, Maroon J C. Trigeminal neuralgia. Current concepts regarding etiology and pathogenesis. *Arch Neurol* 1984; 41: 1204–1207
15. Dandy W E. Concerning cause of trigeminal neuralgia. *Am J Surg* 1934; 24: 447–455
16. Malis L I. Petrous ridge compression and its surgical correction. *J Neurosurg* 1967; 26: 163–167
17. Kerr F W L. Etiology of trigeminal neuralgia. *Arch Neurol* 1963; 89: 15–25
18. Gardner W J, Miklos M V. Response of trigeminal neuralgia to 'decompression' of sensory root: discussion of the cause of trigeminal neuralgia. *JAMA* 1959; 170: 1773–1776
19. Janetta P J. Arterial compression of the trigeminal nerve at the pons in patients with trigeminal neuralgia. *J Neurosurg* 1967; 26: 159–162
20. Adams C B T, Kaye A H, Teddy P J. The treatment of trigeminal neuralgia by posterior fossa microsurgery. *J Neurol Neurosurg Psychiatry* 1982; 45: 1020–1026
21. Hardy D G, Rhoton A L Jr. Microsurgical relationships of the superior cerebellar artery and the trigeminal nerve. *J Neurosurg* 1978; 49: 669–678
22. Haines S J, Janetta P J, Zorub D S. Microvascular relations of the trigeminal nerve: an anatomical study with clinical correlation. *J Neurosurg* 1980; 52: 381–386
23. Dott N M. Facial pain. Proc Roy Soc Med 1951; 44: 1034–1037
24. Smith G W, Miller J M. The treatment of tic douloureux with stilbamidine. *Bull Johns Hopkins Hosp* 1955; 96: 146–149
25. Murali R, Rovitt R L. Are peripheral neurectomies of value in the treatment of trigeminal neuralgia? An analysis of new cases and cases involving previous radiofrequency Gasserian thermocoagulation. *J Neurosurg* 1996; 85: 435–437
26. Härtl F. Ueber die intracranielle injektionsbehandlung der Trigeminusneuralgie. *Med Klinik* 1914; 10: 582–584
27. Tew J M, Tobler W D, van Loveren H. Percutaneous rhizotomy in the treatment of intractable facial pain. In: Schmideck H H, Sweet W H (eds) *Operative neurosurgical techniques.* New York: Grune & Stratton, 1982; 1083–1100
28. Mullan S, Lichtor T. Percutaneous microcompression of the trigeminal ganglion for trigeminal neuralgia. *J Neurosurg* 1983; 39: 1007–1012
29. Abdennebi B, Bouatta F, Chitti M, Bougatene B. Percutaneous balloon compression of the Gasserian ganglion in trigeminal neuralgia. Long term results in 150 cases. *Acta Neurochir (Wien)* 1995; 136: 72–74
30. Håkanson S. Trigeminal neuralgia treated by the injection of glycerol into the trigeminal cistern. *Neurosurgery* 1981; 9: 638–646
31. Burchiel K J. Percutaneous retrogasserian glycerol rhizolysis in the management of trigeminal neuralgia. *J Neurosurg* 1988; 69: 361–366
32. Sweet W H. The treatment of trigeminal neuralgia (tic douloureux). *N Engl J Med* 1986; 15: 174–177
33. Taha J M, Tew J M, Buncher C R. A prospective 15 year follow-up of 154 consecutive patients with trigeminal neuralgia treated by percutaneous stereotactic radiofrequency thermal rhizotomy. *J Neurosurg* 1995; 83: 989–993
34. Hartley F. Intracranial neurectomy of the second and third divisions of the fifth nerve: a new method. *NY Med J* 1892; 55: 317–319
35. Krause F. Resection des trigeminus innerhalb der schadelhohle. Arch Klin Chir 1892; 44: 821–832
36. Tiffany L M. Intracranial operations for the cure of facial neuralgia. *Ann Surg* 1896; 24: 575–619

37. Vengsarkar U S, Mokashi Y B, Bhatjiwale M G. Middle fossa decompression of the trigeminal sensory root for trigeminal neuralgia – a re-appraisal and a modification of the technique. *Br J Neurosurg* 1995; 9: 743–748
38. Dandy W E. The treatment of trigeminal neuralgia by the cerebellar root. *Ann Surg* 1932; 96: 787–793
39. Barker F G, Janetta P J, Bissonette D J, Larkins M V, Dong Jho H. The long term outcome of microvascular decompression for trigeminal neuralgia. *N Engl J Med* 1996; 34: 1079–1083
40. Burchiel K J, Clark H, Heglund M *et al.* Long term efficacy of microvascular decompression in trigeminal neuralgia. *J Neurosurg* 1988; 69: 35–38
41. Barbra D, Alksne J F. Success of microvascular decompression with and without prior surgical therapy for trigeminal neuralgia. *J Neurosurg* 1984; 60: 104–107
42. Sjöquist O. Studies on pain conduction in trigeminal nerve. *Acta Psychiatr Neurol Suppl* 1938;17: 1–139
43. Schvarcz J R. Postherpetic craniofacial dysaesthesia. Their management by stereotaxic trigeminal nucleotomy. *Acta Neurochir* 1977; 38: 65–72
44. Nashold B S Jr, Caputti F, Bernard E. Trigeminal DREZ: caudalis nuclear lesions for the relief of facial pain. Neurosurgery 1986; 19: 150
45. Nashold B S Jr, El-Naggar A, Abdulhak M M *et al.* Trigeminal caudalis dorsal root entry zone: a new surgical approach. *Stereotact Funct Neurosurg* 1992; 59: 45–51
46. Bernard E J Jr, Nashold S Jr, Caputti F, Moossy J J Jr. Nucleus caudalis DREZ for facial pain. *Br J Neurosurg* 1987; 1: 81–92
47. Gorecki J P, Nashold B S Jr. The Duke experience with the nucleus caudalis DREZ operation. *Acta Neurochir* Suppl 1995; 64: 128–131
48. Nashold B S Jr. Extensive cephalic and oral pain relieved by midbrain tractotomy. *Confinia Neurol* 1972; 34: 382–388
49. Kondziolka D, Flickinger J C, Lunsford L D, Habeck M. Trigeminal neuralgia radiosurgery: the University of Pittsburgh experience. *Stereotact Funct Neurosurg* 1997; 66: 343–348
50. Adams J E, Hosobuchi Y, Fields H L. Stimulation of internal capsule for relief of chronic pain. *J Neurosurg* 1974; 41: 740–744
51. Richardson D E. Analgesia produced by stimulation of various sites in the human beta-endorphin system. *Appl Neurophysiol* 1982; 45: 116–122

ADDITIONAL BIBLIOGRAPHY

1. White J C, Sweet W H. *Pain and the neurosurgeon, a 40 year experience.* Springfield: Charles C Thomas, 1969
2. Gybels J M, Sweet W H. *Neurosurgical treatment of persistent pain.* Basel: Karger, 1989
3. Rovitt R L, Murali R, Janetta P J. *Trigeminal neuralgia.* Baltimore: Williams and Wilkins, 1990

Index